AF566773

Cytogenetics

Cytogenetics

Darbeshwar Roy

Alpha Science International Ltd.
Oxford, U.K.

Cytogenetics
280 pgs. | 76 figs. | 17 tbls.

Darbeshwar Roy
Department of Genetics and Plant Breeding
College of Agriculture
Govind Ballabh Pant University of Agriculture and Technology
Pantnagar, India

ALPHA SCIENCE INTERNATIONAL LTD.
7200 The Quorum, Oxford Business Park North
Garsington Road, Oxford OX4 2JZ, U.K.

www.alphasci.com

ISBN 978-1-84265-487-3

Printed in India

To

Dr. G. H. Jones who taught Cytogenetics

Preface

Cytogenetics is now a more important subject in this era of biotechnology and bioinformatics. Biologists are trying to understand the function of cell at molecular level using systems biology approach. Findings from molecular analysis regarding mapping of genes, variation in chromosome number, structural variations(duplication, deficiency, inversion, translocation), chromosome organization, etc can be corroborated with cytological observations. Interphase cytogenetics and molecular cytogenetics will help in the studies of single loci or chromosomal regions in interphase nuclei. Thus the knowledge of cytological and cytogenetic techniques is more relevant today. This book on cytogenetics is the first in the series of books, the second and the third being biotechnology and bioinformatics, respectively. This book deals with structure and function of a cell, study of organization, function, transmission and recombination of genetic material using chromosome during mitosis and meiosis. Mendel's laws of heredity, the deviations from the laws and genetic recombination provide understanding of the transmission of traits from parents to offspring. It is essential to understand the applications of chromosome engineering for improvement of crop plants and the uses of manipulation of cytoplasm and 2n gametes. Cytological techniques described will help in conducting practical and the cytogenetics and genetics problems provided will enhance the analytical skill.

The source for inspiration for writing this book is the superb teaching that the author received from Dr. G. H. Jones at the University of Birmingham, England, while doing Ph.D. during 1980-83.

I would like to express my gratitude to Mr. N. K. Mehra, M/s Narosa Publishing House for proposing to publish manuscript in a series and for his sincere interest and attention. I would like to thank Mrs. Anupama Jauhry, Assistant General Manager (Production), Narosa for her attention to my manuscript.

I would like to thank Dr. S.N. Tiwari, and Dr. V.S. Pundhir, for using their facilities for scanning some figures.

Finally, I would like to thank my wife Veena, daughter Cynthia and son Shishir for their encouragement during the course of writing this series of books.

DARBESHWAR ROY

Contents

Cell (Eukaryotes and Prokaryotes)

1.1 EUKARYOTIC CELL STRUCTURE

The different diverse organisms such as higher plants, animals, several kinds of fungi and algae belong to a class called **eukaryotes** (Greek eu, 'true and karyon,' nucleus) as their nucleus is enclosed within a double membrane called nuclear envelope and they have common features of cell organization which are quite different from another class of organisms called **prokaryotes** (Greek pro, 'before') which include bacteria and other allied forms such as archaebacteria and these have nucleus without nuclear envelope. The hereditary material resides in the cell. The cell contains the nucleus which is surrounded by a double-walled membrane and other cell organelles such as mitochondria, plastids, endoplasmic reticulum, golgi complex, lysosomes, centrosomes are found in the cytoplasm. The boundary of nucleus is a double lipid bilayer termed nuclear envelope which consists of an outer nuclear membrane that is continuous with the membrane system of endoplasmic reticulum and inner nuclear membrane of a different protein composition. The proteins of this inner nuclear membrane plays a vital role in genome organization(Kutay and Muhlhausser, 2006). The portion of a cell's contents enclosed by the nuclear membrane is called **nucleoplasm**. A filamentous meshwork located between the inner nuclear membrane and heterochromatin which provides potential attachments sites for chromatin and cytoplasmic intermediate filaments is called **nuclear matrix/nuclear scaffold**. Nucleoplasm and cytoplasm constitute the **protoplasm**. The aqueous cytoplasm with its dissolved solids minus the cell organelles is called **cytosol**. The nucleus contains the chromosomes which are the seats of mendelian factors(or genes). A plant cell is similar to an animal cell (Figure 1.1) except that an autotrophic plant cell contains plastids and a flowering plant cell contains no centrosome. The nucleus contains one or more nucleoli(singular, nucleolus) associated with

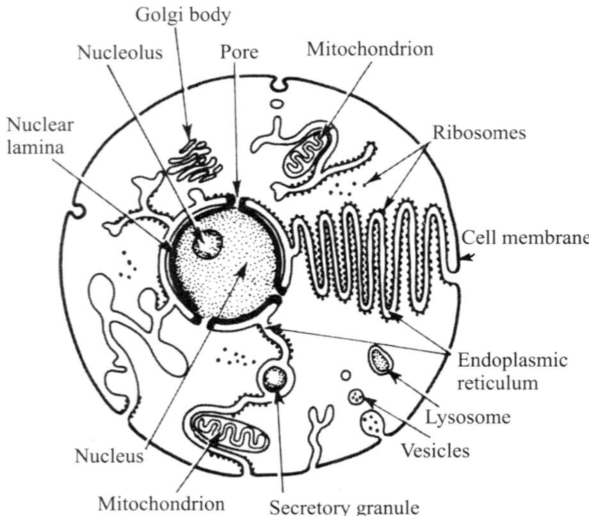

Fig. 1.1 Showing structure of a typical animal cell.

region of specific chromosomes called **nucleolus organizer regions**(NORs, Barbara Mclintock, 1934). Darlington(1936) classified the total hereditary material of the cell into genome and plasmons. **Genome** refers to all the hereditary material of chromosome complement whereas **plasmons** refer to all hereditary material of extra chromosomal complement. The extra chromosomal complement includes plastids, kinetosomes, mitochondria and centrioles. The genes present in the extra chromosomal complement are called plasma genes and thus there are plastogenes, kintogenes, chondriogenes and centrigenes. The sum total of all the plasmagenes constitute the plasmon and the sum total of all the nuclear genes constitutes the genome. The plasmons show non-mendelian inheritance. These are self perpetuating or self duplicating structures and show physical continuity from one generation to another. These cell organelles are not assembled by the cell. They arise by division of the existing organelle. Division of chloroplast and mitochondria are by a process called fission, similar to the division of bacteria.

1.1.1 Extra Chromosomal Cell Constituents with Physical Continuity

Mitochondria-Mitochondria are the power house of the cell. Mitochondria vary in shape, can be spherical, rod like or filamentous. Their shapes vary characteristically from one cell type to another in the same individual and they may even differ in shape within a single cell. Although the chemical composition of mitochondria varies but the principal constituents are protein and lipids, mostly phospholipids. Mitochondria divide transversely. The physical continuity is through the male gametes The break down of fats, carbohydrates takes place in the mitochondria and during this process there is release of energy which is stored in ATP molecule. **Plastids**-Chloroplasts are present in all green plants and are associated with photosynthesis. Chromoplasts are associated with coloring of petals and roots whereas leucoplasts serve as storage of starch, oil or protein. In lower plants(some algae) a small and constant number of rather large chloroplasts are present in each cell. **Centrosome** – It is present in all animal cells and lower plants. The centrosomes can be with or without a central granule(a centriole) which generates astral rays during cell division. Centrosomes play a central role in cell cycle regulation and function as signaling centers. Functional genomics screens in Drosophila have discovered a key regulator of centrosome duplication.Whenever centriole is present each cell possesses a pair of centrioles. **Kinetosomes** – The basal granules or blaspharoplasts or kinetosomes generate cilia, flagella and sperm tails wherever a free swimming motile stage is found.

1.1.2 Other Extra Chromosomal Structures-Kinetoplasts

They have fine structure similar to that of mitochondria but these are not essential for survival. **Microsomes**- They are small structures, rich in RNA, phospholipids and protein. Ultra centrifugation invariably separates microsomes leaving nuclei and mitochondria. Microsome originates as a result of fragmentation of endoplasmic reticulum during the homogenization of the cells. Many ribosomes are attached to the membranous components. In bacteria ribosomes are found free through out the cytoplasm and thus are not associated with any membranous structure. **Endoplasmic reticulum**- These are delicate membranous structures lined with ribosomes(ribonucleoprotein particles) which are the seats of protein synthesis. **Golgi complex**- Golgi apparatus is commonly found in secretory cells(liver, pancreas) and a similar structure occurs in certain plant root cells. They have secretory function. **Lysosomes**- Lysosomes are the sites of special hydrolytic enzymes including acid phosphotase, ribonuclease, deoxyribonuclease, cathepsin, glucuronidase, uricase, etc and are found in animals.

1.1.3 Infective Particles

The extra chromosomal elements such as platids , mitochondria are required for functioning of cells whereas kappa, mate-killer, mu (Siegel, 1952) and metagon characteristics in paramecium (Gibson and Beale, 1962), sigma in Drosophila (a virus particle causing CO2 sensitivity), sex ratio in Drosophila(because of the presence of infective agent, a protozoan spirochaete of the genus *Treponema* and causing production of predominantly or exclusively female) willistinii are examples of extra chromosomal elements that are not normally necessary for cell function but that become part of the heritable characteristics of the individuals. These particles are not only self-reproducing but they are infective to some degree since it is through this means that they are introduced into new hosts. These particles exist either as a parasite or more possibly a symbiont. Kappa, mu and sigma particles depend for their maintenance in the cell on the presence of appropriate genes. They exist independently in the cytoplasm or they are able to attach themselves to the host chromosome and thus they are called **episomes.** The episomes that are virus like infective particles and known to occur in bacteria include lysogeny, sex factor(F^+), colicinogenic factors and resistance transfer factor(RTF). The term 'episome' was coined to denote plasmids like F with both autonomous and passive (integrated) modes of propagation. However, this property did not reflect a fundamental distinction from other plasmids and so this term is no longer considered useful (Lane, 1999).

1.1.4 Sexual Reproduction

System of sexual reproduction is another characteristic of eukaryotes in which there is alteration of haploid and diploid phases in their life cycle. The germ cells (gametes) –eggs and sperms in animals, ova and pollen tubes in flowering plant and morphologically undifferentiated cells in budding yeast are haploid. In case of plants megaspore mother cell and microspore mother cell produce ova and pollen, respectively. In case of animal female oocytes and male spermatocytes which undergo meiosis to form eggs and sperms, respectively. Flowering plants differ from animals in that they have a brief propagation of the haploid phase following meiosis but preceding **karyogamy** (fusion of nuclei) within the pollen tube and the embryo sac of the ovule. Ferns, mosses and liverworts have more extensive haploid phases. In fern the haploid phase is in the form of small and short lived but free-living green prothallus and in mosses and liverworts there is whole green plant with the exception of the spore bearing capsule which is diploid.

Cell division-To understand how the chromosomes are transferred from one generation to the next, how the consistency in chromosome number from one generation to the next is maintained during sexual reproduction and how the chromosomes are evenly distributed between cells of an organism one will have to study mitosis and meiosis.

1.1.4.1 Mitosis

Mitosis comprises of nuclear division (karyokinesis) and cell division (cytokinesis). It is an equational division in which generally there is no change in the chromosome number between the ancestor and the daughter cells. For example, in a plant with $2n = 14$, the daughter cells will also have $2n = 14$. The four different stages of mitosis are: 1. Prophase 2. Metaphase 3. Anaphase 4. Telophase (Figure 1.2). The characteristic features of each of these stages are as follows:

Prophase-The chromosomes become visible due to progressive coiling or condensation. As prophase proceeds and the chromosomes become shorter and thicker the chromosomes appear longitudinally split into two duplicates (called sister chromatids). Nucleolus is seen during interphase and prophase. The end of prophase is marked by the disappearance of nuclear membrane and nucleolus as well as by the division of the centrosome and formation of spindle.

Metaphase-The chromosomes are at their highest level of coiling or contraction and therefore appear to be shorter and thicker than in any other stage. Chromosome at metaphase is of the order of 1000 times shorter than that of prophase. There is movement and arrangement of all chromosomes on the metaphase plate midway between the two poles of spindles. In many plants centrosomes and centrioles accompanying them are absent but the spindles are nevertheless present. In animals spindle fibers are formed as a result of astral rays coming from the two centrioles. Spindle fibers are an array of protein fibers with contractile properties. The mitotic spindles are assembled by microtubule-based motor proteins(Scholey et al., 2003; Compton, 2000). The end of metaphase is marked by splitting of the centromere and separation of chromatids. This is the stage when chromosomes are most easily identified and distinguished and their numbers can be counted.

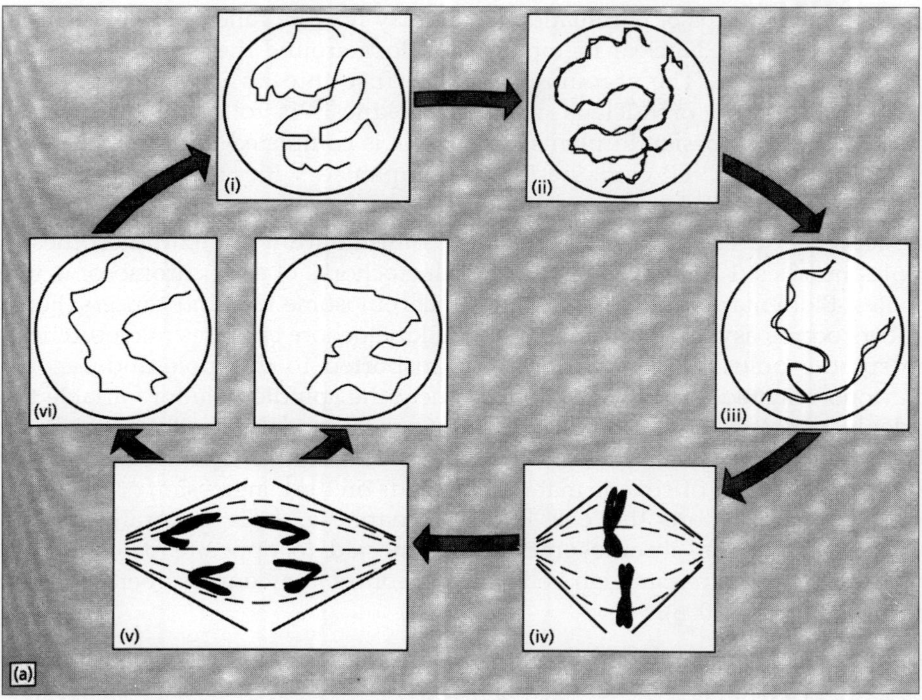

Fig. 1.2 Showing phases of mitotic cycle, (i) G1; (ii) G2; (iii) prophase: (iv) metaphase and (v) anaphase (adapted from Fincham, 1994).

In all eukaryotes a hallmark of functional centromeres-both normal one and those formed aberrantly at atypical loci, is the accumulation of centromere protein A (CENP-A), a histone variant that replaces H3 in centromeric nucleosomes. Centromeres and their associated kinetochores are protein-DNA complexes which mediate spindle mictotubule attachment during mitosis and are necessary to mediate delivery of one copy of each chromosome to each daughter cell. Mammalian centromeres contain megabases of repetitive, a-satellite DNA that is organized into specialized chromatin consisting of nucleosomes in which histone 3 is replaced by the variant CENP-A. Removal of CENP-A reduces the fidelity of chromosome segregation leading to the proposal that centromere-specific CENP-A containing nucleosomes may recruit other centromere and kinetochore components, at least within the context of an array of CENP-A structure required for the formation of the inner kinetochore surface. CENP-A represents the most likely candidate to be the principal epigenetic marker that maintains the identity of centromere after DNA replication in organisms containing multisubunit centromere such as the humans. Maintaining centromere identity during normal cell division may be its most important role but how it does it and even the process by which CENP-A is targeted to centromere remains unclear(Black et al., 2004).

Anaphase-Each centromere splits into two daughter centromeres. There is repulsion between sister chromatids and thus the two sister chromatids separate and move towards opposite poles of the spindle. It is the shortest of all the mitotic stages.

Telophase-The two groups of chromatids which may now be called daughter chromosomes reach the opposite poles. Nuclear membrane develops around the two daughter nuclei and the nucleoli are formed. The chromosomes now fuse into an indistinguishable mass of chromatin. The division of nucleus called karyokinesis is followed by the division of cytoplasm (called cytokinesis). In plants cytokinesis is completed with the formation of cell plate whereas in animal cell cytokinesis is accompanied with cleavage, the indented cell furrow.

Successful cell division requires proper '**biorientation**' of chromosomes whereby mictotubules bundles (K-fibers) connect sister kinetochores of each chromosome to opposite spindle poles. Biorientation errors are linked to chromosome loss and cancers. Formation of sister K-fibers occurs asynchronously and once a kinetochore captures microtubules growing from the spindle pole, the chromosome is transported to this pole and become mono-oriented. Mono-oriented chromosomes remain near the spindle pole for variable times until they suddenly 'congress' to the spindle equator. Current models of mitotic spindle formation postulate that chromosome congression occurs as a result of bi-orientation (Mc Ewen et al., 1997). Stable propagation of genetic material depends on the congression of chromosomes on the spindle equator before the cell initiates anaphase. It is generally assumed that congression requires that chromosomes are connected to opposite poles of the bipolar spindle (bioriented). In mammalian cells chromosomes can congress before becoming bi-oriented (Kapoor et al., 2006).

1.1.4.2 Stages of large scale structural changes in mitosis

There occur changes of chromosome structure during the mitotic cell cycle. There are three stages of large-scale structural changes of chromosomes in mitosis (Tatsuya, 2000). The linkage between duplicated sister DNA is established during S phase and maintained throughout the G2 phase(cohesion). Replication of chromosomal DNA during S phase produces pairs of sister chromatids. In early mitosis (prometaphase), dramatic structural changes occur to produce metaphase chromosomes, each comprising a pair of condensed sister chromatids (condensation). At anaphase onset (metaphase-anaphase transition), a signal is produced to disrupt the linkage between sister chromatids (separation) allowing them to be pulled apart to opposite poles. Although all these three events-cohesion, condensation and separation occur at different stages of the cell cycle, they must be tightly co-ordinated together. Malfunctioning of any of these processes could lead to genetic instability such as aneuploidy and chromosome breakage which are potentially contributing to tumor development.

1.1.5 Meiosis

Meiosis is a reductional division in which the chromosome number is reduced by half. The male and female gametes, sperm and egg or pollen and ova thus formed are haploid which upon fertilization forms zygote or embryonic cell in which the chromosome number is restored to original number (diploid number). For example, a plant with 2n=14 will produce pollens and ova with n=7 which upon fertilization will unite and form zygote with 2n=14. The meiotic division consists of two successive divisions-1st meiotic division and 2nd meiotic division. The first meiotic division is a reductional division by the end of which the chromosome number is reduced by half. A diploid cell after undergoing first meiotic division will give rise to two daughter cells will be in haploid condition. The second meiotic division is an equational division in which the chromosome number remains constant and each of the two daughter cells gives rise to two cells and thus meiosis results in the production of four haploid cells. In other words, sister chromatids move to opposite poles (disjoin) in an equational pattern. The second meiotic division is in fact a mitotic division. The first meiotic division consists of four stages: Prophase I, Metaphase I, Anaphase I and Telophase I. In the interphase I the nucleus appears homogenous. Chromosomes appear as long, loosely coiled threads and it is a period of chromosome duplication.

Prophase I-In prophase I chromosomes shorten, homologous chromosomes come together and pair and duplication becomes evident. The prophase I is divided into five stages, namely, leptotene, zygotene, pachytene, diplotene and diakinesis (Figures 1.3 and 1.4). The characteristic features of these stages are as follows:

Leptotene-The term 'Leptotene' has been derived from Greek word('leptos'=thin). The chromosomes appear longer, thinner(as long threads) with appearance of bead like structures called **chromomeres** along their length but in certain animals chromosomes appear polarized in which the ends of the chromosomes seem to be attached to the nuclear membrane close to the centriole and in certain plant the chromosomes are densely clumped to one side of the nucleus- a condition referred to as **synizesis** (Mc Clung, 1905). Chromomeres are structures resulting from the coiling of a continuous DNA strand. A chromomere or interchromomeric region(or a G- or R-band) comprises on the order of several to tens of loops, depending on the organism and the region of the genome(Zickler and Kleckner, 1999). The two tightly pressed sister chromatids of each leptotene chromosome are bound to a common protein core known as an axial element which appear to hold the meiotic chromosome in a looped configuration. These axial elements become the lateral elements of **synaptonemal complex** (SC) which are joined together by transverse elements and a central element. There is progressive reduction in the size of chromomeres from the centromere. Centromere stains lighter than other parts of the chromosome. Thus in this phase the chromosomes are discernibly individualized appearing thin and thread like and the total array of chromosomes appear as a dense tangle of such threads.

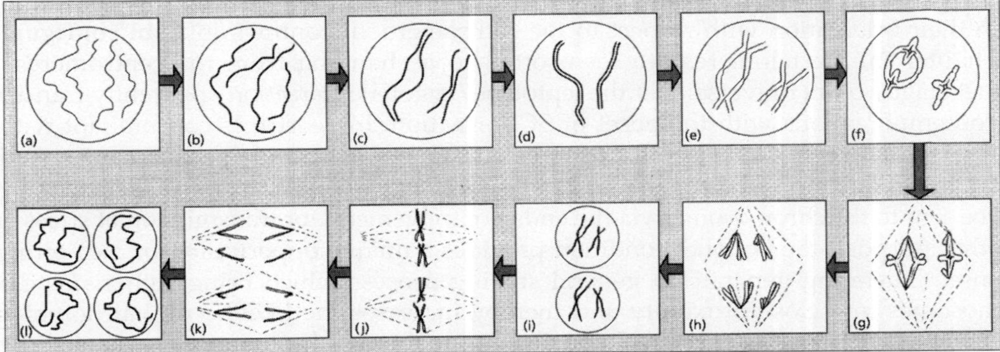

Fig. 1.3 Showing different stages of meiotic cycle, (a) G1; (b) G2; (c) Zygotene; (d) Pachytene; (e) Diplotene; (f) Diakinesis; (g) Metaphase I; (h) Anaphase I (i) Telophase I (adapted from Fincham, 1994).

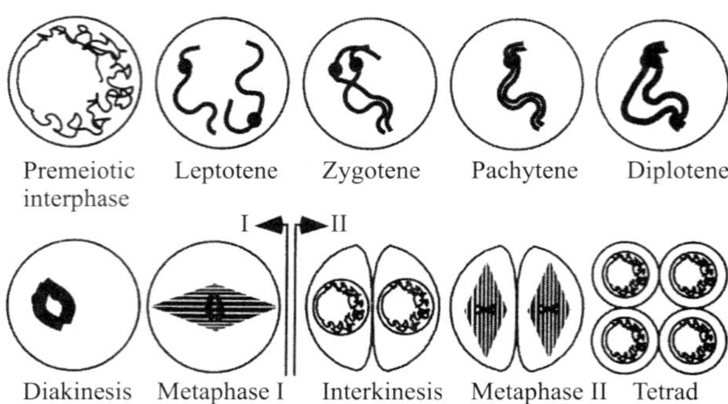

Fig. 1.4 Showing meiosis in plants right from premeiotic interphase. Microsporogenesis leadsto formation of tetragonal tetrads (in other species the four daughter cells form a tetrahedron (adapted from Dawe, 1998).

Zygotene-Zygotene has been derived from the Greek word 'zygos which means pair. It is the stage when pairing of homologous chromosomes (called synapsis) starts. Pairing is specific and is gene by gene. Pairing occurs in a zipper like fashion starting at any or even at several points along the chromosome and proceeding all homologous segments. Pairing is not always completely finished.

Homology identification-Loidle(1990) identified three possible mechanisms- premeiotic association, specific interactions at prophase and random contacts for early stages of chromosome alignment. Of which random contacts initiate synapsis and the mechanism increases the number or efficiency of random contacts is a phenomenon known as the **bouquet stage** (Figures 1.5A & B)). The bouquet stage refers to the clustering of telomeres to a small region of the nuclear envelop during zygotene. In the bouquet, all chromosome ends irrespective of arm length, are directly attached to the inner membrane of the nuclear envelope via special attachment plaques and are strongly polarized (Zickler and Kleckner, 1999). Chromosomes thus loop out from their attachment regions. Chromosomes in bouquet no longer have their centromeres but instead their telomeres facing the centrosome. Bouquet formation requires that centromeres separate while telomeres cluster and chromosomes switch their orientation with respect to the centromere. In contrast, in **Rabl configuration** (see chapter 2), the telomeres are seem often to be hanging from the centromeres. The bouquet stage occurs universally at the leptotene/zygotene transition apparently transiently. It is contemporaneous with the onset of SC formation. In yeast it is concomitant with the progression between double stranded breaks and stable strand-exchange intermediates (double Holliday junction). Bouquet is also observed in plants. When some particular signals must be sent to the chromosomes via the ends, for movement, for transmission of stress or for any other purpose, the bouquet condition provides a unique opportunity for a co-ordinated response. Clustering generates a general stirring process which brings otherwise distant chromosomes into close proximity and thereby increases the likelihood that homologous contacts will occur. Early observations suggested that chromosome synapsed before recombination was initiated but now observations in yeast suggest that early events of recombination may precede synapsis. The visual evidence of homology recognition, is a

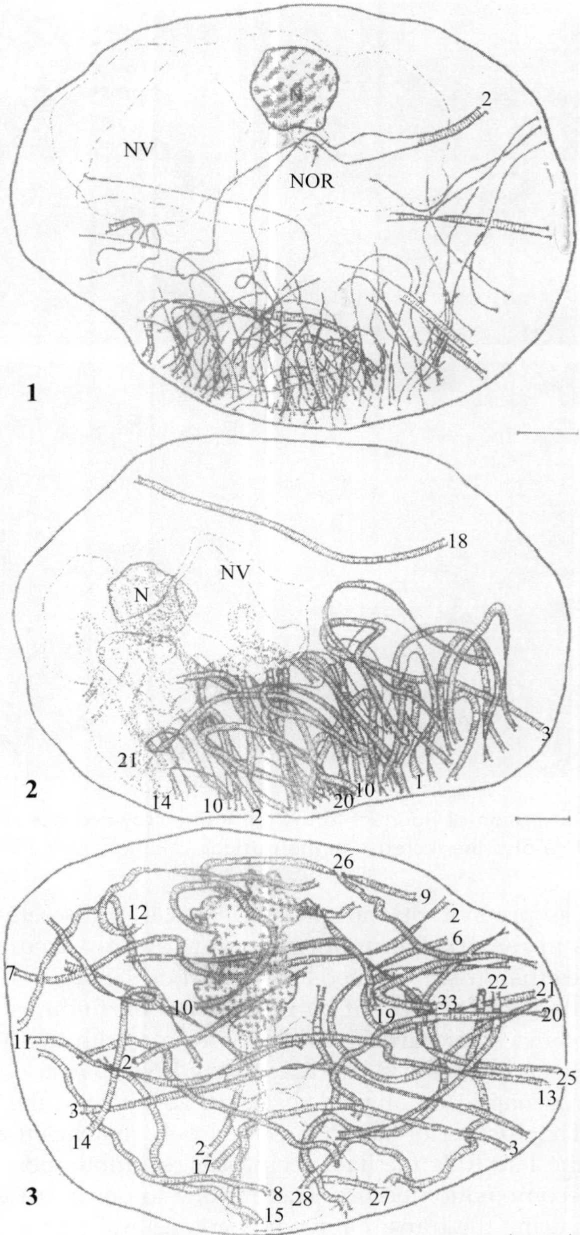

Fig. 1.5A Bouquet formation and release in Bombyx oocytes. (1) Mid-zygotene nucleus (2) Lase zygotene (3) Pachytene, release of the telomere clustering. N = nucleoulus; NV = nuclear vacuole and NOR = nuclear organizer.

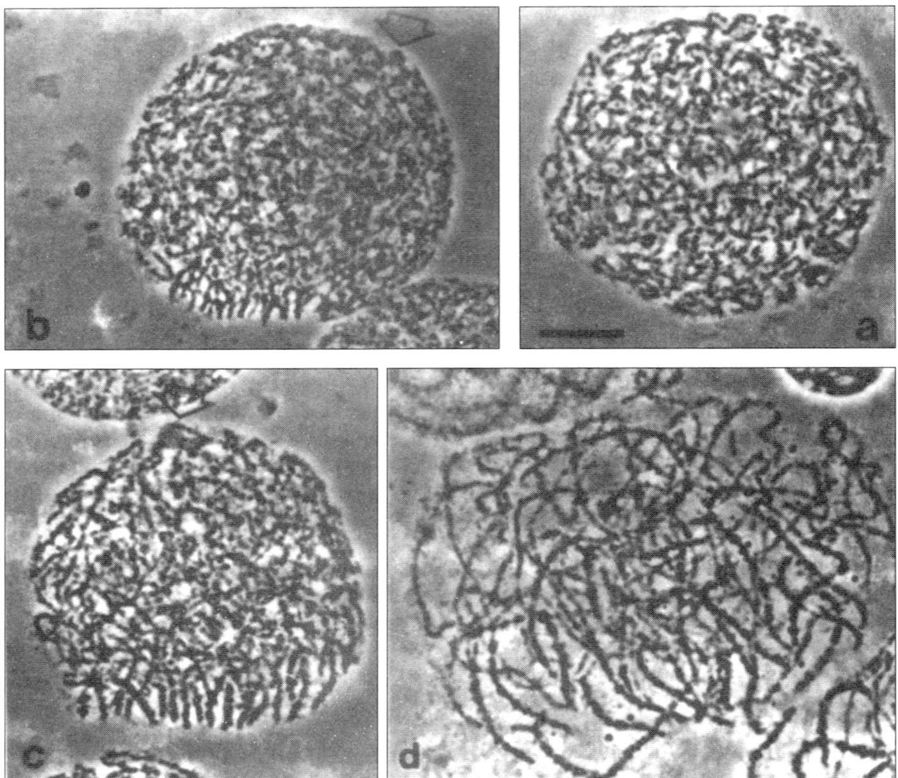

Fig. 1.5B Beautiful illustration of bouquet formation in the crowded nucleus of salamander. (*a-d*) Note also the increase in the nuclear volume (From 156).

phenomenon known as **pre-synaptic alignment**. This is a discrete phase in the pairing process which results in remarkable uniform and apparently accurate alignment of axial elements over distances that greatly exceeds the width of the synaptonemal complex (up to 2.5 μm). This function can be carried out by **recombination nodules** (RNs). Recombination nodules are small proteinaceous particles which associate with SC. There are two types of recombination nodules- early RNs and late RNs. Early RNs have contractile roles by pulling chromosomes from presynaptically aligned distances (2-3 μm) to the proximity required to form the SC(0.3 μm). The number of late RNs very closely matches the number of chiasmata which suggests that the late RNs mediate reciprocal recombination or become associated with the sites where recombination occurs. There is one-to-one correspondence between late RNs and crossovers. During the transition from zygotene and pachytene the majority of the early RNs are either degraded or dissociated from the SC. RNs contain enzymes which mediate meiotic recombination and also provide support to the idea that early events of recombination precede synapsis.

Meiosis involves two rounds of chromosome segregation following a single round of DNA replication. Meiotic cells contain two copies of most chromosomes, one from each parent. After DNA replication each chromosome comprises a pair of sister chromatids held together by cohesion complexes. Sister centromeres attach as a single unit to mictotubules from a spindle pole(Neale and Keeney, 2006). Exchange of chromosome arms between nonsister chromatids yields a chiasma. Dissolution of sister chromatid cohesion along the arms allows the homologues to separate at the first meiotic division. Chromosome must first pair with the correct partner and then become physically connected so that they orient together on the meiotic spindle. Connection is established by the exchange of homologous chromosome arms(the point of CO being called chiasma) plus cohesion between sister chromatids. Exchange uses a specialized pathway of homologous recombination that repairs DNA double stranded breaks. Basically the cell damages its own DNA and then uses the repair system to lock homologous chromosomes together. A central step in recombination involves proteins related to bacterial Rec A that catalyzes the pairing and exchange of DNA strands between single stranded DNA formed at the break and intact, homologous double stranded DNA. Most eukaryotes have two Rec A homologues, the ubiquitous Rad 51 and its meiosis specific counterpart Dmc1(disrupted meiotic cDNA).

Pachytene-Pachytene has been derived from Greek word (Pachus=thick). Pairing of homologous chromosomes is completed and thus bivalents are formed. There is progressive shortening and coiling of chromosomes. The synapsed homologous chromosomes are shorter and thicker and tightly associated. At this stage each chromosome consists of two sister chromatids and each pachytene bivalent (the pair of synapsed chromosomes or a pair of connected homologues) consists of four sister chromatids forming a tetrad chromosome. Electron microscope reveals a protein ribbon called synaptonemal complex (Moses, 1956, 1968; Fawcett, 1956; Westergaard and von Wettstein, 1972) running between each pair of homologous chromosomes. Once homologs are paired, the chromosomes are connected by a specific structure- the synaptonemal complex. The synaptonemal complex consists of a central element and two lateral elements. The lateral elements are in direct contact with the chromosomes. In leptotene each chromosome possesses a single lateral element and in the zygotene the lateral elements of homologues pair up and a central element forms between them to complete the synaptonemal complex (Figure 1.6). These lateral elements bind to silver stain, making it clearly visible under electron microscope. The chromosomes are completely free of the synaptonemal complex by metaphase.

SC normally forms only between homologous chromosomes at pachytene but intimate pairing and SC formation can occur between non-homologous chromosomes. There is no causal relationship between the SC and genetic recombination. SC may be required as a scaffold to house and/or stabilize recombination nodules during the final stages of recombination. Non-homologous pairing can occur during zygotene as well as pachytene and homology is probably not a pre-requisite for SC formation at any stage of meiosis(Dawe, 1998).Pachytene chromosomes are much longer than mitotic prophase chromosomes and have been used in several species to construct rough cytological maps.

Diplotene-Diplotene has been derived from Greek word 'diplos'(=double). Chromosomes further contract and thicken. Separation occurs between homologous chromosomes due to repulsion but chiasmata (Greek singular; chiasma = beam), the visible evidence of physical crossing over becomes apparent. Chiasma refers to a point where homologous chromatids have broken and rejoined cross-wise. Each chiasma involves just one chromatid from each

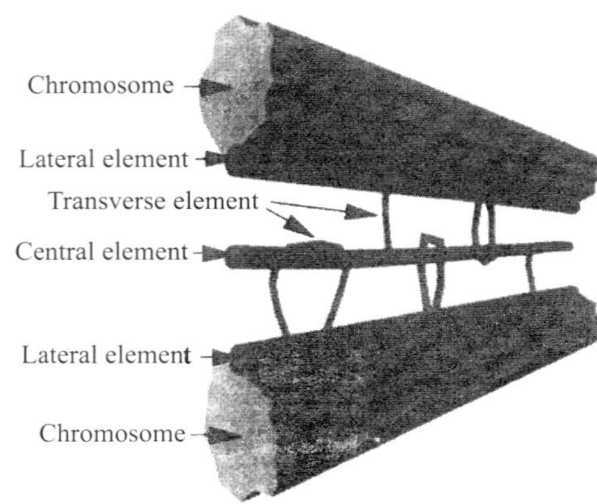

Fig. 1.6 Showing synaptonemal complex in maize, obtained by high voltage electron microscopy followed by Computerized axial tomography (J. Fung et al. unpublished data).

chromosome. As the diplotene progresses the chiasmata move away from the centromere and decreases in number and this process is called **chiasma terminalization**. The terminalization process may be complete, partial or altogether absent. The number of chiasmata depends on the species and the length of the chromosomes. The chromosomes are thus seen split into chromatids except at the centromeres.

Chiasmata which are the result of recombination hold the chromosomes together and two sister kinetochores are observed as a single unit in normal cells. Chiasmata often occur preferentially in subterminal regions in plant and animal chromosomes. If chiasmata do not form (as in case of dy mutant in maize) the homologous chromosomes can segregate in either direction and may arrive at the same pole. If the sister kinetochores are separated (as in case of *pc* mutant in tomato) they can interact with different spindle poles and randomly segregate.A chiasma provides a connection between the homologs that restrains pole ward movement and causes tension at the kinetochores. In the absence of a chiasma the resulting univalent kinetochores do not cause tension and produce a signal which causes the cell to delay anaphase. Chiasma also ensures proper meiotic spindle assembly. In synaptic mutations where recombinations are disrupted, a general failure in bipolar spindle formation is observed (Dawe, 1998)

Diakinesis-Diakinesis is derived from the Greek word, kinesis(=movement). Chiasmata terminalization is complete and thus separation of homologous chromosomes is complete. The bivalents migrate and become evenly distributed. The chromosomes contract lengthwise by a spiraling process and by prometaphase (immediately before metaphase) they are thickened and highly condensed.

Role of meiotic kinetochore-Meiotic kinetochores regulate chromosome movement in plant meiosis and its structure predicts its behaviour. When the two sister kinetochores, comprise a single structure in meiosis I, they usually segregate together and when the kinetochores are visibly separated from each other, they usually disjoin. Ab 10 in maize provides unique opportunity to study meiotic kinetochore, the integrated DNA/protein complex which forms an active centromere (see chapter 2). Mitotic kinetochores of higher eukaryotes are only accessible by immunocytological approaches.

Metaphase I-It is one of the most ideal stages for counting the number of chromosomes in an individual. The bivalents(chromosome pairs) move from haphazard arrangements to an orderly position on the equator of the spindle and held in equilibrium by the equal and opposite force of its two centromeres to the spindle poles. The spindle is formed in prometaphase-metaphase. Chromosome pairing is assessed by squashing meiocytes at metaphase I and classifying the specific chromosome configuration at this stage.

Anaphase I-The bivalents separate with one of each kind of chromosomes moving to opposite poles of spindle. In other words, chiasmata are released and sister chromatids segregate to the same pole. The distribution of members of any given pair is random. **Telophase I**-The bivalents are separated into two groups of n double chromosomes or dyads and cell wall is generally formed. The unique feature of meiosis I is that homologous chromosomes, each containing two sister chromatids, segregate away from each other.

An interphase II stage may or may not occur and the second meiotic division starts quickly after the telophase I. **Metaphase II-** Chromosomes in each cell again line up on the equator of spindle. **Anaphase II**- Each centromere splits and sister chromatids (in comparison to mitosis the two sister chromatids are not necessarily identical) separate and move to opposite poles. **Telophase II**- There are four groups of n undivided chromosomes and cell walls are formed. Each of the four cells has half the number of the original number of chromosomes with one of each kind.

1.1.5.1 Interphase

Interphase refers to that part of the cell cycle(also called resting stage) during which metabolism and synthesis occur without visible evidence of division. In other words, when the nucleus of a cell is not visibly involved in division (mitosis, meiosis), it is referred to as an 'interphase nucleus' or resting nucleus. During the interphase of the cell cycle, chromosomes can not be observed and the mass of DNA protein is collectively known as chromatin. Chromosomes are visible only during particular phases of the cell division cycle in which they are tightly condensed. Interphase is the stage in which a cell prepares itself for cell division (mitosis or meiosis). The interphase can be divided into three phases, namely, **G1**, **S** and **G2.** G is the gap or lag phase between the divisions, i.e. which separates the replication of the DNA (S phase) and the segregation of the chromosomes (**M** phase, mitosis). **GO** is the resting or quiescent phase of the cell cycle.G1, the first gap phase, is the period of interphase before DNA replicated and in this phase RNA and protein synthesis occur; S is the period during which DNA synthesis occurs i.e, when DNA is replicated and thus doubles the amount of DNA in the cell whereas G2, the second gap phase, is the period after DNA replication and before the start of cell division and in this phase although no DNA synthesis takes place but RNA and protein synthesis continue. In other words, G1 is the phase of the cell cycle which occurs after mitosis (M phase) and before initiation of DNA synthesis (S phase). Cells in G2 are discriminated from G1 cells by possessing a double DNA content. During the interphase between meiosis I and II (called interkinesis) there is no DNA replication. The pre-meiotic DNA replication occurs in the leptotene stage of meiotic prophase. Successful completion of G1 which leads to the longest part of the cell cycle is required to progression to S phase. G2 is the phase of cell cycle immediately following DNA synthesis (S phase) and in which cells are tetraploid. This phase lasts for several hours and is followed by mitosis(M phase). Thus, G1- S – G2- M (mitosis)-G1 constitute a eukaryotic **cell**

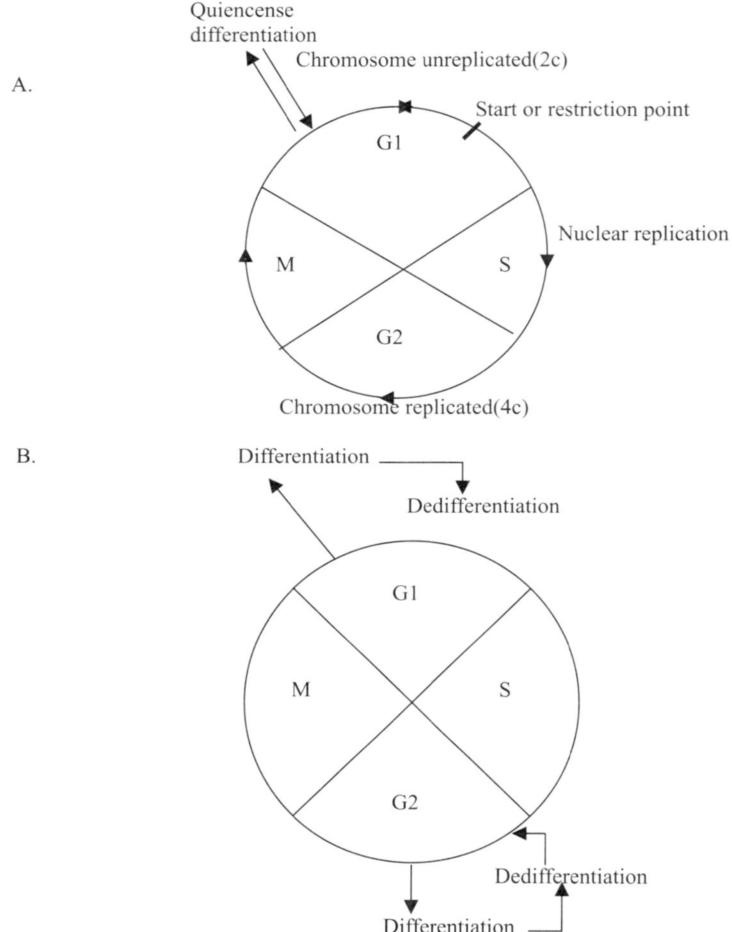

Fig. 1.7 Showing cell cycle in yeast and animals(A) and plants(B).G1/S transition is regulated by the check point called start or restriction point.

cycle as shown in the figure 1.7. Meiosis can be distinguished from mitotic cell cycle beginning with the initiation of premeiotic DNA synthesis. The cell cycle refers to the precisely timed sequence of events occurring during mitotic divisions. Cell division thus occurs in four well defined stages. These different phases can be identified by measuring the DNA content of the cell and by counting the number of chromatids present. In terms of the DNA content the amount of DNA in the cell of G1 stage is 2C where C is the amount of DNA in the haploid gamete. The amount of DNA in the cell of G2 stage will be 4C and those of S phase a DNA content intermediate between 2C and 4C. In terms of the number of chromatids one chromosome in the G1 stage will be made up of a single chromatid (made up from a single double helix DNA molecule) whereas in G2 stage each chromosome will be made up of two sister chromatids (two molecules of double helix DNA molecules). Cells that are noncycling and either quiescent (G0, G1Q) or differentiating (G1D) can be distinguished from G1 cells active in the cell cycle through biparametric analysis of DNA and RNA content. The duration of various stages particularly G1 and G2 are highly variable depending upon the cell types and species. In cultured animal cell the entire process takes about 24 hours. After passing through M and into G1 a cell either continues through another cell division or ceases to divide and enters a quiescent phase(GO) that may last for hours, days or life time of the cell. These terminally differentiated cells withdraw from cell cycles. In plants (less in animal cell) wounding or some other treatment will often cause differentiated cells to dedifferentiate and thus they become meristematic and they re-enter the cell cycle. Where they enter the cycle will depend on where they left it. One of the daughter cells produced by mitosis may not continue in the cell cycle but may enlarge and differentiate. If this occurs before DNA replication then the differentiated cells will have the normal diploid chromosome number and the amount of chromatin but in plants it is not unusual for differentiated cells to occur after DNA replication then the differentiated cells will have more than the diploid quantity of chromatin. Some times chromosomes are further duplicated without cell division and so the differentiated cells are polyploids. Mitosis in which chromosome condensation and segregation takes place within a short period (often about one hour) of the total cell cycle. Most DNA synthesis occurs during interphase(S) over a period of several hours. The CT, the duration of mitotic cycle, varies from 6.0 or 7.8 (*Helianthus annuus*) to 10.5 (*Zea mays*) to 17 to 23 hrs (*Allium cepa*) to 24 hrs (*Vicia faba* or *T. aestivum*) to 32.6 hrs (*Scilla campanulata*). The duration of meiosis ranges from 24 hrs (*T. aestivum*) to 72.0(*Vicia faba*) to 96.0 (*Allium cepa*). In meiosis the meiosis I takes more time in comparison to meiosis II (the mitosis). Thus we see that there is a great variation among different cells in the length of time a cell stays in any phase. The duration of various interphase stages and mitosis or meiosis depends on the genetic factor(the DNA content of the cell) as well as the environmental factors (temperature, etc) and the following generalization can be made (Hof, Vant, 1974).

1. The duration of G1, S, G2, M and CT decreases with the increase in the temperature.
2. The CT(duration of mitosis) increases approximately 0.34 hr while S increases about 0.17 hr per picogram of DNA.
3. Meiosis lengthens with increased DNA content of haploid nuclei.
4. When polyploids are compared with diploids, plants of unrelated species diploids have the longer duration

DNA synthesis occurs in all the stages (G1, S and G2) of interphase in cell organelles.

In both plants and animals, cell size increases linearly with DNA content per nucleus and cell cycle time also increases. The length of meiotic prophase is shorter in polyploids in comparison with their diploid progenitors which is the opposite from what could be expected. The more genomes or chromosomes are present, the shorter the meiotic prophase and hence the time required to sort them. This is because of the reason that the centromere association is occurring prior to telomere association in these species. However, diploid progenitors do not associate their centromeres until meiotic prophase. The centromeres in hexaploid wheat lines lacking the *Ph1* locus associate premeiotically 3 days prior to meiotic prophase. Homologous chromosomes colocalize along their length. However, the telomers of the homologues do not associate as pairs. The meiocytes progress through S phase. The telomeres cluster to form a bouquet. The homologous chromosomes separate along their length and are associated by the centromeres and telomeres. Maize chromosomes do not align along their length during premeiotic interphase. During the leptotene-zygotene transition of meiotic prophase telomeres cluster to form a bouquet and then heterochromatin knobs associate. The homologues intimately associate along their length with the telomeres still clustered. So if centromeres in maize associate premeiotically then chromosome behavior in maize and the timing of the telomere bouquet would resemble that for wheat lines lacking *Ph1* gene. Pairs of homologous chromosomes can be fluorescently labeled in hexaploid wheat and their behavior followed from early floral (anther development) through meiotic prophase. Association of chromosomes implies some conservation of chromosomal structure(including genes) and suggests a high degree of conservation of gene order on the chromosomes of wheat and its wild relatives.

1.1.5.2 Regulation of cell division

In animal cells, growth and synthesis of components required for these phases are regulated by extracellular growth factors and occur mainly in two gap phases, G1 (between M and S) and G2 (between S and M). The major regulatory points in the cell cycle operate at the G1/S and G2/M boundaries which correspond to points of the potential arrest as a consequence of evaluation of external conditions (Dewitte and Murray, 2003). The ordered progression of the cell cycle phases is orchestrated by cyclin-dependent kinases (CDKs) whose activity is controlled by phosphorylation and by association with specific regulatory subunits, the cyclins (Koepp et al, 1999; Murray, 2004; Kastan and Bartek, 2004), the two universal components of the control mechanism. The timing of cell cycle is regulated by a family of protein kinases that act at specific periods of the cycle. The activity of kinase changes in response to cellular signals. Activation of a protein kinase changes the activity of its target proteins by adding a phosphoryl group to a Ser, Thr, or Tyr residue and thus by phosphorylating specific proteins at precisely timed interval these protein kinases co-ordinate the cell division. Phosphorylation of proteins leads to major events of M-phase such as chromosome condensation, cytoskeletal reorganization, nuclear envelope breakdown and cell shape changes. Protein kinases are heterodimers with a regulatory unit, cyclin and a catalytic subunit, cyclin dependent protein kinase (CDK). CDK activity is regulated at multiple levels. CDKs are active when associated with cyclins and regulation of CDK is by phosphorylation. Its activity can, therefore be regulated by the abundance of CDK or cyclin components through controlled transcription, translation, intracellular localization or regulated destruction or by the regulated activity of CDK-activity kinase(CAK). In addition, activity can be inhibited by phosphorylation by WEE1 kinases or the binding of inhibitor proteins [Kip-related proteins (KRP)]. Inhibitors may block the assembly of CDK/cycline complexes or inhibit the kinase activity of assembled dimmers. CDK subunits(CKS) proteins scaffold interactions with target substrates. Levels of cyclin dependent protein kinases vary and thus there is variation in the activities of specific CDKs during cell cycle. There are various forms of cyclin and cyclin is degraded as cells exit M-phase. The activity of Cyclin E-CDK2 is at its peak near G1-S boundary. The amount of cyclin A-CDK2 rises during S phase and G2 phase whereas the quantity of cyclin-CDK1 is at its maximum in M phase. The activity of a cyclin-CDK complex changes during the cell cycle through differential synthesis of CDKs, specific degradation of cyclin, phosphorylation and dephosphrylation of critical residues in CDKs and binding of inhibitory proteins to specific cyclin-CDKs. For high throughput loss of functional analysis of eukaryotic cell-cycle see Roy (2009).

1.1.5.3 Spindle assembly check point

The spindle assembly check points guard the fidelity of chromosome segregation. The cell division is regulated by the spindle assembly check point. Before a cell divides the sister chromatids from all of its chromosomes have to attach to two poles of the mitotic spindle. As long as chromosomes are present, they do not obey this rule and spindle assembly check point is active. This cellular surveillance system inhibits the continuation of cell division and therefore prevents an abnormal number of sister chromatids reaching each daughter cell (Peters, 2007). It requires close co-operation of cell cycle regulatory proteins and cytoskeletal elements to sense spindle integrity. The spindle check point delays cell cycle progression until microtubules attach each pair of sister chromosomes to opposite poles of the mitotic spindle. Following sister chromatid separation, however, check point ignores chromosomes whose kinetochores are attached to only one spindle pole, a condition that activates the check point prior to metaphase. At the molecular level check point proteins Mad2 and BubR1 associate with Cdc20 and inhibit its activity to activate the ubiquitin ligase APC/C. Anaphase initiation is regulated by antagonistic ubiquitination and deubiquitination activities. Ubiquitination by anaphase-promoting complex (APC)drives spindle check point inactivation.

Cell division in plants is controlled by the activity of cyclin-dependent kinase complexes (Dewitte and Murray, 2003). Although this basic mechanism is conserved with all other eukaryotes, plants show a novel feature of cell-cycle control in the molecules involved and their regulation including novel CDKs showing strong transcriptional regulation in meiosis. Plant development is characterized by indeterminate growth and reiteration of organogenesis and is therefore intimately associated with cell division. Plants have a large number of cell cycle regulators that appear to have overlapping and distinct functions. In the proposed model D-type cyclins are primary mediators of the G1/S transition and hence have a major responsibility for stimulating the mitosis. Transcription of the D-type cyclins is activated by extracellular signals and leads to the formation of active CDKA-CYCD complexes. These phosphorylate and hence inactivate the Rb proteins so that it looses its E2F association and hence can no longer block the activation of E2F-regulated genes. E2F is then able to activate transcription of genes involved in S phase and other growth and cell cycle processes. The G2/M transition and mitosis is controlled by the level of CDK-CYCB kinase activity and initiation of CDKs arrests cells in G2 and stabilizes the proprophase band. During G2 levels of transcripts encoding CYCB increase and probably associate with both CDKA and CDKB. Further, regulation of CDK activity by both KRP proteins and inhibitory phosphorylation by WEE1 kinases is likely. Phosphatases therefore probably play an important role in M-phase onset, although a direct homologoues of CDC25 phosphatases that activate CDKs in animals and yeast has not been found. However, as inhibition of serine/threonine-specific protein phosphatases in alfalfa by endothall causes premature, imperfect microtubular reorganizations and activation of CDKB2.1, these phosphatases might be involved in regulating M-phase progression. The protein kinases CK2 shows discrete activity peaks at G1/S and M in tobacco BY-2 cells, blocking its activity during G1 abolished the G2/M check point, resulting in premature entry into prophase and showing links between G1 processes and G2 controls. There is dual control of mitotic exit(Shirayama et al. 1999). The genetic material is duplicated in S phase. Then after a rest phase(G2), it is segregated equally between two daughter cells(mitosis). Cells finally enter another resting phase G1 before the cycle starts over again. Transition to G1 is known as exit from mitosis and is regulated by a protein called Cdc14 which acts as phosphatese, i.e. it dephosphyrates specific protein targets. Another protein Cdc20 is also known to promote exit from mitosis. Key cell cycle transitions-be it entry into S phase or progression through mitosis are mediated by specialized protein degradation systems. These consists of the socalled ubiquitination machinery which attaches a peptide (ubiquitin) onto the protein to be degraded and a protease, the 26S proteasome that degrades this ubiquitin-tagged proteins. Thus mitosis is regulated by the anaphase promoting complex(APC) ubiquitination machinery along with a family of APC activators, Cdc20 and Cdha/Hct1(Prinz and Amon, 1999). The exit from mitosis is ultimately triggered by inactivation of mitotic cyclin-dependent protein kinases(CDKs) which consist of a catalytic protein kinase subunit and a regulatory cyclin subunit.

Cytokinins and auxins are indispensable for maintaining undifferentiated cells in proliferation during in vitro cell culture and are hormones most directly linked to cell proliferation. Because most plant hormones also provoke morphogenetic effects, the cell cycle consequences may be direct or part of the morphogenetic response. Cytokinins have effects on the G1/S and G2/M transitions as well on progression through S phase (Jacqmard et al. 1994). Cytokinins and brassinosteroids induce the expression of CYCD3;1 and the overexpression of CYCD3;1 conferred cytokinin autotrophic initiation and growth of Arabidopsis leaf calli showing that in this system high levels of CYCD3;1 are sufficient to replace exogenous cytokinins.

One must look for clear phenotypes including arrest in G1, G2/M and S, cytokinesis, and DNA replication defects and apoptosis and/or cell death when targeting regulators of these processes.

1.1.5.4 Importance of Meiotic Process

The meiotic process provides the physical explanation for Mendel's laws and for their most important exception, the occurrence of crossing over. According to Mendel's first law, maternal and paternal versions of any given single gene assort randomly. Each of the gametes from a round of meiosis contains either paternal or maternal version of a given gene with two types represented in equal numbers within a total pool and zygotes are formed by two gametes, each drawn randomly from such a pool. Mendels' second law states that the alleles for two different genes determining two different traits segregate independently of one another. This finding can be explained in a simple way if the genes for the two traits lie on different chromosomes. Each bivalent aligns on the meiotic I spindle independently of all other bivalent and in most cases without respect to the parental origin of the component chromosomes. However, if the two genes for the two traits lie on the same chromosome, the segregation pattern observed depends on the frequency of crossing over between the two corresponding loci. In the absence of any crossing over (tight linkage) the maternal and paternal alleles of two traits present on the same chromosome will never segregate from one another, the opposite of Mendel's law. In fact, Mendelian segregation will be observed only if there is a very high crossing over frequency (in case of no linkage) between the two loci and in that case the four possible combinations of the pair of alleles (two parental and two recombinants) will occur with equal frequency. In case of intermediate crossing over (a case of partial linkage) the parental frequency will not be equal to recombinant frequency. This exception to Mendel's second law led to the conclusion that the genetic traits are organized in a linear array(i.e., genes determining different traits are organized in a linear array) and there exists the genetic recombination.

1.2 PROKARYOTIC AND UNICELLULAR EUKARYOTIC CELLS

1.2.1 Bacteria

Bacteria are single celled organism. They vary in shape(round/oval/spherical, rod/ cylindrical, ellipsoidal and spirals) and sizes (1 to 5 u). Strains of bacteria which lack the ability to produce certain proteins/enzymes are called auxotrophs whereas those that can normally produce theses substances are called prototrophs. They are haploid and have a single main chromosome with a few thousands genes plus one to several **episomes** and plasmids. Bacterial chromosome is enclosed in a structure but has no surrounding membrane called the **nucleoid**. Plasmids are highly variable in size and carry 3 to several hundred genes. Plasmids are found in almost all bacterial strains. In many cases they are indispensable and impart resistance to certain environmental conditions (presence of different antibiotics). They are capable of independent replication. The three major types of plasmids that have been extensively studied are : 1. F and F′ plasmids, the sex or fertility factors 2. R′ plasmids(RTF, resistance transfer factor) 3. Col plasmids. All F′ and F plasmids, many R plasmids and some Col plasmids are **transmissible plasmids** and they mediate transfer of DNA by conjugation (see Roy, 2009). With respect to F-plasmid there are four types of bacterial cells. 1. **F^+** cell 2. **F^-** cell 3. **Hfr** cell 4. F′ cell. F-plasmid(sex factor, fertility factor, F-factor) is a large E.coli plasmid(a close loop molecule of DNA) whose presence determines the sex(maleness) of the bacterium. A bacterium cell with an F-plasmid is called F^+ cell and it is a donor or male. A bacterium cell which lacks an F-plasmid is called an F^- cell and it is a recipient cell or female. F-plasmid is a transmissible plasmid and contains *tra* genes and can transfer parts of the bacterial chromosome from donor (F^+ male) to the recipient (F^- female). F consists of four functional blocks and a region of unknown composition. The *tra* operon consisting of 33kb encodes the essential components of donor ability, the mating apparatus and the DNA manipulation enzymes responsible for transmitting F into recipient cells along with the origin of transfer (*ori*T). The ~13kb leading region which follows *ori*T is the first to enter recipient cell. The region at the end of the *tra* operon contains a defunct replicon and seems to have acted as a trap for transposable elements which contribute indirectly to mating as one of them is inserted in the fineO regulator gene causing constitutive expression of the *tra* operon and higher transfer rates and also because homologous recombination with similar elements in the chromosome is the major route of F integration to form Hfr donors. The fourth region contains all the significant determinants of F maintenance. The *rep* and *sop* loci together constitute the basic maintenance system. F plasmid contains three replicative determinants, namely *rep*FIA, *rep*FIB and *rep*FIC but only repFIA is fully functional. It is because this that F is stably maintained in only a narrow range of hosts, E.coli and similar Enterobacteria (Lane, 1999). The *sop* region consists of three elements *sop*A, *sop*B and *sop*C, all essential for partition. F plasmid is a closed double stranded loop DNA molecule, about 100kb in size that replicates autonomously in the bacterial cell and is occasionally integrated into bacterial chromosome by crossing–over. Its properties such as stability and low copy number make it suitable as vector for cloning large fragments of eukaryotic DNA. The integrated plasmid DNA undergoes occasional excision, usually precisely but sometimes carries with it an adjacent segment of the bacterial DNA, the nature of which depends on where the plasmid has been inserted. This modified F is termed F′ and it is transmitted from cell to cell like F itself. When an F-plasmid from an F^+ cell is transferred to an F^- cell with high frequency, the F^- cell then becomes an F^+ cell. Hfr cell refers to any strain of E.coli containing the F-factor integrated into its chromosome (Demerec et al., 1966) and as a consequence the strain is able to undergo recombination at a high frequency. The F-factor determines a mechanism for its own transfer from cell to cell. This mechanism includes the synthesis on the surface of the F^+ cell of a special filament, a pilus which can attach to another cell and an enzyme complex which replicates the F^- DNA. So the F-factor in an ordinary F^+ strain is highly infectious by cell contact but then it is hardly transferred to F^- cells in an Hfr strain. An F′ cell is a bacterium which carries an F-plasmid attached to a fragment of bacterial genome. F′-lac plasmid carrying segments of DNA from the lac operon of the bacterial chromosome elucidated the complementation relationships in E.coli.

R plasmids carry genes conferring resistance to one or a number of drugs, for example, resistance to chloramphenicol, ampicillin and tetracycline, for example, RP4 plasmid of *Pseudomonas.* Col plasmids synthesize chemical, colicin which is toxic to bacteria. For example, Col 1 of E. coli. Plasmids of Rhizobium and Agrobacterium carry the genes responsible for nitrogen fixation and induce crown gall formation in plants. Bacteria divide asexually by simple binary fission which leads to equal division of their genetic material (chromosome and plasmids if any) into two progeny and thus a single bacterium gives rise to a clone of descendents which are genetically identical with their ancestor. Bacteria do not under go mitosis or meiosis, however, recombination occurs through transformation, transduction and conjugation.

1.2.2 Yeast

Most fungi including the filamentous Ascomycetes are haploid with meiosis following immediately after karyogamy. Yeast is predominantly a diploid (2n =4 or 8) unicellular eukaryote. The *Saccharomyces cerevisiae,* the baker's yeast consists of diploid oval shaped cells. It reproduces vegetatively as well as through sexual means. It is able to propagate itself vegetatively in either the haploid or the diploid phase. The vegetative reproduction is through '**budding**'. A small cytoplasmic bud extrudes from the cell wall and there is then division of nucleus and the bud receives one of the two daughter nuclei and it eventually gets separated and exists as a distinct cell. Under unfavourable conditions the cell wall thickens and the diploid cell forms an ascus (plural : asci) and this diploid cell undergoes meiosis and forms four spherical haploid products called ascospores. Two haploid ascospores under favourable conditions form a diploid zygote. Haploid cells are not distinguishably male and female but some yeasts exist in two self-incompatible but mutually compatible mating types (+ and –) and so conjugation between only opposite mating types of ascospores will produce diploid zygote. So long as the different mating types are kept separate the haploid condition can be maintained. Occasionally the nucleus of a haploid ascospore can duplicate itself without undergoing cytoplasmic division and thus forms a diploid cell (**direct diploidization**). Diploid cultures are stable so long as they are well nourished but starvation induces meiosis. There is another species of yeast, *S. pombe* which is being used in experiments. *S. pombe* is characterized by its relatively large chromosomes. The advantage with *S. cerevisiae* is that its mitochondrial genome is dispensible and thus is genetically more tractable.

1.2.3 Viruses

The structure of virus consists of genetic material (DNA or RNA) in the form of a single haploid chromosome which is surrounded by a protein envelope called capsid. The bacteriophages or phages are viruses that specifically infect bacteria. The virus ranges in shape from round (mumps) to rod-like/filamentous (tobacco mosaic virus, M13) to polyhedral with attached tail/Head- and tail ('T-even' bacteriophages) (Figure 1.8) and the size ranges from the vaccinia virus, as large as a small bacterium, to the minute Japanese B encephalitis no longer than some protein-1000 A (mumps virus) to 1μ (T2, T4, TMV). The infection in case of bacteriophage is a three step process : 1. The phage particle attaches to the bacterium and injects its DNA into the cell 2. The phase DNA replicates inside the bacterium with enzymes coded by genes (ori) of the phage 3. The phage's other genes direct synthesis of protein components required for the capsid formation and thus new phage particles are assembled and released by lysis of the cell. The above mentioned three processes constitute the infection cycle. In case of some phage types the lysis of bacterial cell occurs at some 20 to 25 minutes (latent period or incubation period) after initial infection. The bacterial cell wall dissolves and some 150-300 new viral particles are released. In some resistant strains of bacteria although the phage chromosome is inserted into bacterial chromosome but it is in an inactive form (dormant state) and thus retained in the host bacterium possibly for many thousands of cell division. Such viral chromosome is called a **prophage** and the bacteriophages that can be carried in such a passive fashion are termed as **temperate phage** and the bacterium is referred to as a lysogen (lysogenic bacterium). The **virulent** phages of

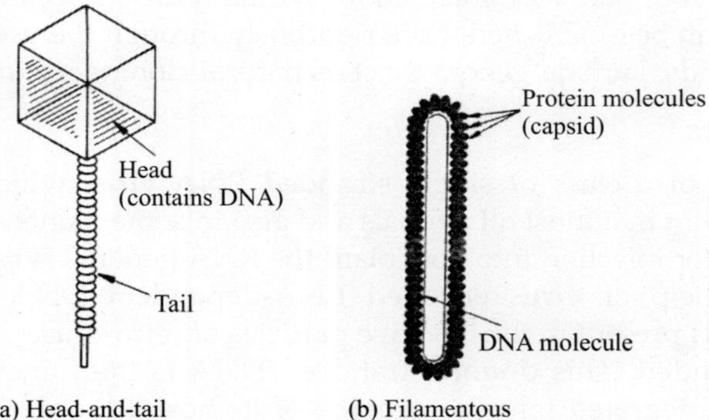

Fig. 1.8 Showing structure of two types of phage (a). Head- and –tail (eg. λ) and (b). Filamentous (eg. M13).

viruses are the ones which cause lysis of the cell upon infection and the phage can be said to be in vegetative or lytic state.

1.2.3.1 Bacteriophage lambda

The temperate phage, bacteriophage lambda(λ) can persist in the host cell in a latent state but certain damaging treatments notable UV irradiation will trigger its lytic proliferation. The studies of the cross, Hfr x F^-(λ) and reciprocal cross, Hfr(λ) x F^- and mapping studies by P1 transduction have shown that λ genome is integrated as a prophage at a specific chromosomal locus between markers *gal* and *bio*(nutritional requirement for the vitamin biotin) in its latent state. It is restrained from excising from the chromosome and multiplying by its own repressor protein. When this lysogenic bacterium is UV irradiated the repressor protein gets destroyed and the λ phage starts its multiplication and causes lysis of the host cell. Sometimes the prophage is excised from the chromosome imprecisely and thus carries with it chromosomal material from one or the other side of its site of integration. Thus they act like specialized transducing agents and specially they are able to transduce the markers *gal* or *bio*.

1.2.3.2 Bacteriophage T4

The *E.coli* bacteriophage T4 is exclusively virulent with no lysogenic option. In the T4 the recombination takes place in a T4 cross, i.e. when *E.coli* cells are infected simultaneously with two different T4 mutants. On lysis the two original mutant types along with wild type and double mutant recombinants are isolated. Recombination is thus brought about by bringing together of haploid genomes through mixed infection, followed by recombination and then segregation of pure recombinant haploids. The different mutant derivatives of T4 differ from the wild-type in the *E.coli* strains that they are able to lyse and/or in the rapidity of lysis, expressed in different sizes of circular clearings(plaques). Mutations can be mapped to some extent on the basis of recombination frequencies. Frequencies of recombination vary widely the maximum being in the region of 40%. The problem with interpretation of recombination frequencies is that they are not of single controlled crosses between pairs of virus genomes but of successive random pair-wise interactions within a mixed population over a period of time. So map order can be established more rigorously through the use of deletion mapping. Further like F′ plasmids, lambda genomes act as natural cloning vectors.

1.2.3.3 Retroviruses

It refers to any one of a class of single stranded RNA virus which infects eukaryotes. Retroviruses are present in almost all animals and also in some plants and they contain RNA rather than DNA. After infection into host plant the RNA genome is reverse transcribed into a cDNA with the help of virus encoded RNA-dependent DNA polymerase (reverse transcriptase) which is present in the infective particles of retroviruses. The reverse transcript is made double stranded. This double stranded cDNA is then uncircularized in the cell nucleus and finally integrated into the genome of its host at border sequences containing long terminal repeats (LTRs) via a mechanism similar to transposition. The integrated viral genome (endogenous retrovirus) is called as provirus. The LTRs contain enhancer and promoter elements and flank the coding region that harbors the *gag*-gene encoding the viral-capsid proteins, the *pol* gene encoding reverse transcriptase and the *env* gene encoding the glycoprotein of the viral lipid coat. The termini of the viral genome carry characteristic R segments, direct repeat sequences of about 10-80 nucleotides in lengths which are flanked by specific regions (U5 at the 5' end, about 80-00 nucleotides long, U3 at the 3'end, about 170 to more than 1000 nucleotides long). Some retroviruses have been engineered to function as vector for cloning of mammalian genes. Many of the eukaryotic transposons are related to retroviruses and their mechanism of transposition includes an RNA intermediate.

Mode of expression of the provirus The entire length of the provirus sequence is transcribed as a single unit which consists of three functionally distinct regions named *gag*(group-specific antigen), *pol*(polymerase, i.e., reverse transcriptase) and *env*(virus envelope protein) respectively according to the types of protein they encode. A portion of this long transcript is processed by splicing-out of the *gag/pol* regions and thus leaving only the *env* domain. The unprocessed fraction contains all three regions. Such types of processing indicate that these different messengers are required in different proportions for the construction of the infective virus particles. The polypeptide chain resulted from the *env* messenger transcript is finally cleaved into two products, SU(surface protein) and TM(transmembrane protein, for penetrating the host cell membrane). The full length messenger transcript (*gag-pol* regions) is translated into a protein called *gag-pol* polyprotein which is ultimately cut into a number of functionally distinct polypeptides. The boundary between the *gag* and *pol* parts of the mRNA is marked in different retroviruses by either a chain termination codon or a reading frame. Thus translation of *gag-pol* mRNA results in the production of *gag-pol* polypeptide and *gag*-only products. The *gag* products are structural proteins and make the four inner components of virus particle whereas the *pol* products are functional proteins and yield a reverse transcriptase and a protease which is involved in polyprotein processing. Thus a typical retroviral genome has a single unit of transcription, two processed transcription products, three primary translation products and eight functionally distinct proteins.

Chromosome and Chromosome Strucutre

2.1 CHROMOSOME

The chromosome is usually referred to as the hereditary material consisting of nucleic acid and protein. In 1943 Avery and Chase first reported that the nucleic acid (DNA/RNA) is the hereditary material. In higher organisms, double stranded DNA is the hereditary material and the whole chromosome is one immense DNA molecule whereas in some micro-organisms including some viruses single stranded DNA seems to be the core of genetic information. All plant viruses contain RNA as the genetic material. Some animal viruses also contain RNA. The chromosomes appear as threads of chromatin under light microscope. **Chromosome domain/territory**- It refers to the space of each chromosome in the interphase nucleus. The individual chromosomes occupy discrete patches which are separated by interchromosomal domains. Within individual territories the chromatin fibre is looped with the late replicating DNA near the nuclear periphery and the early replicating DNA oriented towards the interior. Active genes tend to be preferentially localized to the periphery of chromosome territories and the newly synthesized RNA transcripts are released into interchromosomal domain channels for further processing and transport.

2.2 CHROMATIN

The chromosomal material consists of DNA and protein. Staining properties of the chromosomes in interphase and prophase have shown that some areas of chromosomes are darkly stained (called **heterochromatin**) (Heitz, 1928) whereas others stain lightly or not at all (called **euchromatin)**. In other words, heterochromatin is identified as regions of the genome which stain differently to euchromatin (gene-rich regions). Euchromatin is found at chromosome arms whereas heterochromatin is found at centromers and telomeres. Heterochromatin is compact, gene-poor regions of a genome which are enriched in simple sequence repeats whereas euchromatin contains unique sequences. Heterochromatin is a major component of metazoan and plant genomes (e.g., ~20% of the human genome). Euchromatin is replicated throughout S phase but the heterochromatin is replicated in late S phase. Recombination takes place during meiosis in euchromatin but no meiotic recombination occurs in heterochromatin (Gregwal and Elign, 2007). Heterochromatin thus provides greater compaction than other genomic regions during interphase, lower accessibility than other regions to transcription and recombination machinery and formation of structured nucleosome arrays. Silencing of TE sequences within heterochromatin is probably a genome defense strategy. However, heterochromatin can also have important roles during chromosomal segregation and transcription and epigenetic silencing have been shown to both modulate gene expression and contribute to cis-regulatory sequences. As it can be impossible to clone, heterochromatin is often ignored when calculating the percentage of a genome that has been sequenced. Euchromatin undergoes de-condensation. The heterochromatin can be constitutive, for example, satellite DNA or facultative. The **constitutive heterochromatin** remains heterochromatized during the cell cycle whereas the **facultative heterochromatin** is transitory and it exists in alternate euchromatic and heterochromatic states during the cell cycle. Constitutive heterochromatin is found in the centromeric region, near the telomeres, in the satellite and in the nucleolus organizer region. Constitutive heterochromatin has also been termed as satellite DNA or repetitive DNA(or redundant DNA). The interstitial sites of heterochromatin(knobs) are common in plant genomes and were first described in maize(Mc Clintock, 1951). Corn heterochromatin is the constitutive type which also remains condensed. Under certain conditions, the telomeric heterochromatin (knob) in maize, rye, wheat and Bromus can be activated to function as 'neocentromeres' that form kinetochores during meiosis (Moore et al., 1997). In a typical diploid cell of *Drosophila melanogaster*, 30-35% of the karyotype is heterohromatic and heterochromatic regions are generally located at pericentric or telocentric locations. Heterochromatin is highly enriched in repetitive DNAs. Heterochromatin is molecularly heterogeneous with specific types of satellite and middle repetitive sequences preferentially clustered at certain chromosomal locations. The arrays of satellite DNA are interspersed with and may be interrupted by a more complex sequence that include different types of middle repetitive DNAs. Unlike the single transposable elements (TEs) found in the euchromatin, heterochromatic clusters of TE-like DNAs show little variation in chromosomal location. Hence middle repetitive DNAs may constitute rather stable structural components of the heterochromatin. Heterochromatin in salivary glands consists of two morphologically distinguishable types of heterochromatin termed α and β (Heitz, 1934). α heterochromatin is composed largely of satellite sequences whereas β-heterochromatin consists of middle repetitive and single copy sequences. Some β-heterochromatin sequences are actively transcribed. Single copy DNA sequences from heterochromatin have also been isolated and in several cases correspond to functional genes. Heterochromatin is both necessary for the expression of heterochromatic genes and inhibitory for the expression of eukaryotic genes. Novel euchromatin-heterochromatin junctions can affect the expression of euchromatic and heterochromatic genes located several megabases away, distinguishing higher order chromatin structure from most other regulatory mechanisms (Weiler and Wakimoto, 1995).

Conversion from a euchromatic to a heterochromatic condition results in gene inactivation. In a radiation–induced chromosomal translocation in Drosophila the epigenetic 'on-off' transcriptional states are largely dependent on the position of a gene within an accessible (euchromatic) or an inaccessible (heterochromatic) chromatin environment. This phenomenon of influence of heterochromatin on gene expression is called **position effect variegation**. Position effect variegation results when a chromosomal segment of euchromatin is translocated adjacent to a heterochromatic region (Wakinmoto, 1998). Similarities between position effect variegation in Drosophila and gene silencing in maize mediated by 'controlling elements' (that is TEs) led in part to the proposal that heterochromatin is composed of TEs and that such elements are scattered throughout the genome might regulate development. Using microarray analysis, heterochromatin in Arabidopsis is determined by TEs and related tandem repeats, under control of chromatin remodeling ATPase DDMI (decrease in DNA methylation). Small interfering RNA (siRNAs) corresponding to these sequences suggest a role in guiding DDMI. TEs can regulate genes epigenetically only when inserted within or very close to them. This probably accounts for the regulation by DDMI and the DNA methyl transferase MET1 of the euchromatic, imprinted gene FWA as its promoter is provided by TE-derived tandem repeats that are associated with siRNAs (Lipmann et al., 2004). Thus we see that heterochromatin contributes to genomic stability and chromosome segregation, nuclear organization and gene expression. Heterochromatin participates in RNAi (RNA interference) mechanism (see Roy, 2009 for detail) that epigenetically represses gene and TE expression which may immunize genomes against the invasion and expansion of selfish DNA elements. But difficulties in cloning, mapping and assembling regions rich in repetitive elements have hindered the genomic analysis of heterochromatin.

2.2.1 Formation of Heterochromatin

The formation of heterochromatin has been associated with silencing of gene transcription. But recent evidence suggests RNAi to be involved in targeting and maintenance of heterochromatin and that transcription is required to generate the siRNAs. The silenced form of chromatin- the heterochromatin should not be transcribed but there have been reports of low level transcription in heterochromatic regions and several hundred genes are found in these regions in Drosophila. Various factors implicated in heterochromatin formation are given in the table 2.1 given below.

Table 2.1 Shows factors involved heterochromatin formation in various organisms

	S. pombe	*Neurospora*	*Drosophila*	*Mouse*	*Arabidopsis*
Repetitious DNA	Y	Y	Y	Y	Y
DNA methylation	N	Y	NO	Y	Y
H3K9 methylation	Y	Y	Y	Y	Y
HP1	Y	Y	Y	Y	NO
Small RNAs	Y	NO	Y	Y	Y
Polymerase II	Y	ND	ND	ND	ND
RDR	Y	NO	NO	NO	Y

No-indicates that factor to be absent in the organism

ND- means that the factor is present but its role is unclear

2.2.2 Heterochromatin-euchromatin Transition Zone

The boundaries between heterochromatin and euchromatin may not be precisely defined. The genomes of eukaryotes generally contain heterochromatic regions surrounding the centromeres which are intractable to all current sequencing methods (see Roy, 2009). The heterochromatin in Drosophila consists primarily of simple sequence satellites, transposons and tandem arrays of ribosomal RNA genes. The transition zones between euchromatin and heterochromatin contain many previously unknown genes including counterparts to human cyclin K and mouse Krox-4. There are small islands of unique sequence embedded within heterochromatin–for example, the mitogen–activated protein kinase gene is flanked on each side by at least 3Mb of heterochromatin.

The distribution of the major types of repetitive DNAs on the cytogenetic map of the heterochromatin has been determined by combining chromosome banding techniques and *in situ* hybridization with repetitive DNA.

2.3 PROTEINS

There are two types of protein- histones and non-histones.

2.3.1 Histones

Histones are positively charged and highly basic whereas the **non-histones** are largely acidic and negatively charged. Histones contain 20-30% arginine and lysine. H1 is lysine rich whereas H2A and H2B are slightly lysine rich. H3 and H4 are arginine rich. Nucleic acid is acidic and negatively charged. Histones interact with negatively charged phosphate backbone of DNA by salt bridge. The histones of all higher plants consist of five major proteins, H1, H2a, H2b, H3 and H4 and are present in the ratio of 1 :2 : 2 : 2 : 2. The amount of histone present in the cell is equivalent to the amount of DNA. Histones have got a structural role to play. Together with DNA they produce the structural subunit of chromatin called **nucleosome**. The small ellipsoidal beads called are nucleosomes which are 110 Å in diameter by 60 Å high. Four of the five histones, namely H2a, H2b, H3 and H4 are highly conserved, i.e. they are similar in all higher eukaryotes and thus they are important in chromosome structure. Thus the role of histone has been one of a general and static nature. Histones are absent in prokaryotes. A key characteristic of the nucleosome unit is its inherent dynamic nature. The linker histone H1 has an on/off rate of seconds and the core histones are much more stable with exchange rates measured at orders of minutes (for H2A/H2B) to hours (for H3/H4). Covalent modification of histones by various enzymes especially at the N-terminal 'tails' that extend out of the histone core octamer core can significantly change the stability of the nucleosome unit.

Thus we see that eukaryotes have several histone variants as a result of their altered amino acid composition and this can affect both the structure of individual nucleosomes and the ability of the nucleosomes to form higher order chromatin structure. Moreover, binding of proteins such as linker histones and high mobility group proteins affect the architecture and the degree of chromatin compaction.

2.3.2 Non-histones

Estimates of non-histone protein (N.H.P.) are variable depending upon isolation procedure and cell type used. N.H.P. consists of a large number of very heterogeneous proteins. Some non-histone proteins are present in all cell types while others are tissue specific. Thus composition of N.H.P fraction varies widely among different cell types of the same organism and this variation in amount of non-histone protein and RNA is closely related to physiological activity of the nucleus(Ris, 1977) and thus are the likely candidates for roles in regulation of expression of specific genes or set of genes. Histones- DNA complex may be regarded as the substance on which NHPs work. NHPs affect histone-DNA interaction indirectly vis-à-vis enzymatic modifications. This may also interact directly to mediate histone restriction of RNA synthesis. NHPs also interact with DNA and they may act as specific repressor and as structural protein. NHPs thus seem to be involved in controlling the dynamic aspects of chromosome structure and function.

2.3.3 Role of Proteins in Gene Regulation

Specific proteins access the DNA and read the DNA sequence to either regulate or mediate gene expression. The reading process is impeded by chromatin-the tight packages of DNA and histone proteins required for nuclear compartmentalization of the genome. For gene transcription, opening of the chromatin is a must. Thus chromatin dynamic structure regulates all aspects of DNA metabolism and genome inheritance. On a local scale the positioning of nucleosomes-the fundamental units of chromatin, comprising octamers of histone proteins wrapped by the DNA double strand, affect DNA binding proteins to access the target sequence. On global scale, nucleosome positioning is non-random along the DNA of eukaryotic chromosomes. Promoter sequences are typically deficient in nucleosomes whereas coding regions tend to be nucleosome-rich. A regulatory sequence within the promoter of a gene which is supercoiled around a nucleosome generally is not accessible for trans-acting or regulatory proteins. Once such a nucleosome has been partly dissociated, the regulatory sequence becomes fully available for binding proteins. Several strategies(Ardt, 2007) probably determine such consistent nucleosomal patterns. These include nucleotide sequence and structural features of DNA which disfavour nucleosome formation and active mechanisms such as competition between transcription factors and histones for binding to DNA and enzyme-mediated chromatin remodeling.

Factors governing nucleosome positioning *in vivo* The first factor is the intrinsic DNA sequence preference for the histone octamer which favors or disfavors nucleosome assembly. Use of intrinsic sequence preferences for predicting nucleosome positions across a genome on the basis of DNA sequence has met modest success in that *in vivo* nucleosome positions have been predicted for only a subset of genomic loci. Another factor includes a variety of protein factors that contribute to nucleosome positioning *in vivo*. Principal among these are ATP-dependent chromatin remodeling enzymes which change the positions of nucleosome *in vivo* and *in vitro*. Thus DNA sequence and chromatin remodeling enzymes will provide information on how nucleosome positions are specified.

2.3.4 Chromatin Remodelling

We will see in section below that in the nucleus of eukaryotic cells, DNA is wrapped around octamers of histone proteins to form chromosomes which further assemble into higher order chromatin structure. This type of packaging compacts the entire genomic DNA inside the nucleus. However, it also prevents access to DNA for many regulatory proteins necessary for transcription, replication, DNA repair, recombination and chromosome segregation. Therefore, a dynamic change of chromatin structure (chromatin remodeling) is both essential and fundamental to the function and regulation of these genetic processes. Each protein has both a histone fold domain which mediated the histone-histone and histone-DNA interactions that are crucial for the assembly of the nucleosome core particles and a flexible amino-terminal tail domain which protrudes from the nucleosome. H2A and H2B also have an important carboxy-terminal tail domain. Eukaryotes have several histone variants which as a result of their altered amino acid composition can affect both the structure of the individual nucleosomes and the ability of nucleosomes to form higher order chromatin structure.

Chromatin structure can be changed by enzymes that remodel chromatin or covalently modify histones. The term chromatin remodeling encompasses a wide variety of changes in chromatin structure but can be defined as a discernible change in histone-DNA contacts(Downs et al., 2007). Such changes in contacts can result from repositioning(sliding) of nucleosomes on DNA, removal of part or all of the histone octamer from DNA, an induced change in the accessibility of the DNA in chromatin to protein such as transcription factors (TFs) or nucleases and the exchange of histone variants for core-histones.

An enzymatic mechanism for altering chromatin structure involves the covalent modifications of histones. This is carried out by a wide variety of enzymes including histone acetyltransferase(HATs), histone methyltransferases, protein kinases and ubiquitin- and SUMO-protein ligases. The combination of modifications ('marks') produced by these enzymes have been proposed to constitute a code ('histone code') that regulates downstream processes like transcription, DNA repair and apoptosis (Jenuwein, 2001).

There are two distinct classes of enzymes that are implicated in these processes. 1. ATP-dependent chromatin remodeling enzymes (eg. SWI2/SNF2 ATPase) (Flaus and Owen-Hughes, 2004; Isw2-a remodeling factor in yeast (Ardt, 2007)) and histone-modifying enzymes which catalyze the chemical modification of histone tails(eg. acetylation, methylation and ubiquitination) (Strahl and Allis, 2000). Usually these enzymes act with many other associated subunits in the context of large multiprotein complexes. Moreover, diverse complexes are not functionally independent but rather co-operate to regulate chromatin structure(Narlikar et al., 2002).

2.3.4.2 Role of non-enzymatic protein modules or domains in chromatin remodeling

Many conserved non-enzymatic protein modules or domains have been identified and characterized among these large remodeling complexes such as the SANT domain (Aasland et al., 1996; Boyer et al., 2004), chromodomains (Koonin et al., 1995; Brehm et al., 2004), bromodomains (Dhalluin et al., 1999) and PHD finger (Bienz, 2005). These non-enzymatic domains contribute greatly to chromatin remodeling is based on the following findings.

1. These domains commonly bear histone binding activity specifically to different modification states of histone tails. For example, bromodomains selectively bind lysine-acetylated histone, chromodomains bind lysine-methylated histone and SANT domains bind unacetylated histone.
2. These domains mediate interactions among remodeling proteins and other regulators like transcriptional factors. For example, SANT domains in diverse protein can

interact with TFSp1, the Brf1 subunit of RNA polymerase and several nuclear receptors.

3. Some domains, for example, chromodomains possess DNA binding and RNA binding activities

Thus non-enzymatic domains determine complex interactions among re-modeling proteins, histones, DNA, RNA and other regulators which facilitate the assembly of multiprotein complexes and target substrate recognition during remodeling (Zhang et al., 2006). These are discussed in detail below.

2.3.4.3 Histone code

It refers to various post-translational modifications of histone protein at a given time which are recognized by other proteins involved in chromatin modeling and transcriptional regulation. For example, acetylation of 13 different lysine residues in all different core histones (H2A, H2B, H3 and H4), methylation of lysine and arginine residues on histones, H4 and H4, phosphorylation and ubiquitinylation of all histones are such codes. Histones can be post-translationally modified by acetylation, methylation, phosphorylation, poly (ADP) ribose polymerization or reduction (See Roy, 2009 for more on role of histone).

Histone acetylation—It refers to the enzymatic transfer of acetyl group from acetyl-CoA to some amino acids of certain histone molecules. Acetylation especially of serine residues at the N-terminus of H1, H2A and H4 may occur during their synthesis and is irreversible. Other acetylations especially of the N-terminal lysine residues of H2A, H2B, H3 and H4 may facilitate repulsion of histones from the phosphate backbone of DNA in nucleosome because of introduced negative charges. This induces conformational changes in DNA which may be a pre-requisite for gene activation.

Methylation pattern—Transcriptional control of gene expression depends on DNA accessibility which is epigenetically regulated by histone modifications. There is specific distribution of methylated side chain residues in histones within the chromatin of eukaryotic cells which is continuously changing during the life cycle of a eukaryote. Methylation predominantly occurs on lysine and arginine residues in histones, H3 and H4. The methylated lysine (lysine residue 4 and 9(K9) in H3 and lysine 20 in H4 are methylated) is the only binding site for heterochromatin protein, HP1 which is associated with the silent heterochromatic region of a genome. Heterochromatin is defined by distinct post-translational modifications on histones such as methylation of histone H3 lysine 9(H3K9) which permits heterochromatin protein HP1- related chromodomain proteins to bind. Heterochromatin is frequently found near CENP-A chromatin which is the key element of kinetochore assembly. However, the primary signals specifying the site of CENP-A chromatin and thus the kinetochore assembly are unknown. The RNAi-directed heterochromatin flanking the central kinetochore domain at fission yeast centromere is needed to promote CENP-A^{cnp1} and kinetochore assembly over the central domain. The H3K9 methyltransferase Clr4(Suv39); the ribonuclease Dicer which cleaves heterochromatin double stranded RNA to siRNAs; Chp1- a component of the RNAi effector complex(RNAi-induced initiation of transcriptional gene silencing, RITS) and Swi6(HP1) are required to establish CENP-A^{cnp1} chromatin on naïve templates. Once established, CENP-A^{cnp1} chromatin is propagated by epigenetic means in the absence of heterochromatin. The modification of lysine 4 of histone 3(H3K4 methylation) is a well characterized feature of transcriptionally active genes, showing that the action of histone methyltransferase on H3K4 marks genes for transcriptional activation according to specific tissues and temporal patterns. In Drosophila, human and fission yeast, kinetochores are embedded in heterochromatin. Fission yeast centromeres have two distinct domains: outer repeats that flank the central kinetochore domain composed of innermost repeats and central core DNA. Heterochromatin containing Swi6(HP1) bound to histone H3 dimethylated on lysine 9(H3K9m2) forms over the outer repeats whereas CENP-A^{cnp1} replaces histone H3 in central domain (Diego Folco et al., 2008).

In animals an arginine-rich highly basic proteins called **protamines** replace histones in sperms heads and package the DNA into an extremely compact form. Protamines contain up to 50% of arginine residues and are clustered. Protamines can be grouped into mono, di or triprotamines depending upon the presence of only one type (e.g. arginine), two types(e.g. arginine, lysine) or three types (arginine, lysine, histidine) of basic amino acids.

Covalent modification of histone tails can itself influence the state of chromatin compaction(Shogren-knaak et al., 2006). However, the main mechanism by which these modifications influence chromatin structure is by controlling the binding of non-histone effector proteins to chromatin, some of which(e.g. remodellers) have the capacity to change histone-DNA contacts. Such effector proteins have modules that recognize specific histone modifications. For example, bromodomains bind acetylated lysine residues whereas chromodomains, tudor domains and plant homeodomain(PHD)-finger domains bind distinct methylated residues.

Double stranded breaks (DBS, Takahashi and Kaneko, 1985) cause chromatin reorganization as shown by altered access of nucleases to damaged DNA. DBSs have been shown to induce a local decrease in the density of chromatin fibre, in addition to changing the position of nucleosomes and causing the eviction of histone octamer. These remodeling events can make the lesion more accessible to 'damage sensor' proteins by exposing features of the nucleosomes such as constitutive histone modifications that are normally concealed in the unperturbed cells. In addition to remodeling chromatin and unmasking histone modifications buried in the histone fold domains, DBSs can also elicit post translational modifications on protruding histone tails.

2.4 DNA AND ITS PACKAGING

The chromosome contains a single giant DNA molecule which extends from one end through the centromere all the way to the other end of the chromosome. Chromosomal DNAs are much more longer than the cells in which they are packaged. The length of E. coli DNA chromosome is 1.7mm whereas the length of typical E. coli cell is 2μm and thus the DNA molecule is 850 times larger. Animal cells are 5 to 30 μm in diameter whereas plant cells are 10 to 100 μm in diameter. In human one of the smaller chromosome has a contour length(length of double helix) of about 30 nm and the largest human chromosome is about 8cm in length. The combined length of DNA is about 2m in human cell when the combined length of all 46 chromosomes in only about 200 μm. DNA molecule of about 10^5 μm long is found inside the cell nucleus of diameter 5 to 10 μm. They are folded or coiled in a very organized and systematic manner not yet fully understood. DNA compaction in a chromosome can be about 1000 fold or even greater. The chromosome at metaphase is of the order of 10000 times shorter. In the **first level of contraction (packaging)** DNA is coiled in the form of a double helix in which both strands of DNA coil around an axis and further coiling of that axis upon itself produces DNA supercoiling (see Roy, 2009). In the **second level of contraction** the DNA which is 20 Å or 2 nm in diameter is wrapped around histone core (an octamer of histones consisting of 2 molecules each of H2A, H2B, H3 and H4 which forms the core of a nucleosome) (Kornberg, 1974) to form a series of nucleosomes of 10 nm in diameter. Nucleosome is the basic repeating unit of chromatin. DNA wraps twice around an octamer of histone proteins as a left handed superhelix. Studies have shown that 200 bps segment of DNA(1.75 turns of super helical DNA) and histone core constitute the nucleosome. Out of 200 bps segment of DNA 146 bps segment is tightly linked to the protein core(which consists of eight histones, two copies each of H2a, H2b, H3 and H4) and the remaining 56 bps serve as **linker** molecule between nucleosomes. The complex of 146 bps of DNA with histone forms a bead and thus bead plus the connecting DNA(**linker DNA**) is called nucleosome (Figure 2.1A). Linker DNA of variable lengths connects two adjacent nucleosomes in eukaryotic chromosome. The whole structure thus looks like **'beads-on-a string'** form of chromatin. Linear arrangements of beads in a partially extended form(beads-on-string) was reported by Olins and Olins(1974) in chromatin. The nucleosome is interspersed by regions associated with sequence specific non-histones. The DNA wound around nucleosome core(H1 histone) provides about 7 fold compaction of DNA. This is the

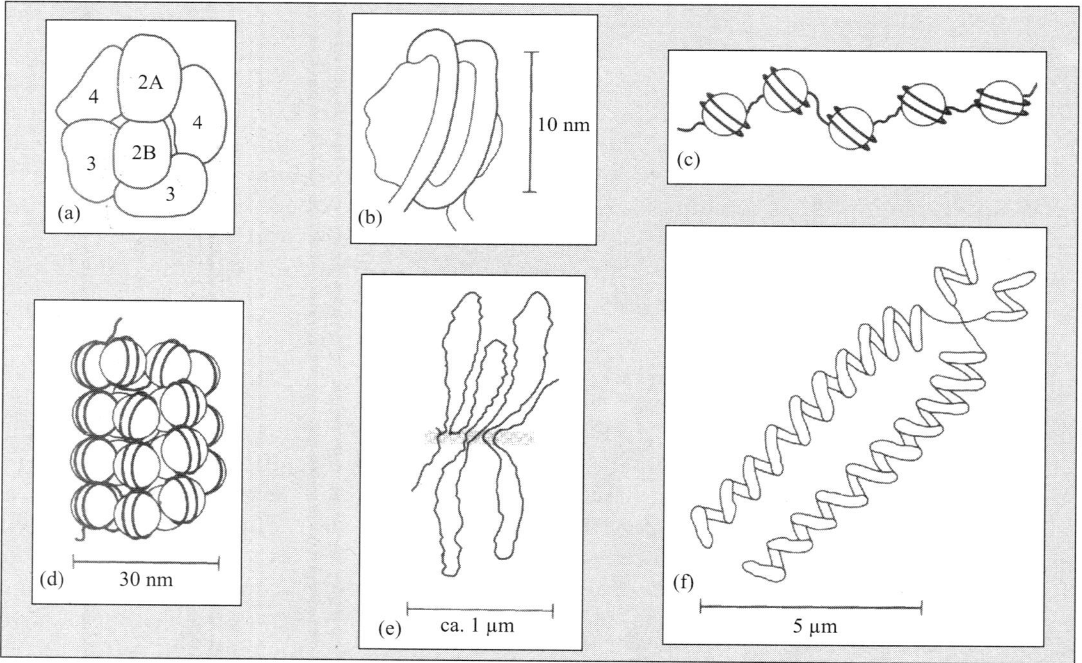

Fig. 2.1A Showing chromosome structure. (a) Protein core of a nucleosome (b). Binding of the DNA duplex around each Nucleosome string (d). chromatin fibre (e). Looping of the chromatin fibre (f). Helical folding of the chromosome scaffold at metaphase I.

fundamental level of organization which is best understood. The linker length varies from 10 to 80bp and variability of linker length may be related to the regulatory functions that control the particular locus. Thus the first level of organized compaction of nuclear DNA plays a structural role and has a compaction ratio of ~5-to-10 fold (ie., from 2-nm diameter of naked superhelical DNA to about 11nm). Further, it has a functional role in that it creates a barrier for accessibility for nuclear enzymes such as TFs to locate their target site of action.

In the **third level of contraction** the superhelical nucleosomal chains are held together by histone H1 in so called solenoids (corresponding to a 30nm chromatin fiber). This fifth histone, histone H1, stablizes this 30nm fiber form of chromatin by binding to the linker DNA and to the nucleosomes. Linker DNA especially at the entry and exit site of DNA in a nucleosome are bound to H1 histone. Nucleosomes thus appear to be organized into a structure called **chromatin fiber** (30 nm or 300Å or 0.03 μm in diameter) and which results from the helical packing of the nucleosome string) which provides ~ 100 fold compaction of DNA and in the **fourth level of contraction** some 20-200 kb of nucleosomally and/or solenoidally packed DNA (Finch and Klug, 1976) is arranged into loops which are anchored at the nuclear membrane or the nuclear scaffold by non-histone protein, i.e. the chromatin fibers are themselves folded to provide ≥ 10,000 fold compaction needed to fit a typical eukaryotic chromosome into a cell nucleus. In other words, the nucleosomal structure is then organized in chromosomal loops and such loops are known to form solenoids(Daban and Bermudex, 1998). Solenoids in turn form the chromosomal fibers which is visible by microscopy. The tightly folded chromatin fiber constitutes the chromatid which is ~ 600 nm in diameter and the two sister chromatids constitute one chromosome. Although higher levels of organization are not yet understood but certain regions of DNA are associated with a nuclear scaffold (chromosome matrix). In some cases the chromatin fiber (30 nm in diameter) forms loops. Each loop consists of ~ 75000 bps segment of DNA. Loops (active regions of the genome) often contain groups of genes with related functions. Six such loops of DNA (which form a rosette) are associated with a nuclear scaffold. The chromatin loop is the region of chromosomal DNA located between two contact points of DNA with a protein framework within the nucleus, the so-called nuclear matrix. These contact points are marked in the genomic DNA as Matrix/Scaffold Attachment Regions(S/MARs). Association of DNA with this nuclear matrix is a pre-requisite for transcription of DNA. The nuclear scaffold appears to contain large amount of histone H1 and topoisomerase II, the protein which is essential for DNA replication (unwinding of DNA during replication) and DNA packaging. Thirty such rosettes form a coil and ten such coils form a chromatid. Thus DNA compaction in the chromosome looks like involving coils upon coils upon coils. In plants the chromatin fibre forms a rosette-like structure whereas in higher eukaryotes chromatin forms interconnected mega-based domain. Other models suggest looping of the fibre from a central backbone (Figure 2.1B). The two chromatids are held together at one point called centromere. Centromere is a sequence of DNA that functions as an attachment point for proteins that link the chromosome to mitotic or meiotic spindle and thus ensures equal and orderly distribution of chromosomes to daughter cells.

2.4.1 Role of Nucleosome in Gene Expression

Nucleosome, the central unit of chromatin, can modulate the read out from DNA in at least three ways (Baylin and Schuebel, 2007).

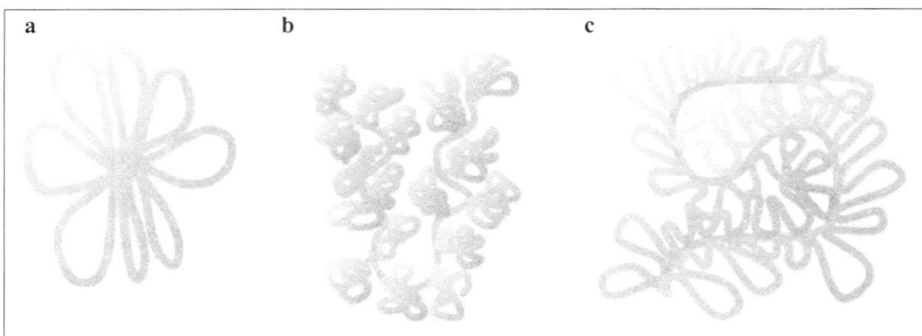

Fig. 2.1B Showing higher-order organization of chromosome territories.(a) showing formation of a rosette-like structure by chromatin fibre in plants; (b) showing chromatin fibre to be forming interconnected domains and (c) showing looping of fibre from a central backbone.

1. Nucleosome can be physically rearranged on DNA by chromatin remodeling proteins. Generally, greater the distance between nucleosomes and so the 'openness' of chromatin, the higher the likelihood that such regions of DNA will be transcribed.
2. Many nucleosomes can be compacted into higher order aggregates to form 'closed' chromatin or heterochromatin thereby preventing transcription. The balance between open and closed parts of the genome facilitates proper gene expression patterns in given cell types and also prevents unwanted gene expression
3. There is complex interplay between enzymes that can modify particular amino acids in the histone component of the nucleosomes and those that reverse the modifications. The modifications or histone 'marks' interact with proteins that bind to and interprete them. The marks were initially seen as a 'histone code', the idea being that a restricted number of them would specify the 'on or off' state of RNA production from DNA(i.e. gene transcription). But it has now been recognized that constituents of chromatin and nucleosome structure, position and modification are highly complex and a balance between these factors marks an individual gene, group of genes, for various levels and states of expression and thus there is no simple 'on-off code'.

2.4.2 Genome-scale Identification of Nucleosome Positioning

Sequence-dependent nucleosome positioning implies both rotational and translational settings of the DNA on the histone surface. Rotational setting appears to depend on the existence of periodically distributed short ologonucleotide tracts which deflect the DNA helical axis (the so-called bending) or are more easily compressed in major or minor groove (bendability). A distribution of these tracts in phase with the double helix periodicity may then leads to permanent or easily induced unidirectional curvature of the DNA and would lower the energy needed for its wrapping around the nucleosome (Travers, 1985). The basis for translational setting is more obscure. It may result from the pecularity of the DNA of not being smoothly bent on the histone surface but distorted in various ways at specific locations. Its particular sequences along the DNA are conformationally more flexible than their preferential localization at these sites of higher DNA distortions would be expected to decrease the free energy of nucleosome formation. The nucleosome positions actually observed probably result from a complex interplay between rotational and translational constraints.

Nucleosome formation can be inhibited by DNA supercoiling when the DNA sequences undergo supercoiling-induced structural transitions. G+C-rich alternating purine-pyrimidine stretches may flip to the Z-conformation while inverted repeats may extrude cruciforms and both of these transitions have been shown to prevent nucleosome reconstitution. Two more transitions such as slippage loops which may occur in the directed repeats and the B-H transition in which triple stranded together with single stranded structures can form from homopurine-homopyrimidine sequences, are likely to hinder nucleosome reconstitution.

Positioning of nucleosomes has been implicated in the regulation of gene expression in eukaryotic cells. Thus high resolution measurements of nucleosome positions over genome-wide scale would enhance understanding of chromatin structure and function. To measure nucleosome positions on a genomic scale, Yuan et al. (2005) developed a DNA microarray method to identify nucleosomal and linker DNA sequences on the basis of susceptibility of linker DNA to micrococcal nuclease(S1). In this method, nucleosomal DNA was isolated, labeled with Cy3 fluorescent dye (green) and mixed with Cy5 labeled total genomic DNA (red). This mixture was hybridized to microarrays printed with overlapping 50-mer oligonucleotides probes tiled every 20bps across chromosomal regions of interest.

A graph of green : red ratio values for spots along the chromosome was expected to show nucleosomes as peaks about 140 bps long or six to eight microarray spots surrounded by lower ratio values corresponding to linker regions. Further, HMM (Hidden Markov Model) was used to determine nucleosome/linker boundaries. HMM used observable signals (that are hybridization values of the tiled probes) to infer hidden states(nucleosomal and linker states) responsible for generating the signal. Well positioned nucleosomes should cover ~140bps or six to eight probes and have a high green : red ratio whereas stretches of ≥ 9 probes were classified as 'fuzzy' or delocalized nucleosomes. Linkers are expected to have lower ratios and may have variable length. The model thus calculates the probability that a given probe on the array corresponded to nucleosomal, fuzzy nucleosomal or linker DNA and identifies the most likely nucleosome position. They found majority of the identified nucleosomes (65 to 69% of the nucleosomal DNA) to be well positioned. The chromatin organization at pol II promoters consisted of a nucleosome-free region (NFR) about 200 bps upstream of the start codon flanked on both sides by positioned nucleosomes. The NFRs are evolutionary conserved and were enriched in poly(dA) or poly(dT) sequences globally. Poly(dA-dT) stretches incorporate poorly in nucleosomes because of their relative rigidity. Many occupied TF binding motifs were devoid of nucleosomes thereby suggesting nucleosome positioning to be a global determinant of TF access. Functional TF motifs were similarly found ~100 to 500bps upstream of the start codon. NFRs are transcriptional start sites. Nucleosomes are prevented from association with promoter regions either by sequence characteristics such as poly(dA- dT) elements or by nucleosome eviction by recruited proteins and nucleosomes are subsequently well positioned between nearby nucleosome free regions because of structural constraints imposed by packaging short stretches of sequence with nucleosomes.

In chromatin, NFRs known as nuclease hypersensitive sites are thought to represent the 'open window' which allow enhanced access of crucial resident cis-acting DNA sequences to trans-acting factors. NFRs have pronounced sensitivity to nuclease cleavage or chemical modifications. NFRs are two orders of magnitude more sensitive than regions in bulk chromatin. Hypersensitive sites generally represent a minor(ca. 1%) but highly selective fraction of the genome.

2.4.3 Genome-wide Analysis of Isw2-dependent Chromatin Remodeling

Use of high resolution tiled microarrays was made to map the positions of nucleosomes and defined how their positioning is altered in a mutant strain. Whitehouse et al. (2007) sought to discover Isw2 target across the yeast genome by identifying nucleosomes whose positioning is changed in mutant strain. They purified nucleosomal DNA from both wild-type and mutant yeast and hybridized each to high resolution tiled microarray that cover the entire yeast genome with 5bp spacing. Nucleosome-sized peaks were found across the genome. Arrays of positioned nucleosomes generated a periodic signal whose maxima and minima were separated by the nucleosome repeat length of 165bp. Comparison of wild-type nucleosome positions with those from a mutant revealed many sites of altered nucleosome positioning. After having established the locations of chromatin changes, ChIP in conjunction with tiled microarrays was used to determine whether Isw2 was acting directly at these loci. They found that nucleosome positioning by Iws2 to be directional and resulting in compact chromatin adjacent to sites of transcriptional activity. Loss of Isw2 led to nucleosome positional changes and inappropriate transcription, resulting in generation of non-coding transcripts.

2.5 SISTER CHROMATID COHESION

Eukaryotic chromosome is characterized by sister chromatid cohesion. Cohesion plays an important role in chromosome segregation as well as post replicative repair of double strand breaks. It is mediated by a large ring-shaped complex, cohesion and its associated proteins, Pds5 (Huang et al., 2005; Meluh et al., 2002). Cohesin between sister chromatids is believed to be produced only during ongoining DNA replication by an obligate coupling between cohesion establishment factors such as Eco1(Ctf7) and the replisome. Thus the current model is that the cohesion generation can only occur in the context of DNA replication and Eco1 merely allowing the replisome to slide through the cohesion ring in S-phase (Lengronne et al., 2006). However, in case of budding yeast Onal et al. (2007) observed cohesion to be generated by an Eco1 dependent but replication independent mechanism in response to double strand breaks in G2/M. Cohesin has two functions- a cohesion activity and a conserved acetytransferase activity which triggers the generation of cohesion in response to the double strand break and the DNA damage check point. As double strand break occurs not only on broken chromosomes but also one unbroken chromosomes, the DNA damage check point through Eco1 can be said to provide genome-wide protection of chromosome integrity.

Figure 2.2 shows procedure for studying the various histones present in the nucleosome whereas figure 2.3 shows procedure for demonstrating the nucleosome phasing.

2.6 STRUCTURE OF A TELOMERE

The heterochromatic state of the telomeres, their gene-less nature and their ability to silence transcription of inserted subtelomeric reporter genes (which is known as telomere position effect) indicate that telomeres are transcriptionally silent. Telomeres consist of long stretches of tandem repeats of short sequence elements, typically 5-8bp in length that are rich in G

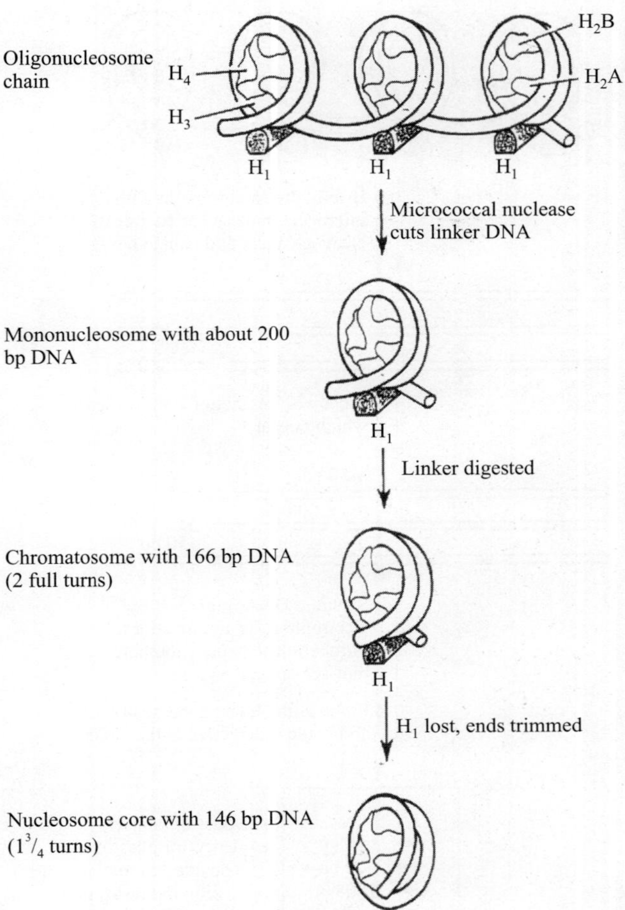

Fig. 2.2 Showing relative positions of the various proteins in the nucleosome determined through the digestion of Chromatin with micrococcal nuclease (adapted from Richmond et al. (1982).

residues on the strand oriented in 5′ to 3′ direction toward the end of chromosome. Mammalian telomeres comprise tandem arrays of duplex 5′-TTAGGG-3′ repeats. The G-rich strand (G strand) extends beyond the complementary C-rich strand and terminates as a single stranded 3′ overhang as shown in Figure. In the chromosome, the double stranded part of the telomeric DNA loops back on itself, forming a lariat structure as shown in Figure. The single stranded terminus (the 3′ G strand extension) invades the double stranded DNA (duplex telomeric repeats) forming a D loop (displacement loop). Thus the telomeric terminus is masked and protected from enzymes which act on the DNA ends. The double stranded part of the telomere is bound to a multiprotein complex called shetrin which prevents telomeres from being recognized and processed as double stranded break. In plants, telomeres are mostly composed of many tandemly repeated copies of basic oligonucleotides

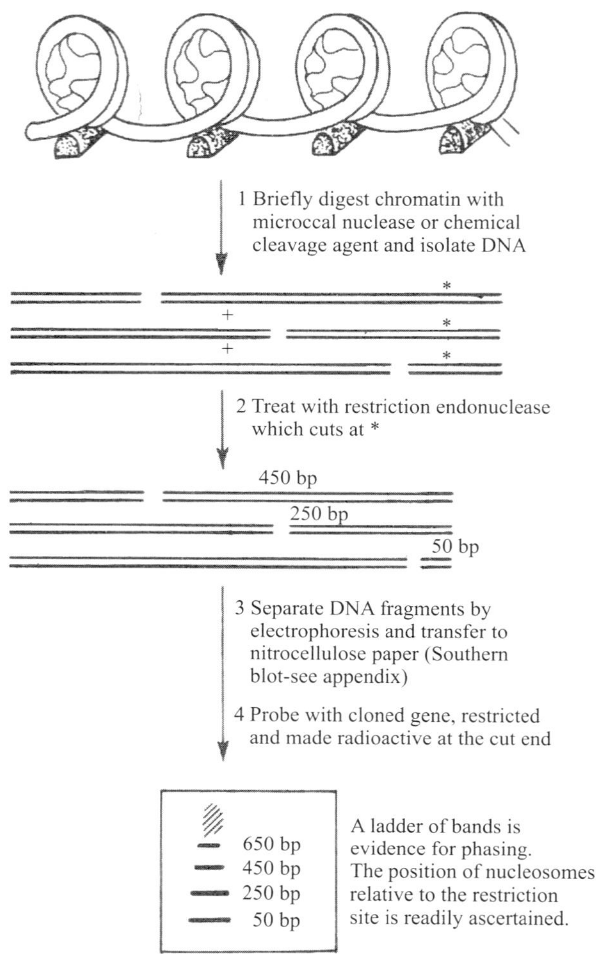

Fig. 2.3 Showing method of study of nucleosome phasing through digestion with micrococcal nuclease to produce Core particle DNA which can be cloned and sequenced or Mapped with restriction endonucleases (adapted from Adams, Knowler and Leader, 1986).

of the sequence $(TTTAGGG)_n$-characterized by the G-rich strand running in the 5′ to 3′ direction and C-rich strand running toward the centromere.

Among cereals the locations of centromere and telomere sites have been conserved at the gross level. However, in mammals, the locations of centromeric sites have not been found to be conserved across species. Obligate bivalent formation in polyploids is achieved by initiating chromosome pairing at two sites. Analysis of autopolyploids and allopolyploids suggest that centromeres and telomeric regions are the two sites. Watanabe suggested that there could be two sites under independent and functionally different genetic control and pairing at one site always precedes the pairing at the other.

2.6.1 Telomere Capping

Duplex telomere-binding proteins bind along the length of the telomere repeats whereas a specialized telomere-binding protein bind the D loop at the junction of the lariat. A series of specific proteins assemble at the ends of chromosomes to form a complex which protects telomere from being degraded. The tandemly reiterated telomeric DNA repeats bind a set of sequence-specific DNA binding proteins amongst which telomerase itself and other structural and catalytic proteins which include TRF2, RAP1, Ku and others. A series of transposons forms an integral part of the telomeres of chromosomes of *Drosophila melanogaster* and replace the conventional telomeric repeats. Telomeric transposons resemble retrotransposons, i.e. they have poly(A) stretches at the 3′ end and internal sequences with homology to *gag* regions of the retrotransposable elements(and retroviruses). One of these transposons is the 6kb HeT-A element and the other is less abundant 10kb TART. Telomeric transposons transpose only to chromosome ends and are responsible for telomeric maintenance.

The length of telomere is species specific. For example, telomeres may be 40-150kbp long in mice and humans whereas they are relatively short in Arabidopsis (only 2-4kbp). In animal systems, telomeres are reported to shorten at the rate of about 5-10kbp per cell generation and when telomere lengths falls below a critical level, senescence and cell death follows. In other words, telomeres equip normal human somatic cells with a cellular clock which determines their replicative life span besides providing chromosome stability during mitosis and meiosis. Telomere length is maintained by an enzyme called telomerase. In animals, telomerase activity is high in differentiating cells but the activity decreases and is undetectable in differentiated and mature cells. However, in plants, telomerase activity is high in proliferating cells but is lost in differentiating cells and is undetectable in mature cells. Plant's telomeres are structurally similar to those in other eukaryotes and have a consensus repeat sequence, -TTTAGGG-. Barley telomeres may have several thousand such repeats. There are conflicting reports about shortening of telomeres in plants.

Functions of telomere - First, the telomeres preserve a chromosome's linear integrity and block illegitimate recombination of the chromosome ends. Secondly, it facilitates complete replication of the extreme end of chromosomes and interact with different proteins to repress the expression of adjacent genes and thus shows telomeric position effect. Finally, it maintains the three dimensional architecture of the chromosomes within the nucleus and are involved in the attachment of chromosomes to the nuclear membrane.

2.7 CHROMOSOME MORPHOLOGY

Karyotype refers to chromosome size, morphology and number of a chromosome complement of an individual (Battaglia, 1952). Chromosomes vary in size, location of centromere (kinetochore or primary constriction) and the presence or absence of satellites (that are attached to proximal portion of chromosome arm through thin stalk structure called **secondary constriction**). **Long and short arms**- The regions either side of the centromere, a compact part of a chromosome, are known as arms. When the centromere is at the tip of the chromosome the chromosome is classified as telocentric chromosome. Such chromosome has got only one arm. In the **acrocentric** chromosome the centromere is near the end of the chromosome and as a result one arm of the chromosome is very short. When the centromere lies in an approximately central position, the chromosome is classified as **metacentric** chromosome and in such chromosome both arms are of equal length. Sorghum chromosomes are metacentric. In the **submetacentric** chromosome the centromere is nearer to one than the other resulting in two arms being unequal, i.e. one short arm and one long arm (Figure 2.4). Certain chromosomes have secondary constriction which forms nucleolus during interphase and thus secondary constriction is also called **nucleolus organizer region**. As a general rule, the number of nucleoli corresponds to the number of secondary constrictions. Each species usually has at least one homologous pair of nucleolus organizer chromosomes. Nucleoli are observed in metaphase. They breakdown in metaphase and are reformed in telophase. The **ideogram** is the diagrammatic representation of the gametic chromosomes of a given species and is constructed for comparison of karyotype (chromosome constitution) of one species with those of the other species. The distinguishing characteristics of karyotypes are; (i). the basic chromosome number (ii). form and relative sizes (length) of different chromosomes (iii). number and size of satellites and secondary constrictions (iv). absolute size (length) of chromosomes and (v). distribution of nuclear materials with staining properties, that is, euchromatin and heterochromatin (in terms of area, volume and density and C-banding of metaphase chromosome). Great variation in mean chromosome size and DNA content exists among plant species with the same haploid number of chromosomes belonging to the same family (up to 36 fold difference) has been observed (Stebbins, 1971). Higher content of DNA can be due to large number of structural or due to large amount of non-sense or repetitive sequence. The pachytene chromosomes of maize (10 pairs), tomato (12 pairs) and barley (7 pairs) as seen when acetocarmine stained anther smears are examined under the microscope are good examples of striking differences in the gross appearance of the chromosomes of different species. Meiotic pachytene chromosomes are ideal for studying chromosome morphology. Chromosomes are numbered in order of decreasing length from 1 to 10 in maize, 1to 12 in the tomato, 1 to 7 in barley with one being the longest chromosome. The two arms of human chromosomes separated by the centromere are called p(petitáe=small) arm and q(=queue) arm. Regions within the chromosome are numbered p1, p2, …..and q1, q2, …. outward from the centromere. Subsequent digits show subdivision of bands. For example, certain bands on the q arm of human chromosome 15 are labeled 15q11.1, 15q11.2, 15q12.

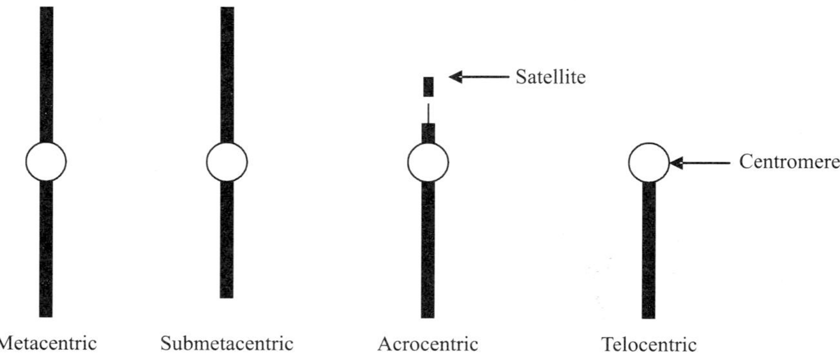

Fig. 2.4 Showing types of chromosome depending on the position of centromere.

Originally bands 15q11 and 15q12 were identified and subsequently 15q11 was divided into 15q11.1 and 15q11.2 The satellited chromosomes with nucleolar organizer regions may be exceptions. Chromosome 6 in maize is shorter than its number would indicate and it is always found associated with the nucleolous. In barley the satellited chromosomes are numbered as 6 and 7 (not in order of decreasing length). The chromosomes comprising the haploid or gametic number can be distinguished on the basis of following morphological criteria.

(i) the relative lengths
(ii) distinctive chromomeres patterns (number, size and distribution of chromomeres of each chromosome)
(iii) deep-staining knobs in characteristics positions
(iv) location of centromere (determining arm ratios)
(v) the degree of **heteropycnosis** in the chromomeres adjacent to the centromeres and
(vi) the nucleolus reorganizer region with nucleolus attached.

The **knobs** are darkely stained bodies whose position and number are constant for a particular species but their number, position and size vary between different races of the same species (Mc Clintock, 1930; Longley, 1939; Brown, 1049). Knobs remain in a relatively contracted condition through out the mitotic cycle (Morgan, 1943) and have been found not to be associated with any vital function. Knobs can be used as cytological markers. The knobs in maize chromosomes differ in number and size in different stocks (Figure 2.5). Chromomeres are beadlike structures found along the length of chromosomes and chromomeres can be resolved into coils. Chromomeres are areas of tightly folded and coiled chromatids. The bands are also tightly packed areas on the chromosomes. Chromomeres may vary in size and position. It is both large and small in maize and of uniform dimension in rye. It can vary in position and in case of tomato the chromomeres are terminal. There are extensive heterochromatic, darkely stained regions in the chromosomes of all stocks of tomato. The chromosomes of barley lack these markers (heterochromatic regions, knobs) which are helpful in distinguishing and recognizing the individual chromosomes. Diakinesis is one of the most ideal stages for counting the number of chromosomes of an individual. Also, mitotic metaphase is the stage when chromosomes are most easily identified and distinguished and can be counted but at this stage heterochromatic regions are usually indistinguishable from euchromatin. Previously the morphological features that could only be measured were the total length of the chromosome and the position of the centromere but they are much more easier to resolve under the microscope with the aid of various stains. Chromatin or DNA can be distinguished from cytoplasm and nucleoli or RNA in the interphase through Feulgen reaction. When a cell is treated with a warm acid and a chemical called Schiff's reagent there is development of reddish purple stain- a characteristic of DNA. The best general-purpose stains such as orcein and hematoxylin bind to chromosome indiscriminately and they can make distinction between chromosomes only if they differ appreciably in size or in position of centromere. The centromere is recognized at metaphase stage as a constriction where the chromatids are held together. Further these stains can also reveal nucleolus organizers (rDNA loci) and at least blocks of heterochromatin and chromomeres during prophase stages of nuclear division.

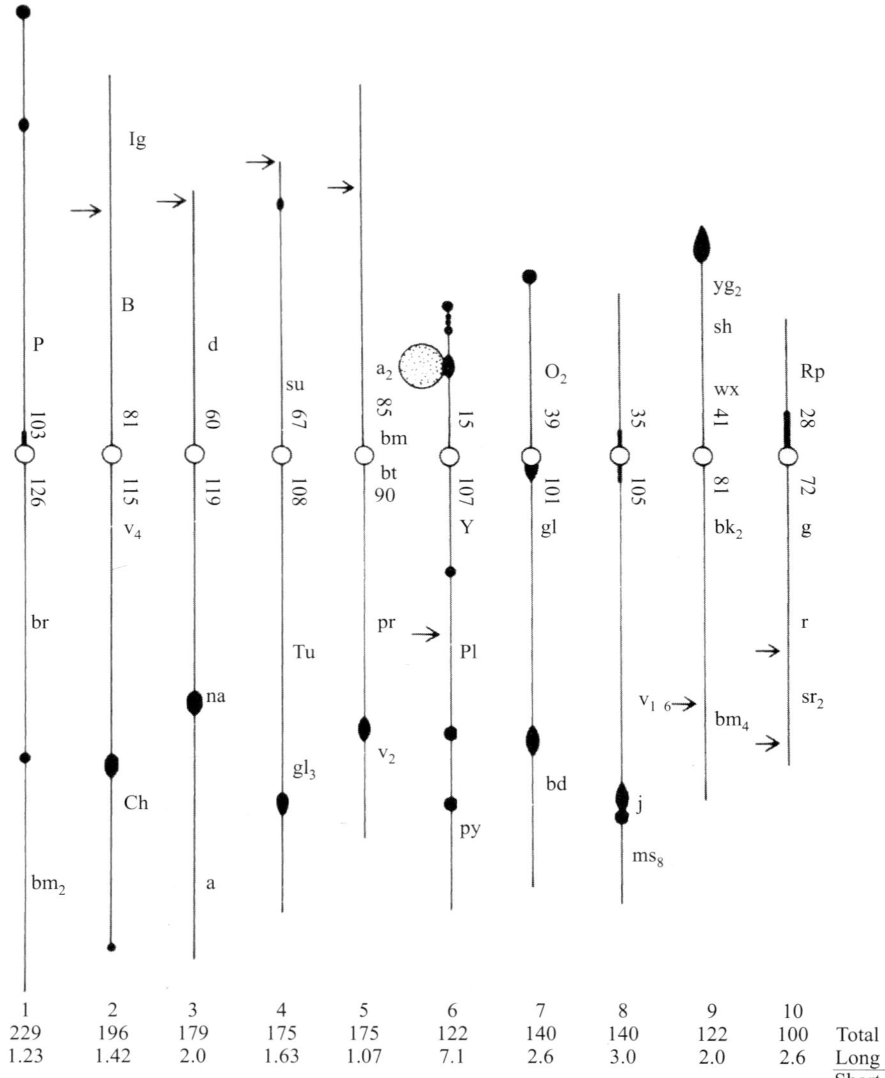

Fig. 2.5 Showing pachytene morphology of maize chromosomes. Lengths are relative based on a length of 100 for chromosome 10. Arrows indicate additional knobs (adapted from Burnham, 1962).

2.8 CHROMOSOME BANDING TECHNIQUES

Genes identified by Mendel were entirely abstract entities whereas chromosomes are physical objects and the binding patterns are their visible landmarks. The development of chromosome banding techniques took around 1970 (Vosa, 1976) and often each chromosome has a unique pattern of bands. Earlier attempts to identify individual chromosomes or genomes relied on chromosome morophology alone, using overall size and the position of various markers such as centromeres and secondary constrictions. Chromosome banding techniques are used for the visualization of light and dark bands in pro-metaphase and metaphase chromosomes. They allow to design a specific banding pattern to each chromosome and depending upon the specific staining technique several band structures can be discriminated. At metaphase stage of cell cycle chromosomes can be banded by a variety of stains in combination with chemical or enzymatic treatments. Banding patterns have been used as a means of recognizing individual chromosomes and under most favorable conditions allow points with about a 10Mb separation to be distinguished. Banding pattern will lead to the production of **cytogenetic map**. The association of disease phenotypes with abnormalities of chromosome banding formed the original basis of gene mapping to chromosome or chromosomal regions. The banding pattern of a particular chromosome is highly specific and constant and is used to distinguish similar sized chromosomes in the karyotype. Banding patterns are used in gene mapping and to detect chromosome aberrations such as partial duplication. The molecular basis of a few of the banding methods(described below) is known to involve base composition. Many correlations have been made between banding and other aspects of genome function and organization. Chromosome banding is an evidence for a regional organization. A **band** is defined as a part of a chromosome that is clearly distinguishable from its adjacent segments by appearing darker or lighter. Each chromosome has a banding pattern which allows it to be recognized easily (Figure 2.6). In other words, the chromosomes are visualized as consisting of a continuous series of light and dark binds with no **interbands**. The banding patterns are consistent and reproducible features of chromosomes. They do not vary from tissue to tissue or change during the course of development. The inherent organization responsible for banding is not random but highly ordered. The **major bands** are those readily identifiable in preparations showing only minimal information and each is given a capital letter starting with A at the centromeric(proximal) end. Each major band is preceded by the number of chromosome thus 5A and 6B. The diagnostic characteristics indicated by 'A' are those seen in fluorescent metaphases of fair technical quality whereas those indicated by 'B' are usually visible only in cells of good quality. A **region** of a chromosome is defined as that part between major bands. In good G-banded preparations, some of the major bands can be subdivided into **minor bands** whose visibility depends on the quality of the banding. Each minor band is given a number within the lettered group, e.g. 5A1, 6B2. In all cases the region consisting of the centromeric heterochromatin is designated as A1. As banding techniques improve, it is highly likely that even the minor bands will be resolved into **sub-bands** (Yunis et al., 1981) such as 5B1.1., 5B1.2 etc. Bands are referred to as **'landmarks'** and are used to describe the first band in each subdivision of a chromosome. A chromosome landmark is defined as a consistent and distinct morphological feature that is an important diagnostic help in identifying a chromosome. Landmarks include the ends of the chromosome arm, the centromere

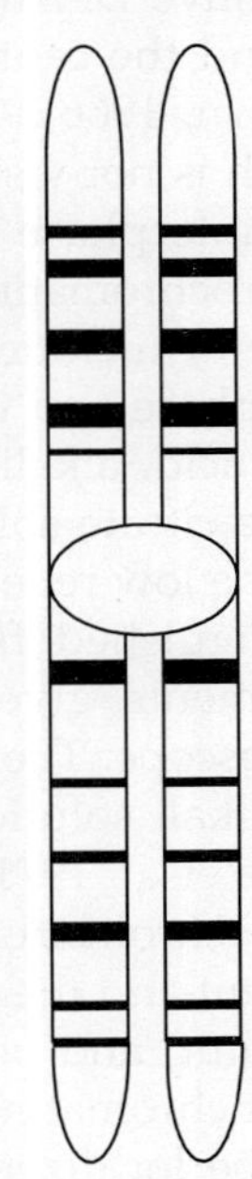

Fig. 2.6 Showing a banded chromosome.

and certain bands. Classification of banding patterns in human chromosomes follows the terminology adopted in 1971 at the Paris conference on Standardization in Human Cytogenetics(Hecht et al., 1974). As constitutive heterochromatin is naturally differentiated by darker staining during interphase and prophase but it is not visible in the mitotic metaphase as the chromosome stains uniformly and so a series of special chromosome banding techniques with use of microscope have been developed to make heterochromatin visible and thus different metaphase chromosomes can be distinguished. Translocations/ inversions can be distinguished by different banding patterns but the paracentric inversions, deletions and duplications which do not significantly change the overall chromosome structure go undetected. Chromosome banding techniques thus help in the identification of individual chromosomes and chromosome segments. The distribution of bands is distinctive between species and also between individual chromosomes within species. The widespread distribution of bands is particularly useful for identifying small chromosome segments. Further, it can also be used to study the inheritance of particular chromosomal segments in practical plant breeding. The four main banding techniques depending upon the specific staining techniques are as follows:

2.8.1 C-banding

C-banding technique arose as a modification of the *in situ* hybridization technique which selectively stains the constitutive heterochromatin (Arrighi and Hsu, 1971). This chromatin remains condensed throughout interphase and mitosis and consists of satellite or highly repetitive DNA(Hsu, 1975). C-banding demonstrates constitutive heterochromatin presence around the centromere, telomere and in the nucleolus reorganizer region(in the satellites) (Brown, 1966). Centromere refers to a highly specific, functional part of the chromosome which is not visible in mitotic chromosomes. Euchromatin stains lightly or not at all during the interphase and prophase while heterochromatin stains darkly in these stages. Heterochromatin is an extremely useful marker for chromosome mapping in the pachytene stage of meiotic prophase. Facultative heterochromatin may or may not be condensed in interphase and they may or may not stain darker. The different methods include treatment with acid, alkali or higher temperature which denatures the DNA and incubation at 60° in saline-citrate solution renatures the constitutive chromatin, the highly repetitive DNA but not the low repetitious and unique sequence DNA. Thus constitutive heterochromatin can be distinguished from euchromatin through differential staining reaction. These different treatments give rise to a characteristic banding pattern visible on fluorescence light microscope. The current techniques of C-banding are based on treating chromosomes with a hot alkali solution, a hot saline solution and staining with Giemsa(Arrighi and Hsu, 1971; Yunis et al. 1971; Sumner, 1972). **Giemsa C-banding** is a means of defining the location of heterochromatic regions in condensed metaphase chromosomes. It is the only technique used in plant and the reason is that the constitutive heterochromatin is much more variable in size (amount) and position in chromosomes of plants and thus are spread whereas constitutive heterochromatin is localized in animal chromosome and if C-banding is used then banding will be localized around centromere. In the human karyotype, C-bands are observed as constitutive heterochromatin at three typical locations, namely, centromere, secondary constriction sites of chromosome 1, 9 and 19 and distal end of the Y chromosome (Jack et al. 1985). Giemsa-stained centromeric regions are referred to a C-bands (Paris Conference, 1972).

C-banding allows one to distinguish each chromosome of barley genome which consists of meta or submetacentric chromosomes of approximately equal length. Relative chromosome length at metaphase was found not to be a reliable parameter for identifying individual chromosome. Different varieties of wheat show variation in the size, number and distribution of C-bands and N-bands. The B-genome chromosomes and chromosomes 4A and 7A are the most readily identifiable as they tend to show larger and more numerous C-bands than other, A- and D genome, chromosomes. The B-genome has a greater proportion of C-banded chromatin than even the rye genome. Although the largest bands or groups of bands (**major bands**) are near the centromere there are many **interstitial bands** (bands located in the interchromomeric regions) and about half of the chromosome arms have telomeric bands. At least 19 of the 21 wheat chromosomes of the haploid set can be identified. The wide spread distribution of bands is particularly useful for identifying small chromosome segments (Flavell et al., 1987). Meiotic and mitotic banding patterns are similar but many of the smaller mitotic bands are absent from the meiotic chromosomes. C-bands of rye occur mostly but not exclusively at or near the telomeres. The smaller interstitial sites facilitate the identification of individual rye chromosomes (Flavell et al.,1987). C-banding involves depurination of DNA in the chromosomes. The chromosome preparations are pretreated with Hcl (0.2M) for several minutes (sugar-phosphate backbone remains intact), exposed to barium hydroxide, $Ba(OH)_2$ (0.7M) (Sumner, 1972) or NaOH(0.7M) (Arrighi and Hsu, 1971) for several minutes (denatures the DNA which aids solubilization), then incubated in warm 2XSSC(SSC= 0.15 $moll^{-1}$ NaCl, 0.015 $moll^{-1}$ trisodiun citrate) before staining with Giemsa-type stain (Gill and Kimber, 1974; Gustafson and Smith, 1977 Iordansky et al.,1978 and Seal and Bennett, 1981). SSC incubation renatures whereas alkali treatment does the denaturation. The warm salt breaks the sugar-phosphate backbone and allows DNA fragments to dissolve. Staining relies on the retention of intact nucleoprotein complexes in the highly compact heterochromatin. Resolution can be increased by (i) using chromosomes isolated early in metaphase when the contraction of chromatin is incomplete and (ii) using microscopic technique known as **epiillumination** in which the chromosomes are viewed from the same side as it is illuminated. This technique thus provides enhanced contrast between stained and unstained bands. The DNA sequences in constitutive heterochromatin are highly repetitive and may include satellite DNA sequences. The distribution of satellite DNA sequences in chromosomes varies from species to species and can even provide information which is specific to a particular individual. The limitations with these banding techniques (C-banding or N-banding) are that the constitutive heterochromatin which is differentially stained by these techniques is heterogeneous in composition in terms of DNA sequences it contains and thus different C-bands on the same or on different wheat chromosomes may not have the same DNA composition. This structural heterogeneity may be partially responsible for the sensitivity of banding patterns to variations in techniques and experimental conditions particularly the **minor bands** which are also important in the chromosome identification can vary with the environmental conditions. But there is close correlation between the site of C- and N-bands and the presence of long, tandem arrays of short repeating sequences in the chromosomal DNA. DNA found around centromere and separating as satellite appears as **major C-bands**. **Polymorphic bands**- Certain chromosome regions differ in size and/or appearance between homologues and individuals. All intensely fluorescent regions are polymorphic and include the centromeric region of human chromosome 3. Size variation is largely due to variation in size of the intensity of fluoresceing Q-band. The Q-band may be completely absent in very small Y chromosome. Y chromosome may differ in size(up to a factor of around 2-3) between unrelated individuals. (Evans, 1973).

2.8.2 G-banding

G-banding usually involves a pretreatment of the chromosomes, followed by staining with Giemsa. Every chromosome within a complement can be identified and deviation from standard type in chromosome structure can be detected. In this technique the chromosomes are treated with a warm Nacl solution/Sodium citrate(SSC) (0.15 M Nacl + 0.015 M Sodium citrate, Sumner, et al. 1971) incubation or with diluted trypsin(Wang and Fedoroff, 1972), urea (Shiraishi and Yosida, 1972), Nacl-urea (Kato and Yosida, 1972) or protease (Berger, 1971; Dutrilaux et al.,1971; Sea bright, 1972) and then with Giemsa stain. In all the various G-banding treatments involve proteolytic enzyme or protein denaturants and thus it is thought that the protein components of chromosomes may be implicated in G-banding mechanism. Trypsin-Giemsa banding is still the most commonly used technique. Giemsa is a Romanowsky type dye mixture consisting of various dyes (methylene blue, Azure B, Azure A, Azure C and thionin) and eosine (Comings, 1975). These dye components bind solely to DNA and not to chromosomal proteins. It shows alternate regions of heavy staining (called G-bands) and light staining (interbands). Giemsa banding could be a result of either stronger chromatin condensation or alteration of the histones and other proteins or change of both DNA and proteins (Hsu, 1973; Ross and Gormley, 1973; Mc Kay, 1973, Ruzika and Schwarzcher, 1974). Although G-banding is the most perfect banding technique in plants G-banding has been used to distinguish human metaphase chromosomes (Figure 2.7). G-banding has not been used in plants because of the reason that plant chromosomes contain much more DNA in metaphase then vertebrate chromosomes of the same length. Banding of chromosomes with Giemsa and staining heterochromatic patterns with fluorochromes like quinacrine mustard can provide phylogenetic value. Several other flurochromes can be used singly or in sequence to band cereal chromosomes.

In case of human chromosomes the chromosomes are fixed with methanol and acetic acid (Carnoy's reagent) which removes the vast majority of the histone proteins. The non-histone protein remains behind and contributes most to the banding patterns. Fixation in formalin prevents the development of G-bands. In formalin fixation, there is presence of histone and basic amino acids. Incubation of methanol-acetic acid fixed chromosomes in a saline solution for one or more hours before Giemsa staining is a standard procedure in most G-banding techniques. G-bands are observed where DNA is accessible to thiazine dyes. Pre-treatment of chromosomes is by enzymatic digestion using a proteinase such as trypsin or a mixture of proteinases such as pronase or high pH can be used which damages protein content of the chromosomes. Staining chromosomal preparations with Giemsa-Leishman's stain or Wright stain result in dark bands which alternate with light bands in prophase and metaphase chromosomes.

Features of G-bands - In terms of chromatin structure G-bands show the presence of A-T rich sequences and absence of potentially active promoter elements (CpG islands). Further, chromatin is late replicating, condenses early in mitotic and meiotic prophase and is not DNase I sensitive. G-banding is most useful in analyzing vertebrate chromosomes. In human

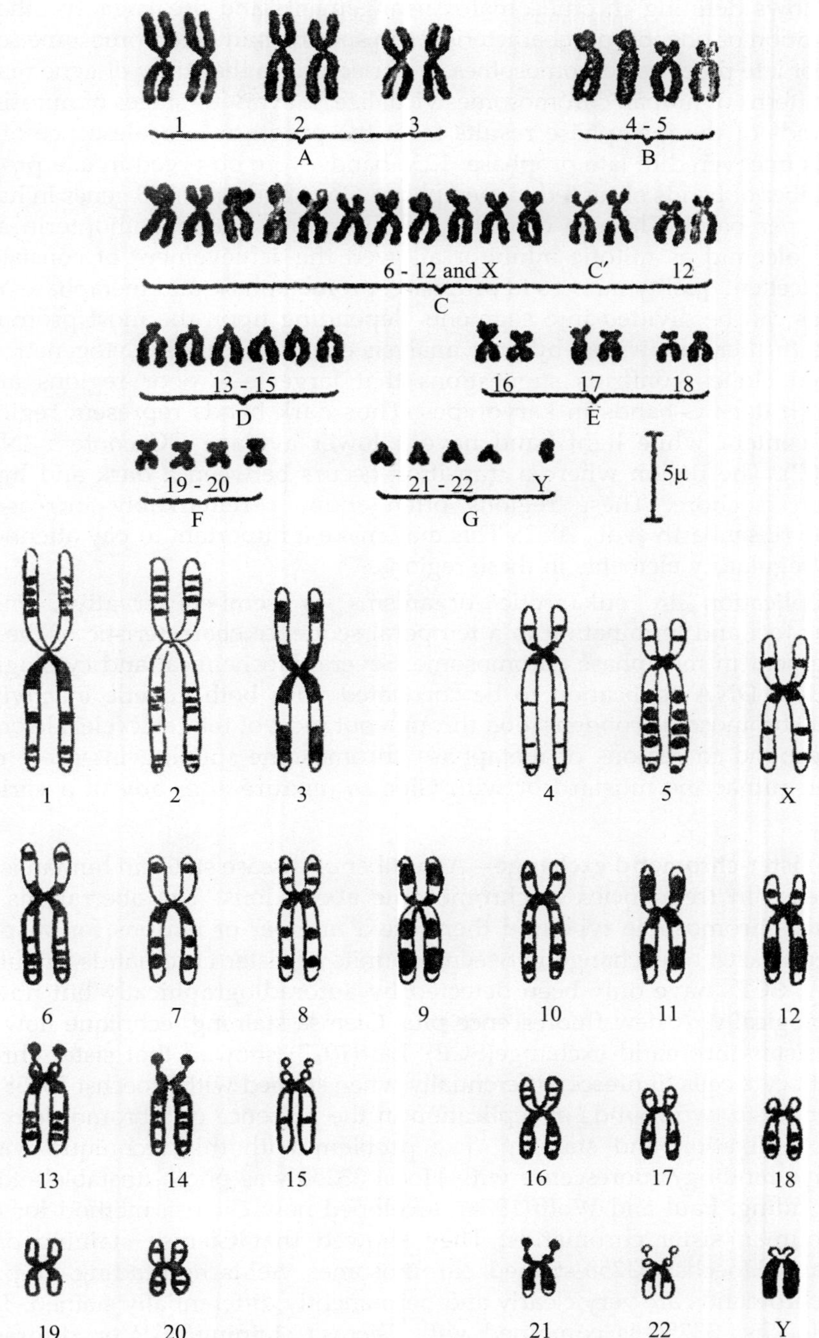

Fig. 2.7 Showing identification of human chromosomes through use of Giemsa banding technique.

G-banding allows defining chromosomal rearrangements and breakage. In other words, it allows localization of phenotypic characteristics to specific minute chromosome segments. So examination of late prophase chromosomes is carried out in the study of gene mapping. The G-banding patterns of human chromosomes visualized at various stages of mitosis show that the major bands of the metaphase results from the progressive coalescence of numerous smaller bands uncovered in late prophase. 1256 bands were observed in late prophase, four times the number of bands observed in metaphase. Considering 30,000 genes in human, there are 23 genes per band. The use of cell synchronization with amethopterin and a brief exposure to colcemid as mitotic inhibitor allowed the achievement of consistently large numbers of excellent quality mitoses in prophase, prometaphase and metaphase(Yunis, 1976) Chromosomes can be divided into segments depending upon the most prominent bands which can be further subdivided by finer analysis of visible bands. Cytogenetic analysis of the sequenced clones confirms suggestions that large G-C poor regions are strongly correlated with dark G-bands in karyotypes. Thus dark bands represent regions of high average GC content while light band have a lower average GC content (Nimura and Gojobori, 2002). The region where a transition occurs between a dark and light band is known as an isochore. These regions often show a remarkably increased rate of recombination (Eisenbarth et al., 2002). This may make it important to pay attention to genes and possible regulatory elements in these regions.

DNA replication in eukaryotic organisms is semi-conservative, initiating at discontinuous foci and terminating in a temporal sequence characteristic of the location of the DNA segment in metaphase chromosome. Several biochemical and cytological studies have found late DNA replication to be correlated with both genetic inactivity and the persistence of chromosome condensation through out most of the cell cycle. Heterochromatic regions correspond to regions of metaphase chromosome staining intensely either with quinacrine or quinacrine mustard or with Giemsa mixture after any of a variety of pre-treatments.

Detection of sister-chromatid exchange—A number of disease states in human is because of marked increase in frequencies of chromosome aberrations. The aberrations are of the chromatid and chromosome type and there are a number of reasons for suspecting that increased frequencies of exchange between subunits of sister chromatids might also exist. Until recently, SCEs have only been detected by autoradiographically but now it can be detected cytologically. A new fluorescence plus Giemsa staining technique now makes the detection of sister-chromatid exchange(SCE). Latt(1973) showed that sister chromatids in human lymphocyte cells fluoresce differentially when stained with Hoechst 33258, if cells are allowed to undergo two rounds of replication in the presence of 5-bromo deoxyuridine(5-BrdU) prior to fixation and staining. The problem with this technique was that the differential dye binding/fluorescence with Hoest 33258 was photo-unstable and subject to rapid image fading. Paul and Wolff(1974) developed new Giemsa method for differential staining of human sister chromatids. They showed that Giemsa staining of 5-Brd U-substituted and Hoechst 33258-stained chromosomes yields non-fading preparations in which sister chromatids are very clearly and permanently differentially stained. Thus in that technique, Hoechst 33258 is combined with Giemsa staining(FGP or **fluorescent plus Giemsa**) to make permanent Giemsa-stained cytological preparations of the differentially labeled chromatids.

2.8.3 R(reverse)-banding

R-banding pattern is just the reverse of G-banding pattern in that the lightly stained G-bands become darkly stained when the chromosomes are incubated in buffer, NaH_2P04(1M) at 88° (~86°C) using a pH of 4-4.5 followed by staining with Giemsa staining which melts more A-T DNA regions to single stranded or acridine orange. Partially single stranded DNA binds the stain much less than the fully double stranded DNA and so R-bands represent the more G-C rich regions between the G-bands. If the pH is changed to 5.5-5.6 the R-banding pattern changes to G-pattern (Dutrillaux and Le jeune, 1971; Schested, 1974). The Kpn family of repeated sequences is enriched in dark bands produced by G-banding whereas the Alu repeat family is enriched in the opposite set of bands, R bands.

2.8.4 Q-banding

Q banding(Caspersson et al., 1968) is obtained with a variety of stains including quinacrine. Chromosomes stained with quinacrine(Q) (the antimalarial drug Atebrin) (Caspersson et al. 1970) or quinacrine mustard(QM) dye or Hoechst 33258[2-[2-(4-hydroxyphenyl)-6-benzimidazolyl]-6-(1-methyl-4-piperazyl)-benzimidazol.3HCL] shows dim and bright bands (Q-bands) under U.V. light and every chromosome pair of human chromosomes could be identified by this fluorescent characteristics. Number, size, intensity and distribution of the bands are specific for each pair and the banding pattern is consistent between different cell types within an individual. The two chromosome fluorescent stains, quinacrine and 33258 Hoechst, have been considered as specific for A-T-rich regions in chromosomal DNA. However, the fluorochromes do not give qualitatively identical staining with certain types of heterochromatin and a general association of chromosome fluorescence brightness with A-T-richness is till in doubt. Q-banding is used for the specification of chromosomes and detection of gross rearrangements, deletion and other abnormalities. Q-banding is used with chromosomes of plants. Quinacrine binds preferentially to DNA regions rich in A-T base pair. Quinacrine fluorescence is enhanced by AT base pair and quenched by GC base pairs with the degree of quenching being related to the GC content of the DNA. Thus regional variations in base composition along the chromosome are the primary basis for Q-banding. G-bands are very similar but not identical to Q-bands and thus on the whole Q-bands coincide with G-bands while R-banding creates the reverse banding patterns, i.e. positively stained G and Q bands correspond to negatively stained R- bands and vice versa. It has been used for studying mice and human chromosome. The relationship between the various types of bands designated as 'Q'-, 'G'-, 'R'- and 'C'- bands in mammalian systems and differentially staining regions in plant chromosomes is not clear.

The G-, Q- and R- banding techniques are particularly important in human chromosome characterization and these have been proved useful visible landmarks and they correlate to some extent with chromatin function. G(or Q) bands have some characteristics of heterochromatin and have less histone acetylation and less number of genes located within them. On the other hand R-bands have relatively more genes and have more histone acetylation. These two groups, G, Q and R of bands also differ in overall amount of long interspersed elements(LINEs) and short interspersed repeat elements(SINEs). LINEs and SINEs are retrotransposons. LINEs occur widely in eukaryotes, especially in some mammals and L1 element of mouse is an example of this. The element contains two open reading frames overlapping by 14 bases and has poly-A tail at 3' end. The upstream ORF does not code anything recognizable but the downstream encodes a protein almost similar to reverse transcriptases from various sources. The element thus has spread like polymerase II transcript. It is scattered throughout the genome and is found where an insertion could be present without disrupting an essential function. Similarly, human Alu element is an example of a SINE which is present in about half a million copies. It shares a great deal of similarity in nucleotide sequence with an abundant small RNA called 7SL. Like LINE elements, Alu has an A-rich sequence at its 3' end and has spread as a RNA polymerase III.

2.8.5 Miscellaneous Bands

H-banding- H-bands have been obtained by either prolonged cold treatment(0°C) (Darlington and La Cour, 1938) or acetic orcein-HCl treatment (Yamasaki, 1956) without prior chilling of plants.

T-banding Higher temperature, for example, 87°C at a pH of 6.7 and Giemsa staining will result especially in staining of terminal regions of chromosome (Dutrillaux, 1973).

N-banding It was used to demonstrate the presence of nucleolus reorganizer (Matsui and Sasaki, 1973). These are found only on a few chromosomes and which are the sites of rRNA genes. In the N-banding technique a hot acid rather than alkali is given to the chromosomes before staining (Gerlach, 1977 and Endo and Gill, 1984). In wheat variety Chinese Spring N-banding pattern is similar to the C-banding pattern but only 16 pairs of wheat chromosomes can be identified by N-banding as many chromosomes lack centromeric and minor bands shown by C-banding. Both C- and N-banding techniques differentially stain constitutive heterochromatin regions and usually allow specific chromosome identification in wheat.

2.8.6 New Banding Techniques

New banding techniques such as biotin labelling which allows the high resolution banding of chromosomal sites containing specific DNA sequences using non-radioactive probes can be used for greater reliability and specificity. These new techniques besides improving chromosomal identification, may enable the identification and location of many gene loci.

In plants the most classical cytological approach to visualizing chromatin is via DNA-binding dyes such as Giemsa stain, aceto-carmine and aceto-orcein. These dyes have relatively low sensitivity for smaller chromosomes and they do not provide much information about the particular DNA sequences. Their reproducible staining patterns for particular chromosomes enable karyotyping and characterization of relatively large changes in genome organization such as rearrangements and breakage of chromosome arm.

DAPI(4′,6-diamidino-2-phenylindole) DAPI, a water soluble fluorescent dye which shows an absorption maximum at 359nm(UV region) and an emission maximum at 461nm (blue color, marks the DNA blue) can be used with both live and fixed cells. It associates with the minor groove of double stranded DNA preferentially at AT rich regions. DAPI unevenly stains chromosomes as AT rich sequences are differentially distributed over their lengths. Repetitive DNA such as perincentric heterochromatin and NORs is usually highly condensed and contains predominantly AT-rich sequences and thus DAPI fluorescence intensities are much higher in these regions than other euchromatic regions. Highly stained regions in interphase chromosomes correspond to 'chromocenters' where heterochromatin exists in a highly condensed state near the centromeres. Assigning these chromocenters is possible using hybridization techniques with BAC clones that are mapped to regions of the genome adjacent to the NORs or neighboring the centromeres. DAPI is used to identify basic chromosome organization in many organisms including plants. The banding dye pair chromomycin A_3 (CMA) and DAPI has been shown to be useful for cytofluorometric DNA base content determination (Schweizer, 1976). With these dyes CMA and DAPI, DNA base composition in individual mitotic chromosomes of Drosophila has been determined. Chromomycin A_3 (CMA) and the closely related mithramycin(MM) are the two DNA-binding guanine-specific antibiotics.

The procedures for different banding techniques are described in chapter 15.

2.8.7 Correlation Between Base Composition and Chromosomal Bands of Metaphase Chromosomes

In human genome, 17-20% of the chromosome complement consists of C-band or constitutive heterochromatin and 80% euchromatic component. The euchromatic component is visible into G-, R- and T-bands. These cytogenetic bands have been presumed to differ in their nucleotide composition and gene density. T-bands are the most G+C and gene–rich. G-bands, the Giemsa-positive bands are formed exclusively by G-C poor isochores. That G-bands are made up of G-C poor DNA was shown by experiments in which G bands were stained with quinacrine, an A-T specific fluorochrome. Bernardi (2000) defined euchromatin as long stretches of DNA of differing base composition, termed **isochores** and denoted L, H1, H2 and H3 which are greater than 300kb in length. Light(L) isochores are G+C poor and thus have gene concentration very low whereas heavy(H) isochores fall into three G+C rich classes(H1, H2, H3) and have higher concentration of genes. In human H2 and H3 isochores are 20-fold more gene-rich. T-bands are formed mainly by isochores of H_2 and H_3 families. R′ bands, defined as R bands exclusive of T bands, comprise both G-C rich isochores (mainly the H_1 family) and G-C poor isochores in almost equal amounts. **Compositional heterogeneity of R′- R′** bands comprise at least three sets of bands, T, T′ and T″ which have a different average base composition as given below in the Table 2.2.

Table 2.2 Shows isochores and corresponding gene concentration.

R bands	*Isochores*	*Gene concentration*
T-bands($H3^+$)	$\mathbf{H_3}$ + $\mathbf{H_2}$ + H_1 + L	+ + + +
R′ bands		
T′ bands H3)	H_3 + H_2 + H_1 + L	+ + +
R″ bands($H3^-$)	H_1 + L	+ +
G-bands(L)	H_1 + **L**	+

Bold-type indicates prominent isochore family; light or GC-poor(L_1 & L_2); Heavy or GC-rich (H_1 and H_2); very GC-rich(H_3)

2.9 OTHER TYPES OF CHROMOSOMES

2.9.1 Lampbrush Chromosomes

The lampbrush chromosomes are the longest chromosomes found in nuclei of certain amphibian oocytes and observed during prophase of meiosis. They are found in the primary oocytes from invertebrates and vertebrates. In frog, *Rana temporaria*(2n=36) the length of each chromosome ranges from ~800 to 1000 μ whereas in case of newt, Triturus viridescens (2n=22) the chromosome length ranges from 350 to 800 μ. The longer size of chromosome can be attributed to uncoiling of the chromosome and also because of very large amount of DNA that the amphibian contains. The lampbrush chromosomes containing the lampbrush nucleolar organizer may produce as many as 1000 nucleoli per cell whereas the normal nucleolar organizer produces only one nucleolus. In the diplotene stage of prophase I of oogenesis the homologous pair is held together by chiasmata. Each chromosome is made up of two chromatids and one chromatid consists of one DNA double helix and thus there are four DNA double helix one in homologous pair. Each pair is having constant pattern of chromomeres. The two major structural components of lampbrush chromosomes are the loops which extend laterally from the chromosome axis and the chromomeres localized on the axis. From chromomeres of each chromosome arises paired of lateral loops that give the chromosome a brush like appearance (Figure 2.8). Each loop is a segment of one chromatid. Each loop or pair of loops is identifiable because of its unique size and character. Some loops are asymmetric in that they are thin at one point of origin from the centromere and thick at the other point. Variously shaped fibers or stalked granules project outwards from the loops. The lops may vary in size (from 1 to over 100 μm corresponding to 3 to more than 300 kb) and shape from chromosome to chromosome. Loops are transcriptionally active regions of single chromatid. Lampbrush chromosomes serve an excellent example of correlation between gene structure and its function. The lampbrush are the structural correlate of transcription of specific set of chromosomal genes.

2.9.2 Polytene Chromosomes

In salivary glands of fruit fly, *Drosophila melanogaster* and some other species polytene chromosomes are found. They are giant chromosomes (the largest chromosome known) having cable like structures with many DNA strands (polytene). They are 100 to 150 times larger than the ordinary mitotic chromosomes. Their unusual length and their banding pattern make them very suitable for chromosome identification and gene localization. Polyteny is a result of endomitosis which itself is a process of chromosome replication (duplication) without karyokinesis and cytokinesis. The DNA molecules do not segregate after replication but remain together during several rounds of cell cycle. After ten rounds of replication there will be more than 1000 DNA double helix molecules lying along side each other with their specific sequences (genes) matched. These interphase chromosomes contain more than 1000 strands of chromatin precisely aligned to produce a characteristic and highly reproducible banding pattern. The cross bands are called chromomeres which are packed with chromatin and interbands regions (achromatic). These polytene chromosomes correspond in linear structure with other chromosomes of the same species. The length of polytene chromosomes in the late larval (3rd instar) stage of Drosophila is ~1180 to 2000 μ,

Lampbrush chromosome

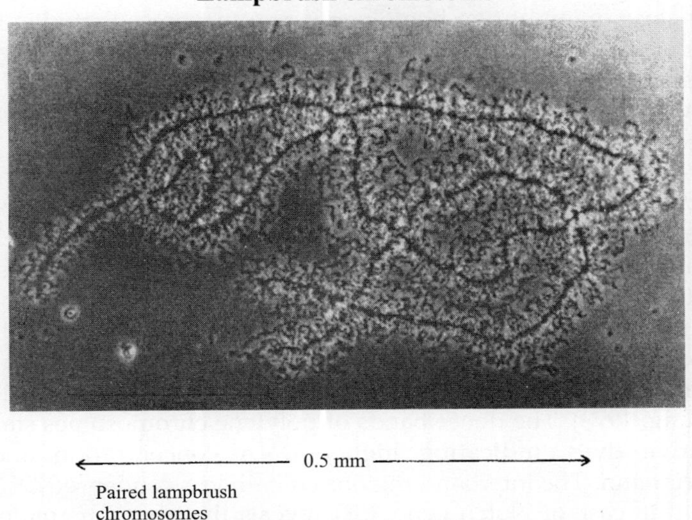

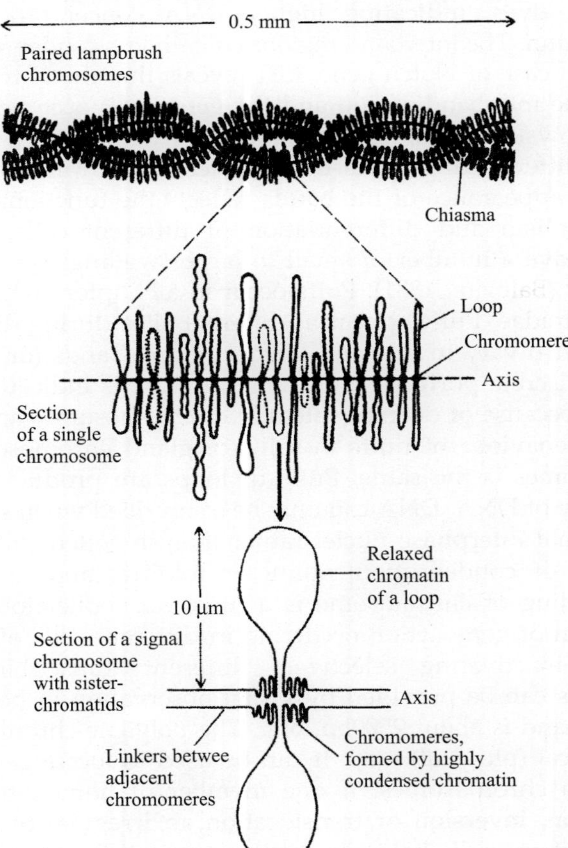

Fig. 2.8 Showing lampbrush chromosome (adapted from Kahl, 2004).

over 100 times the length of somatic chromosome (7.5 μ). Each polytene chromosome really represents a very close union of two homologous chromosomes that give the appearance of only one single chromosome. Chromatids are held together by somatic synapsis (pairing). Bands on the giant chromosomes are associated with gene loci and there is thus a relationship between linearly arranged gene and their banded structure. Polytene chromosomes are useful in correlating cytological and genetic map. The positions of chromosomal inversions, deletions and other rearrangements in particular mutant strains can be determined by analysis of polytene chromosomes by use of light microscopy. Moreover, genetic and molecular maps can be aligned using *in situ* hybridization(ISH) to place cloned DNA segments on the polytene chromosome map. Some of the bands correspond to less than 10kb of DNA with the average approximately 5000 visible bands being about 25kb. Several studies indicate numerical relationship between chromosome band and gene capable of mutating to recessive lethality to be one to one or at any rate not more than one to no more than two. In other words, these bands correlate with the number and positions of the known loci(Judd and Young, 1973). The dense bands of polytene chromosomes stain more intensely with DNA-sensitive dyes, indicating higher DNA concentration and hence greater compaction of chromatin. The interband regions contain less condensed DNA and chromatin as has been shown in case of Notch gene. ISH reveals that the 5′ regulatory regions of the Notch gene lies in the interband region and the gene itself occupies a single cytologically defined band (Rykowski et al., 1988). Banded structure and sizes of the polytene chromosomes may differ during larval development and in different tissues of the same organism. Changing appearance of the bands reflects the functional activities of different genes in the metabolism and differentiation of different cells. These salivary gland chromosomes may have a number of small to large swellings (puffs). **Balbiani rings** are extremely large puffs (Balbiani, 1881). Puffs occur in all Diptera whereas Balbiani rings are restricted to Chironemidae. Puffs appear at different sites during different stages of larval development. Puffs also vary in size and further not all bands (or genes) are involved in puffing. Puffing of discrete portions of the chromosome is indicative of the specific gene action. Puffs appear because of opening laterally of chromosomes and RNA is produced at the puff sites. The behaviour of puffs in salivary gland chromosomes and loops in the lampbrush chromosomes is the same. Puffs or loops are producing RNA as a result of transcriptional activity of DNA. DNA can only be transcribed when stretched out as in puffs or loops and in normal interphase nuclei rather than in coiled state. There is correlation between the degree of condensation(coiling or folding) and the fuctional activity of chromosomes. Uncoiling of chromosome is a must for replication or transcription. The differential expression of gene action occurring in different cells of the same organism is brought about by inactivating selectively different transcribing elements through condensation and this can be provided by visual observation in polytene and lampbrush chromosomes. Each band is about 2500bp long. The polytene chromosomes can be used to establish the cytological (physical) map. It can be used to locate genes and to identify the structural changes in chromosomes. If one member of homologous pair is altered by deficiency, duplication, inversion or translocation an irregularity occurs in pairing and cytological irregularities will help in recognizing different kinds of chromosome modifications and to identify their locations on the chromosome. The absence of a band on a chromosome indicates lack of specific gene action. Identification of small irregularities might go unnoticed in small chromosomes. Polytene chromosomes have also been reported in plants.

Intercalary heterochromatin - There exist specific Drosophila polytene chromosome regions forming frequent contacts with the near centromeric and telomeric heterochromatin and with each other as well (so called ectopic pairing). Kaufman (1939) termed these regions as 'intercalary heterochromatin'. The sites of intercalary heterochromatin have been studied according to the following criteria.

1. Tendency to break (weak points)
2. Ectopic pairing
3. Late replication
4. Existence of repeats (in X and 2R) including those enriched with A-T bases. Highest correlation coefficients were found between weak point behavior, late replication and ectopic pairing. These sites also show higher rate of chromosome aberrations- a characteristic of centromeric heterochromatin.

2.9.3 B-chromosome

The normal chromosome is called A chromosome and the sex chromosome is designated as X or Y chromosome. There is another type of chromosome called B-chromosome. It is a supernumerary element with no vital function. It may be present or absent in different members of a species. B-chromosomes are considered relics of chromosomes that were once involved in the ancestry of plants or it may be parasitic. That is why B-chromosome is called accessory, supernumerary, diminutive, inert, parasitic or ghost chromosome. B-chromosome contains large heterochromatic regions as well as euchromatic segments. **Effects of B-chromosome:** The effects of B-chromosome are: 1. It is detrimental when B-chromosomes are present in unusually high numbers. 2. It affects recombination among A chromosomes by changing the distribution of chiasmata and may 'unlock' genetic regions in which Xta seldom, if ever, occurs and thus it increases genetic variation.It may preferentially enhance recombination in proximal rather than distal chromosomal regions.It has the ability to influence recombination in certain regions of the genome (Hanson, 1961, 1962, 1969). It does so by increasing crossing over with a decline in interference. B-chromosome showed dosage effect with higher numbers giving higher rates of recombination (Nel, 1973) and it is assumed that a number of genes spread along the chromosome affect crossing over (CO). B-chromosome increases the rate of crossing over during microsporogenesis. Further, it does not increase CO in the regions of structural or knob heterozygosity. A considerable number of organisms have B-chromosomes (Battaglia, 1964). It is present in rye (Jones and Rees, 1967) and grasshopper (John and Hewitt, 1965) and shown to change Xta distribution or frequency. Some studies indicate that B-chromosomes confer a selective advantage on individuals which carry them. B-chromosome behaves normally during the development of female gametophyte but is transmitted in excess numbers through the pollen (Roman, 1947, 1948).

2.9.3.1 Mechanism of B-chromsome accumulation

The frequency of B-chromosome increases through modifications of centromeric behaviour. Non-disjunction and preferential fertilization of egg constitute the B-chromosome accumulation mechanism. B-chromosome undergoes non-disjunction regularly at the second pollen mitosis. It occurs in almost 98% of the pollen grains and is seldom observed at frequencies below 50%. Generative nucleus with one B-chromosome produces sperm with zero B's and 2B's and it is the sperm with 2B's which fertilizes egg approximately two-thirds of the time and the polar nuclei one-third of the time and thus preferential fertilization results in an increase in the number of B-chromosome in the progeny in comparison to that present in the parent. Rate of preferential fertilization is constant. The B-chromosomes influence the process of fertilization and they do so by conferring a competitive advantage on sperm in fertilization of the egg (Carlson, 1970b). The genetic regulation of non-disjunction shows that the genetic factor involved in non-disjunction lies on the long arm of B-chromosome (Carlson, 1970a). The sites could be proximal and distal regions of B-chromosome controlling non-disjunction (Roman, 1947, 1949) but then multiple sites could exist between the centromere and the distal tip of B-chromosome which are necessary for non-disjunction. **Molecular study**: Buoyant density and renaturation kinetics showed no difference in DNA attributable to B-chromosomes. Redundant (rapidly annealing) DNA from B-containing plants cross-reacted completely with DNA from zero –B plants which suggested that both B and A chromosomes are closely related which appear to be true for redundant sequences (Chilton and Mc Carthy, 1973). Randolph (1941) concluded that B's and A's do not have genes in common which probably relates to unique rather than repetitive sequences

2.10 STUDY OF INDIVIDUAL CHROMOSOME

The analysis of whole genome can be made by study of either individual chromosomes separately or chromosome fragments. Individual chromosome can be obtained by sorting of chromosomes by a technique called 'Fluorescense Activated Chromosome Sorting (FACS) (Langlois et al., 1982). In this technique spindle arrested (using colchicines or colcemid) cell of metaphase is used for separation of chromosomes. Sorting of chromosomes is based on size and overall base composition of chromosomes. The chromosomes are first stained with a mixture of two dyes that fluoresces in different colors in response to different wavelengths (one in the visible and other one in the U.V light range). One of the dyes binds preferentially to A-T rich region whereas the other binds to G-C rich region. The stream of stained chromosomes in liquid suspension is converted into a spray of droplets, each of which is so small that it supposedly contains one chromosome. These droplets are passed through the laser beams adjusted to the wavelengths required to excite the fluorescence of the two dyes and each droplet is monitored individually. The two colors of fluorescence are recorded using spectrophotometer. As each chromosome differs with respect to I. the amount of fluorescence which is dependent on the size of chromosome and II. the ratio of two colors which varies according to the base composition of DNA, so each chromosome produces a characteristic combination of signals and thus each chromosome can be distinguished. The fluorescence detection system can be adjusted to recognize a particular kind of chromosome and activate an electric field to deflect droplets containing that chromosome in the collection tube. The human chromosomes have been distinguished following FACS using two fluorescent dyes, chromomycin A3 and Hoechst 33258 which bind preferentially of GC and AT DNA regions, respectively. For more on separation of chromosomes see Roy (2009).

2.10.1 Isolation of Chromosome Ends

There is a technique for the isolation of telomeric and subtelomeric regions of eukaryotic chromosomes. In this technique the DNA is treated with terminal deoxynucleotidyl transferase which adds adenosine nucleotides to its ends to form a poly(A) tail. The polyadenylated DNA is then restricted with Sau3 AI which neither cuts in the TTAGGG repeats of the telomere nor in the subtelomeric repeats and thus fragments much larger than bulk of genomic fragments are generated. These larger fragments contain telomeres and are isolated by oligo(dT) affinity chromatography. After isolation the fragments are separated by agarose gel electrohpresis, blotted and hybridized to radiolabeled (TTAGGG)n mer n = 25-40. Lengths of telomere repeat in different eukaryotic species are given below in Table 2.3.

Table 2.3 Shows eukaryotic species along with their length of telomere repeat.

Species	*Length*
Yeast (S. cerevisiae)	300bp
Mouse	50kb
Human	10kb
Arabidopsis	2-5kb
Cereals	12-15kb
Tobacco	60-160kb

2.11 CHROMATIN ORGANIZATION

Interphase chromosomes are generally described as very long uncoiled strands of nucleoprotein, each one randomly scattered and presumably able to occupy any position within the nucleus but then there are exceptions to random arrangement of interphase chromosomes(Fussell, 1975).

Chromatin is not randomly organized within the interphase nucleus but occupies distinct **territories**. Rabl and Boveri suggested that chromosomes in plant cells occupy distinct domains throughout the interphase which reflects their mitotic orientation. In other words, each chromosome is confined to a discrete region referred to as a chromosome territory (Figure 2.9).

Chromosome territories can be visualized using chromosome-specific fluorescent probes. Modern imaging techniques allow all chromosomes in a cell to be visualized simultaneously.

Boveri(1909) showed that chromosomes maintained relatively fixed positions in the nuclei of Ascaris eggs. Telomeres are attached to the nuclear envelop on one side of the nucleus and centromers to the opposite nuclear side and this type of orientation was termed **Rabl configuration(Rabl, 1885)** (Figure 2.10). In other words, centromeres are clustered at one pole of the nucleus while telomeres are left hanging in the other half. This configuration develops as a result of the anaphase movement of the centromeres towards the spindle poles and persists during the subsequent interphase. In human gene- rich chromosomes concentrate at more internal nuclear region whereas the gene-poor chromosomes are located towards the nuclear pheriphery(Croft et al., 1999). DNA constituting an individual chromosome in a eukaryotic nucleus occupies a discrete intra nuclear domain and thus

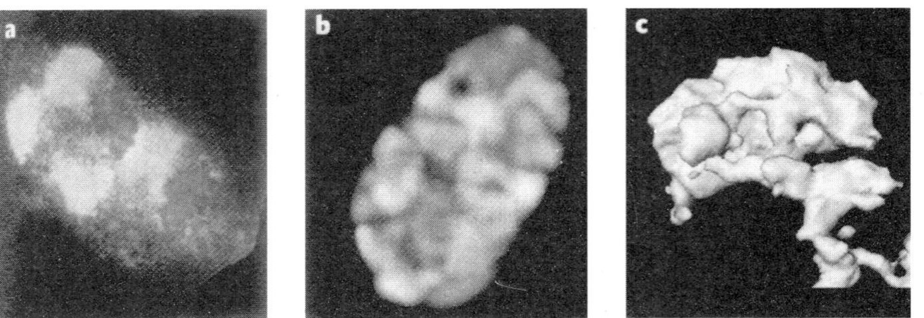

Fig. 2.9 Showing spatial organization of chromosomes in the nucleus.(a) shows visualization of territory of specific chromosomes(12, 14 15) of mouse; (b) shows visualization of all chromosomes of a human fibroblast and (c) showing chromosome territories to be not solid entities.

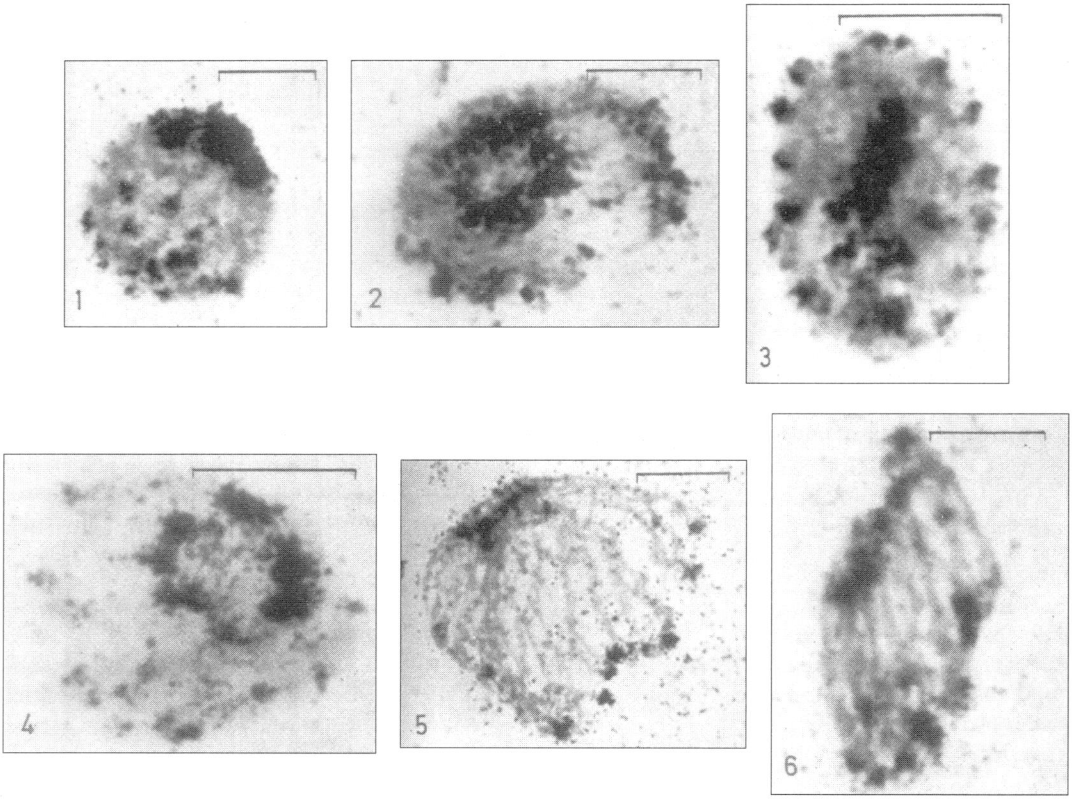

Fig. 2.10 Showing chromosomes in G1, S, G2 and early prophase of Allium cepa root tip nuclei to be oriented in the same position as telophase chromosomes. Chromosomes in higher eukaryotes frequently maintain their telophase orientation from the end of telophase, during interphase and well into the next prophase. The nucleus was treated with ^{3}H-thymidine for 30 minutes. Figures 1 and 2 show interphase nucleus, 3 and 4 show early prophase, 5 shows prophase and 6 shows late prophase nucleus(adapted from Fussell, 1975).

permits the direct quantitation of gene or chromosome numbers given an appropriately specific probe.

Chromosome arrangements are also specific to the cell and tissue type and can change during processes like differentiation and development. Chromosomes 18 and 19 in human tend to occupy a peripheral and internal position, respectively. Chromosomes occupy non-random radial positions relative to the centre of the nucleus. Changing the relative arrangement of genome regions to bring them into close proximity is functionally relevant for the gene activity and for the formation of chromosomal translocations. What is required is to identify the molecular mechanisms responsible for the repositioning of single genes, genome regions and whole chromosomes within the nuclear space and to determine how these mechanisms respond and contribute to physiological cues such as stimulation through signaling pathways. We have seen that placing a gene near a block of heterochromatin represses expression of that gene. Gene loci may also be controlled in a similar way by interactions with particular regulatory regions on the other chromosomes. Interchromosomal interactions may be involved in differentiation-specific gene activation in immune T-cells and in gene imprinting. The relative positioning of DNA sequence elements to one another and their physical interactions is emerging as highly significant for cellular functions. Changing the relative arrangement of genome regions can bring regulatory regions into proximity with otherwise distant genes to control their function.(Meaburne and Misteli, 2007). Chromosome painting technique (see Roy, 2009) has revealed that each chromosome in interphase nuclei is not entangled with others but occupies a discrete territory. Because of abundance of dispersed repetitive DNA sequences throughout the genome in most plant species chromosome painting in plants has been limited to β-chromosomes which exist only in limited species.

Mendel's Laws of Heredity

3.1 LAWS OF HEREDITY

Gregor Mendel formulated two laws of heredity, law of segregation and law of independent assortment in 1865 which were rediscovered in 1900 by de Vries, Tschermak and Correns. Mendel postulated that a **character** is determined by hereditary unit called factor which was later termed gene by Johannsen in 1909 and which occurs doubly in a diploid individual. **Trait** or character(contraction of the word characteristic) refers to one of the many details of structure, form, substance, or function which make up an organism and it develops as a result of interactions of genes with itself and with the environment(P = G + G X E + E). The **gene** is the physical entity transmitted from parent to offspring that influences hereditary traits. The set of genes present in an individual constitutes its genotype and the physical or biochemical expression of the **genotype** is called the **phenotype.** Genes can exist in different forms and these alternative forms of a gene are called **alleles.** Each gene occupies a position(**locus)** along the length of a chromosome and the difference between alleles segregating at meiosis becomes a **marker** for the gene**. Gene designation-** Genes have often been named after the effects of the mutant allele by means of which they were first identified: e.g., w for white eye in Drosophila, *sh* for shrunken endosperm in maize and *gal* for inability to use galactose in yeast. In Mendel's work, only letter symbols were used which were derived from their descriptive names, for example, S for the dominant factor smooth seed and s for the recessive factor wrinkled seed, Y for yellow cotyledon and y for green cotyledon, T for tall plant and t for short or dwarf plant, etc.. At that time there was no idea which factor represents the normal appearance of the organism or is the most commonly found. In Drosophila, genes have sometimes been given capital or lower case initials according to whether the originally identified mutant allele was dominant or recessive(e.g., Abd-A and abd-B). Further genes were often assigned arbitrary letters with capital and lower case initial letters to distinguish dominant and recessive alleles, for example, A/a, R/r determining pigmentation in maize. **A diploid** individual carrying identical alleles of a gene is called **homozygote**(adjective **homozygous**) and a diploid individual carrying two different alleles of a gene is called **heterozygote**(adjective **heterozygous**).

3.1.1 Law of Segregation

Considering a gene determining plant height in case of pea, *Pisum sativum* the diploid individual will carry two alleles(either TT or Tt or tt) for that gene at the gene locus. Here TT, Tt or tt is genetic constitution of the individual(**genotype**). The individual with either TT or tt will be homozygote and Tt individual will be called heterozygote. Here TT, Tt and tt are the genotypes of individuals and the phenotype of TT and Tt genotypes will be tall and of tt genotype will be dwarf. When the cross, Tall(TT) x Dwarf(tt) is made the F1(first filial generation) receives a single factor from each of the two parents differing in a particular trait(the plant height). Only one parental trait appears in the F1, i.e. there is a simple dominance and which subsequently segregate or separate in the gametes and the parental traits will appear in the F2(second filial) generation as can be seen below. The F1 produces two types of gametes, T and t.

♀ \ ♂	T	t
T	TT	Tt
t	Tt	tt

In the F2 generation the genotypic ratio will be 1:2:1 whereas the phenotypic ratio will be 3:1.

In case F1 is testcrossed to recessive dwarf parent, tt then in the test cross progeny the phenotypic ratio of dominant tall and recessive dwarf will be 1:1 as shown below. The genotypic ratio will be 1:1 as well. The F1 will produce two types of gametes- T and t in equal proportion(50:50) as shown below.

♀ \ ♂	T	t
t	Tt	tt

The gene action can be complete dominance, partial dominance, co-dominance or overdominance. When the F1 shows the phenotypic effect of one allele and thus the heterozygote is phenotypically similar to the dominant homozygote it is called **complete dominance**. In the above example T is dominant over t, i.e., F1 will be tall. If the heterozygote is exactly intermediate in appearance between the two homozygotes it is a case of **partial dominance**. In case of **codominance** the heterozygote shows the distinct effects of both alleles. In case of **overdominance** the F 1 is more extreme in appearance than either of the two parents.

3.1.2 Law of Independent Assortment

The different characters are determined by different factors and in his second law of heredity Mendel postulated that the segregation of factors at one locus is independent of segregation at other locus. In other words, every characteristic is inherited independently of every other characteristic. When two parents, one having yellow cotyledon(YY), round seed(SS) and other having green cotyledon(yy) and wrinkled seed(ss) are crossed the F1 is yellow cotyledon and round seed as Y is dominant over y and S is dominant over s When this F1 is selfed, it produces yellow cotyledon round seed, green cotyledon round seed, yellow cotyledon wrinkled seed and green cotyledon wrinkled seed in the ratio of 9:3:3:1 in the F2 generation as shown in the figure below. The F1 the double heterozygote will produce four types of gametes, YS, Ys, yS and ys.

♀ \ ♂	YS	Ys	ys	ys
YS	YYSS	YYSs	YYSs	YySs
Ys	YYSs	YYss	YYss	Yyss
Ys	YYSs	YYss	Yyss	Yyss
ys	YySs	Yyss	yySs	yyss

When the double heterozygote F1 is crossed to the double homozygous recessive parent, yy ss(green cotyledon, wrinkled seed) the genotypic and phenotypic ratio will be 1:1:1:1. The double heterozygote will produce four types of gametes in the ratio of 1:1:1:1 whereas the recessive parent(the female) will produce only one type of gamete, ys, as shown below.

♀ \ ♂	YS	Ys	yS	ys
ys	YySs	Yyss	yySs	yyss

In the presence of epistasis deviations from the 9:3:3:1 will be observed in the F2 generation progeny and so will be the deviation in case of linkage between genes.

3.1.3 Testing Goodness of Fit

We have seen above that in case of monohybrid and dihybrid cross various ratios such 3:1, 1:1, 9:3:3:1 and 1:1:1:1, etc. are observed in the F2 generation. Now when we conduct experiments we count the number of individuals falling into distinct classes and then we ask the question whether or not the observed numbers in these classes conform to the theoretical ratios. The chi-squared (χ^2) test is used for assessing goodness of fit of the data given in the table below.

Single factor segregation					
Phenotypic classes	I	II			
Observed number	O_1	O_2			
Expected number(F2 population)	E_1	E_2			
Expected number(Test cross population)	E_1	E_2			

Two factor segregation					
Phenotypic classes	I	II	III	IV	
Observed number	O_1	O_2	O_3	O_4	
Expected number(F2 population)	E_1	E_2	E_3	E_4	
Expected number(test cross population)	E_1	E_2	E_3	E_4	

The chi-square (χ^2) is calculated as

$$\chi^2 = \Sigma(O- E)^2/E$$

The chi-square is thus calculated by summing the squares of deviations from expectation divided in each case by the expected number, i.e.

$$\chi^2 = (O_1 - E_1)^2 + (O_2 - E_2)^2 \text{ in case of single factor segregation}$$

and

$$\chi^2 = (O_1 - E_1)^2 + (O_2 - E_2)^{2+} + (O_3 - E_3)^2 + (O_4 - E_4)^2 \text{ in case of two factor segregation}$$

The number of degrees of freedom, n equals the one minus the number of classes(N-1) which is one in case of one factor segregation and 3 in case of two factor segregation. The method of chi-square is not used if any of the expected number is 5 or less than 5 and instead the formula is redefined taking into account Yates' correction for continuity.

These two laws were not true biological laws as it was later observed that most contrasting traits do not show simple dominance and further different factors on the same chromosome tend to be inherited together(showing linkage). Blixt (1972) reported that genes for seven pairs of alternative Mendelian traits in garden pea were located on four chromosomes, 2 genes (yellow vs green cotyledon and white vs purple-violet seed coat and flower) on chromosome number 1, 3 genes (tall vs dwarf, axial pods vs terminal pods and full vs constricted pod) on chromosome number 4 and 2, one on each chromosome. The gene for round vs wrinkled seed is found on chromosome number 7 whereas the gene for green vs yellow color of unripe pod is found on chromosome number 5.

3.2 TYPES OF GENES/CHARACTERS

Considering the genetics of traits there are two types of traits-**qualtitative** and **quantitative traits**. Qualitative traits refer to those Mendelian traits(simple traits) which show discrete variation and are under control of one or a few genes and each gene having large effect. Quantitative traits are those traits which show continuous variation and are under control of many genes(polygenes), each having small, equal and additive effects.

3.3 CHROMOSOME THEORY OF HEREDITY

Whether the Mendelian factors were transmitted to the progeny as such or they were related to some other cellular structure was not known. In other words, what was the physical basis of inheritance was not known until then. How this hereditary material which was finally recognized as DNA or RNA is physically arranged within an organism and what are the methods of transfer of hereditary material from one generation to the next became clear later on. During 1865-1900 understanding about cell, nucleus and chromosomes was well advanced and the significance of meiosis and fertilization had been demonstrated by Beneden (1883), Strasburger(1884), Boveri (1890), Hertwig (1890 and Sutton (1902-03). Sutton visualized chromosome in the nucleus as the physical carriers of Mendelian factors and thus saw a physical basis of inheritance. Sutton and Boveri (1902-3) proposed the chromosome theory of heredity and drew a parallelism between the behaviour of chromosome and Mendelian segregation of factors (or genes), i.e. correlation between gene transmission and chromosome transmission. The zygote receives maternal set of chromosomes through egg and paternal set through sperm and thus the somatic nucleus contains pairs of sets of chromosomes(homologous chromosomes). Further different chromosome pairs orient at random on meiotic spindles and there is thus independent assortment of non-homologous chromosomes, thus accounting for the physical basis for the qualitative and quantitative aspects of Mendelian segregation. The principles underlying chromosome theory of heredity are as follows.

1. Somatic cells contain pairs of homologous chromosomes and there is separation of homologous chromosomes as a result of which the chromosome number is reduced from diploid to haploid and further there is exact length wise division of chromosomes allowing for equal distribution of the linearly distributed factors(the genes) to the daughter cells during meiosis. Thus we can see the correspondence between segregation of factors and the disjunction of chromosomes. It further points to that Mendel's paired factors are physically present in the homologous chromosomes.
2. The assumed physical existence of the chromosomes in the nucleus between mitosis gives the genetic continuity necessary for the organs of heredity.
3. Nucleus controls the working of cells wherein each chromosome or chromosome pair plays a definite role.
4. The equality of chromosomes of the fusing germ cells(produced as a result of meiosis) corresponds to equality of male and female in heredity

Mendel's laws have been found working on many plants, animals and fungi. The information from fungi is very valuable as in many of the fungi, especially those of Ascomycete it is possible to isolate and characterize the products of meiosis(haploid gametes) in their original groups of four (tetrads). In case of bread mold, Neurospora all of the products(ascospores) of a single meiotic event are linearly ordered in the ascus(sac) in the same sequence in which the chromatids were on the meiotic metaphase plate and as the meiosis is followed by a mitotic division and so in the mature ascus each of the four chromatids is represented by a pair of ascospores in tandem order. The mature fruiting body(perithecium) contains over 100 asci, each containing eight ascospores. The recovery

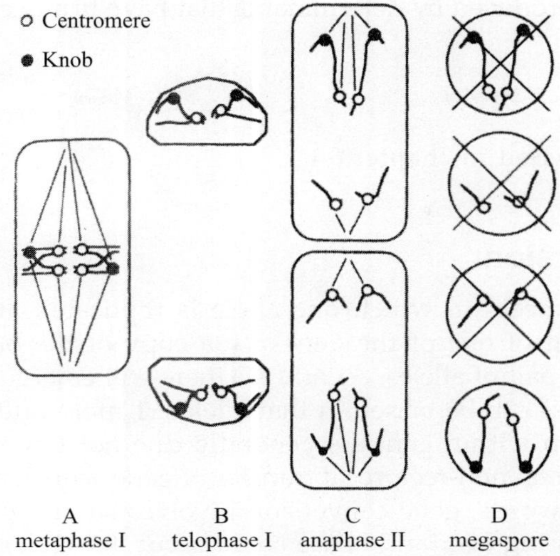

Fig. 3.1 Showing the Rhoades model for meiotic.

and study of all the four products from a meiotic event is called **tetrad analysis**. Here each of the haploid products is isolated and grown separately which germinates into new fungal cell(fungal mycelium composed of interwined filaments called hyphae) and thus phenotype of each gamete can be recorded and analysed. If there is crossing over between centromere and the gene in question then the meiotic products will be in the ratio of 2: 2: 2: 2 but if no crossing over occurs then the linear ratio will be 4:4. Thus whether or not the crossing over has occurred can be ascertained. In case of plants the phenotypes of the gametic products can not be known.

3.4 DEVIATION FROM MENDELIAN SEGREGATION

Use of non-mendelian segregation is used as a criterion of extrachromosomal heredity but this non-mendelian segregation can be produced by determinants that are present either in chromosome or extrachromosomal element. According to the principle of classical mendelian genetics of Mendel(1866) allelic information is stably inherited from one generation to the next resulting in predictable segregation patterns of different alleles. Although several exceptions to this principle are known they all represent specialized cases that are mechanistically restricted to either a limited set of specific genes(for example, mating type conversion in yeast, Klar et al., 1979) or specific type of alleles(for example, alleles containing transposons(Mc Clintock, 1950) or repeated sequences(Gondo et al., 1993). HTH –organ fusion gene in Arabidposis can inherit allele-specific DNA sequence information that was not present in the chromosomal genome of their parents but was present in previous generation. This seems to be a general mechanism for extragenomic inheritance of DNA sequence information. The genetic restoration events are the result of a template directed process which makes use of an ancestral RNA sequence cache(Lollel et al., 2005). Non-Mendelian segregation that can be produced by determinants that have firmly established chromosomal basis are :

i. **Gene conversion**
ii. **Somatic mutation**
iii. **Aneuploidy(discussed in chapter 6)**
iv. **Paramutation**.

3.4.1 Gene Conversion

Gene conversion is a process in which one allele is replicated at the cost of another. It involves the replacement of one of the genes by a copy of the other in a non-reciprocal process. Considering the pair of alleles, A1 and A2 there will emerge two copies of A1 or two copies of A2. This process can be biased in that allele A1 more often converts A2 than vice versa. The recombination within genes is generally due not to crossing over as such but rather to **conversion events**- non-reciprocal transfer of gene sequence from one chromatid to its homologue. In other words, gene conversion involves similar process to recombination but need not lead to crossing over and hence could occur between similar genes at different chromosomal locations without leading to infertility. Gene conversion tends to occur in the immediate vicinity of crossovers but exceptions are there and in Drosophila up to 70-80% of conversions and up to 50-70% in fungi occur without crossing over. In fungi, high rates of conversion are characteristic of markers at one end of a gene with the rate of conversion typically falling with distance from that site. Those conversion events suggested the presence of genetic elements(initiators or recombinators) that impose a high rate of recombination(conversion and crossing over) upon their neighborhood. In case of meiosis in ascomycetes(the fungus that produces ascospores) the ratio may be 6 : 2 or 5 : 3 instead of 4 : 4 or 2 : 2 in the tetrads. Mutations occurring in somatic cells are called somatic mutation.

3.4.2 Paramutations

Paramutation refers to a mitotically and meiotically heritable change in gene expression that does not involve alterations in DNA sequences. Paramutation is a heritable epigenetic modification induced in plants by cross-talk between allelic loci. It was first observed in maize and subsequently in a variety of plants such tobacco and tomato. It is a heritable epigenetic change in the phenotype of a 'paramutable' allele initiating interaction in heterozygotes with a 'paramutagenic' allele form of locus. In other words, paramutation is an allele dependent transfer of epigenetic information which results in the heritable silencing of one allele by another(Chandler and Stam, 2004). Paramutation at the b1 locus in maize is mediated by unique tandem repeats that communicate in *trans* to establish and maintain meiotically heritable transcriptional silency(Stam et al. 2002). The mop 1(mediator of paramutation) gene is required for paramutation and mop 1 mutations reactivate silenced mutator elements(Lisch et al. 2002). Plants carrying mutations in the mop 1 gene also stochastically exhibit pleiotropic developmental phenotype. Paramutation is meiotically stable and inherited in the absence of the inducing allele. In animal species, the closest observation of paramutation is the changes in DNA methylation profiles directed by allele locus and this phenomenon has been described as transvection or paramutation like effects.

Somatic hyper mutation During an immune response, B cells located in germinal centers of the lymph nodes and spleen mutate the variable region of their immunoglobulin(Ig) genes at high rates by the process of somatic hypermutation(SHM). SHM is critical for the generation of high affinity antibodies and efficient immune responses. The reaction is started by the deamination of C to U in Ig genes by activation-induced deaminase(AID). These U's are detected by either uracil DNA glycosylase(UNG) and processed by the base excision repair pathway or by the Msh2/Msh6 dimer and processed by the mismatch repair pathway. Aberrant repair by these pathways in conjunction with error-prone polymerases, especially polymerase η leads to a characteristic pattern of mutations in which A residues are targeted approximately twice as frequently as T residues on the non-transcribed top strand. The origin of this asymmetry remains unknown(Unniraman and Schartz, 2007).

3.4.3 Meiotic Drive

Meiotic drive (MD) refers to preferential transmission of a particular allele(or chromosome) of a heterozygous pair to the progeny. It occurs in natural populations of fungi, plants, insects and mammals. The SD(segregation distorter) locus in chromosome II of *Drosophila melanogaster* has two alleles, one wild type and one a distorter of the wild type. In homozygous wild type produces normal segregation but the heterozygous male(SD/SD$^+$) showed deviation from 1:1 segregation. The gene required for distortion is called Sd which suffered a gain of function mutation which deleted 234 amino acids at the C- terminus. The functional SD product is a truncated version of the nuclear transport protein, RanGAP activator of the Ras- related nuclear regulatory protein, Ran. Duplication of whole chromosome or parts of chromosome may occur in meiosis resulting in irregularities in the gametes and thus only a few sperms contain the normal allele. The phenomenon of hybrid dysgenesis is now known to be caused by transposable P factors in Drosophila.

Meiotic drive describes the outcome of any genetic system which serves to increase the transmission of a chromosome or chromosomal segment by distorting normal Mendelian segregation. Chromosome 10, known as Abnormal 10(Ab 10) causes the preferential segregation or meiotic drive of knobs and loci linked to knobs. Rhoades(1952) proposed a model to explain the meiotic drive. A pre-requisite for MD is that a plant be heterozygous at one or more knobs loci. As knobs are widely separated from a centromere, recombination occurs between knobs and centromeres so that sister chromatids become heteromorphic for the presence of a knob. The spindles than interact directly with knobs to form neocentromeres which cause the knobs and closely linked genes to lie very close to the spindle poles at telophase I. The knobbed chromatids maintain their pheripheral cellular location until metaphase II. Anaphase II segregates the knobbed chromosomes to the outermost cells of the linear tetrad. Finally, as only the basal cell develops into the megagametophyte, knobbed chromosomes are preferentially transmitted. The Figure 3.1 shows the distinct stages in meiotic drive in maize. The recombination between the centromeres and knobs(A), the knobs form neocentromeres and are pulled to the poles(pulled ahead of the true centromere) at telophase I. The knob location is maintained through metaphase II and extreme neocentric activity during anaphase II causes visible extension of chromosome arms(chromosome arms stretched from the metaphase plate all the way to the spindle). All knobs that recombine with the centromere ultimately lie at one of the outermost megaspores and only the basal megaspores survive to form a female gametophyte.

Meiotic drive does not occur in the male as the tetrad is tetragonal (four sided) and all of the products of meiosis produce gametes. Although in maize neocentromere activity of knobs occurs in both male and female meiocytes but meiotic drive is limited to the female. However, there is no direct evidence for the most critical component of the model that neocentromere formation is required for the meiotic drive(Dawe and Cande, 1996).

Meiotic drive has been found in maize, man, Drosophila and Neurospora. In case of Drosophila segregation distorter and the t-haplotypes of mouse, one homolog is preferentially transmitted as it carries a drive locus, or multiple drive locus which can inactivate the sperms carrying the other homolog. As a consequence a pair of genetic markers which would normally be passed to progeny in a ratio 1:1 can be found in ratios as high as 99:1. In maize there is no gametic inactivation and meiotic drive is less pronounced, distorting test cross ratios anywhere between 1:1 and 3:1 depending upon the genetic distance between marker and knob. The preferential segregation is not limited to markers on chromosome 10 but it can be detected with any gene linked to a knob as long as Ab 10 is present. Knobs are found in distal locations on arms of maize chromosomes. In most strains of maize the knobs are inactive and lag behind the true centromeres at anaphase.

3.5 WILD TYPE VS MUTANT ALLELE

The fully functional form of the gene found in the wild(most commonly found) or normal organism is called wild-type allele. There may be many different wild-type, fully functional alleles in an organism, difficult to distinguish without molecular analysis. Wild type alleles are also designated as + superscript(e.g., w^+ , vg^+) in Drosophila, by a capital letter in maize(e.g., Sh,R,A),in yeast by capitalization of the whole symbol(e.g., GAL). Wild type homozygotes can be written as +/+ or just + and heterozygotes for the single gene difference such as vestigial and wild type are symbolized as +/vg. The different wild type genes can be discriminated by using the symbols such as vg+ and H^+ or $+^{vg}$ and $+^{H}$. The wild type/ mutant heterozygote for two gene differences(double heterozygote) is symbolized as vg/+, H/+. **Mutant allele**- A form of gene phenotypically observed and presumed to have derived from the wile-type by spontaneous or induced mutation. Mutant alleles are more or less functionally defective and are usually recessive to the wild-type. In most cases recessive mutants are designated by small letters(e.g., vg for vestigial- winged in Drosophila) and dominant mutations by capitalizing the first letter(e.g., H for hairless flies). When different genes mutate to give the same or similar phenotypes they are often given the same symbol followed by an identifying number such as Sh1, Sh2 in maize and GAL1,GAL2,etc., in yeast. Finally, mutant alleles are usually distinguished by different superscripts, sometimes descriptive of the phenotype of homozygote(e.g., w^a, $w^{ch.}$$w^e$ for apricot, cherry and eosin eyes in Drosophila . In yeast it is more usual to number mutant alleles sequentially, without the use of superscripts, e.g., gal1-1, gal1-2, etc.

3.6 MULTIPLE MARKER GENETIC STOCKS FOR LINKAGE STUDIES

A stock with at least one genetic marker for each chromosome greatly increases the efficiency of the linkage test but one marker per chromosome is not sufficient. Two marker per chromosome, properly placed, may effectively test the major portion of each chromosome for linkage(Carter et al., 1951). A genetic stock with a marker on one arm of each chromosome can be supplemented by a second stock with markers for the arms not marked by the first one. Another possibility is the construction of several stocks, each of which marks both arms of two or more chromosomes. The major problem with this type of method is that the genetic marker itself may affect the expression of the character under investigation(e.g., disease resistance trait). Another problem is that the genetic markers may not be available for certain chromosome arms. For example, in case of maize short arms of chromosomes 4, 5 and 8 are poorly marked or have no markers(Burnham, 1966). Thus character being investigated must be distinguished easily from those in the marker stocks.

3.7 SEX-LINKED INHERITANCE

For sex-linked inheritance see Chapter 4.

Linkage and Genetic Recombination

4.1 LINKAGE

Linkage is the association of genes located in the same chromosome. Such a group of linked genes forms a linkage group. Thus each chromosome can be called a linkage group. The number of linkage groups corresponds in number to the haploid chromosome set. In other words, the number of linkage groups equals the number of pairs of chromosomes. So in case of maize with 2n=20 there are 10 linkage groups. Genes falling into different linkage groups, i.e., present on separate chromosomes show independent segregation and those in the same group show linkage, i.e., combinations of alleles segregating from meiosis tend to be the same as those that entered meiosis. In other words, linked genes tend to remain together but they are not always inherited as single units and even linked genes generally show some frequency of recombination ranging from near zero to a maximum of 50% depending upon the particular genes involved. Recombination between linked genes is a result of occurrence of crossing over between chromatids which can be observed at the diplotene stage of meiosis. Linkage is detected when the proportion of recombinant progeny falls significantly below 0.5(or 50%). F is the frequency of recombinant progeny for any given pair of markers and in the absence of chromatid interference this will not statistically exceed 0.5 irrespective of the number of crossing overs that occur between these loci.

4.1.1 Phases of Linkage

There could be either **coupling** or **repulsion phase** linkage. Considering two loci, A and B, each with two alleles, A and a, and B and b, linkage between either two dominant or two recessive alleles, say AB or ab can be said to be in coupling phase linkage whereas linkage between alleles, say Ab or aB can be said to be in repulsion phase linkage.

4.2 RECOMBINATION

Genetic recombination includes homologous genetic recombination, site specific recombination and DNA transposition. Genetic recombination between homologous chromatids involves breaking and re-joining of DNA. The genetic information is thus exchanged through homologous recombination. This exchange of DNA is also called crossing over which is observed with light microscope. C.O. takes place in the germ line cell. Each cell undergoes two rounds of cell division without an intervening round of DNA replication. Meiosis begins with replication of DNA during the prophase of first meiotic division. The resulting copies remain associated at their centromere and referred to as sister chromatids. Thus two homologous chromosomes exist as 2 pairs of sister chromatids. C.O. links two pairs of sister chromatids together at points called chiasmata (singular chiasma) and thus all four sister chromatids are linked which is essential for the proper segregation of chromosomes in the subsequent meiotic cell division. Chiasmata are the points of reciprocal chromatid exchange or can be said to be the sites of physical exchange between non-sister chromatids. Thus what is seen is that there is actual exchange of segments between paired chromosomes. C.O. occurs at the chromatid level. The chromosomes are divided into chromatids at the time of crossing over and each exchange involves only two of the four chromatids. Chromatid exchange produces two crossover and two non-crossover chromatids whereas C.O. at chromosome level produces all C.O. chromatids. The recovery of C.O. and non C.O. chromatids from a bivalent has demonstrated that C.O. occurs at the chromatid level. In Neurospora all four products of meiosis can be recovered in the ascus (tetrad analysis) providing a complete picture of recombination. C.o. is not an entirely random process and hot spots have been identified on eukaryotic chromosomes. The crossing over is facilitated by the formation of a structure called **synaptonemal complex** which is present throughout the period of synapsis. (Moses, 1968; Westergaars and von Wettstein, 1972).

4.2.1 Synaptonemal Complex

It consists of two lateral elements and a central element. In leptotene each chromosome possesses a single lateral element which probably is attached between sister chromatids. In zygote the lateral elements of homologous chromosomes pair up and a central element forms between them to complete the synaptonemal complex. In the deplotene when the homologous chromosomes start separating out the complex is shed and by diakinesis this structure is terminalized and finally disappears. In most organisms formation of synaptonemal complex is a pre-requisite to crossingover and in general, the absence of synaptonemal complex is associated with no crossing over. Whether synaptonemal complex plays a role in synapsis or crossing over is not certain. Synaptonemal complex has been observed in maize in regions of non-homologous pairing (Gilles, 1973 and Ting, 1973) and further even in the absence of synaptonemal complex C.O. is taking place in males Drosophila (*D. ananassae*) although the rate of C.O. in males is only one-third the rate found in females. Synaptonemal complex is attached to the nuclear envelope in a number of organisms. Generally in animals in pachytene all chromosomes are attached by their telomeres to the nuclear envelope but this is not the case in plants. **Synapsis**- So far the evidence is that the pairing does not begin at the centromere and there is existence of preferred sites of pairing initiation. Moens (1969) reported that formation of synaptonemal complex begins near the nuclear membrane in *Locusta migratoria* and once the pairing is initiated secondary sites of pairing are established along the length of chromosome and gene to gene (a zipper like) synapsis between sites completes pairing. Synapsis in corn begins in subterminal regions and never in the vicinity of the centromere (Burnham et al. 1972).

4.3 PREFERENTIAL PAIRING

Pairing tends to occur within genomes and not between genomes so bivalent formation is favoured at the cost of quadrivalents. Preferential pairing also occurs within a species. Chromosomes which are more closely related in structure or racial origin tend to pair to the exclusion of less closely related homologues. C.O. gives rise to recombination between genes and recombination frequency between any two genes is roughly proportional to the distance between the genes. Genetic recombination is used for determining the relative positions of and distances between different genes. The recombination helps in DNA repair in case of DNA damage, promotes orderly segregation of chromosomes during meiosis II and enhances genetic diversity.

4.4 GENETICS OF RECOMBINATION

Genetic recombination does not occur uniformly in the genome. Numerous elements of different types and from widely different organisms have been found to stimulate homologous recombination (HR) in nearby genetic intervals. In some cases the molecular basis of such differences in recombinogenicity are understood. In E. coli the chi sequence, GCTGGTGG alters the activity of the Rec B C(D) protein to reduce exonuclease and enhances helicase activities, leading to an increased frequency of HR in nearby intervals (Myers and Stahl, 1994). The *X* is present in *E.coli* about once per 5kb. It is present in *S. cerevisiae* at a somewhat lower frequency. It is λ's double chain break site, *cos*, that appears to be analogous with indentified eukaryotic recombinators. The *cis* acting genetic element Cog in Neurospora imposes a high rate of recombination on genes in its neighborhood. M26 creates a recombinator in *S. pombe*. The M26 site in *Schizosaccharomyces pombe* ATGACGT binds a specific heteromeric protein and appears to be a preferred site for DNA cleavage. HR is also stimulated in regions which are near an origin of replication, perhaps because of the nicks or single stranded segments which are associated with replication initiation recombination. Also, transcription has been seen to increase HR in yeast, in mammalian cells and inconsistently in bacteria. The mechanism linking transcription and recombination is unknown but it has been proposed that transcription might render the DNA more accessible to recombinases or that topoisomerase cleavages enhanced by the unwinding associated with transcription create structures which promote recombination (Buzina ans Shulman, 1996).

Genetic recombination depends on both enzymatic and *cis*-acting elements. HR in other systems has been found to be stimulated by transcription and replication as well as by seemingly recombination specific elements. Several methods using either homologous (hit and run or bait- and -switch) or site-specific (lox.Cre or frt/Flp) recombination can be used for this purpose in constructing recombinant immunoglobulin loci. ***Ori*** is a specific sequence of a replicon at which DNA replication is initiated. *Ori* C refers to a sequence of a replicon at which chromosome replication is initiated. For example, the *E.coli Ori* C region spans 0.245kb and contains consensus sequences for replication initiation process. **Replicon** is a segment of DNA under control of one adjacent replication initiation locus and behaving as an autonomous unit during DNA replication (e.g. plasmid, bacterial chromosomes). The number of replicons per organism varies but roughly correlates with evolution in eukaryotes (yeast(500), Drosophila(3500), Xenopus(15,000), Mus(25,000), *Vicia faba*(35, 000). The size of replicon varies from 4.7Mbp(in *E.coli*), 40kb (yeast and Drosophila), 150kb (mouse), 200kb (*Xenopus laevis*) to 300kb(*Vicia faba*).

4.5 Mechanism of Genetic Recombination

Homologous genetic recombination involves genetic exchanges between any two DNA molecules (or segments of the same DNA molecule) which share an extended region of nearly identical sequence. In other words, it can involve any two homologous sequences. Homologous recombination occurs between two homologous chromosomes but is can occur between a chromosome and an extra chromosomal element provided the later carries a region with complete or nearly complete sequence complementarity. Although a number of models were presented by different workers, the Holliday's model is widely accepted. The steps involved in the **'Double strand break repair model'** for genetic recombination are as follows.

1. **Homologous chromosomes align** Homologous means that the linear order of genes may be the same on both chromosomes but base composition of some of genes may differ slightly, for example, a gene may contain different alleles on the two chromosomes. A double strand break occurs in one of the two homologues by the action of endonuclease which leads to generation of 5' and 3' ends whereas the other homolog remains intact (see Figure 4.1a). In the nicking the phosphodiester bond is broken leaving a free 3' OH and a free 5'P.
2. **Start of degradation of 5' and 3' ends by exonucleases**. 5'ends are degraded more in comparison to 3' ends resulting in smaller 5' - 3' strands (see Figure 4.1b).
3. The exposed 3' ends pair with its complement in the intact homolog and the other strand of this duplex is displaced (see Figure 4.1c)
4. The invading 3' end acts as a primer and is extended by DNA polymerase and this is followed by branch migration and eventually generation of a DNA molecule with two crossovers called Holliday intermediates. In this step when the template strand pairs with two different complementary strands, a branch is formed at the point where the three complementary strands meet. The branch migrates when a base pair to one of the two complementary strands is broken and replaced with a base pair to the other complementary strand. This process can move the branch in either direction but this branch migration stops where ever one of the otherwise complementary strand has a sequence non-identical to the other (Figure 4.1d). DNA replication now replaces the DNA missing from the site of the original double strand break. Thus the two double stranded DNA molecules are joined together through reciprocal cross over involving one strand of each DNA molecule.
5. Cleavage of the Holliday intermediates (crossovers) by nucleases generates either of the two products. In product set2 the DNA on either side of the region undergoing repair is recombined (Figure 4.1e).

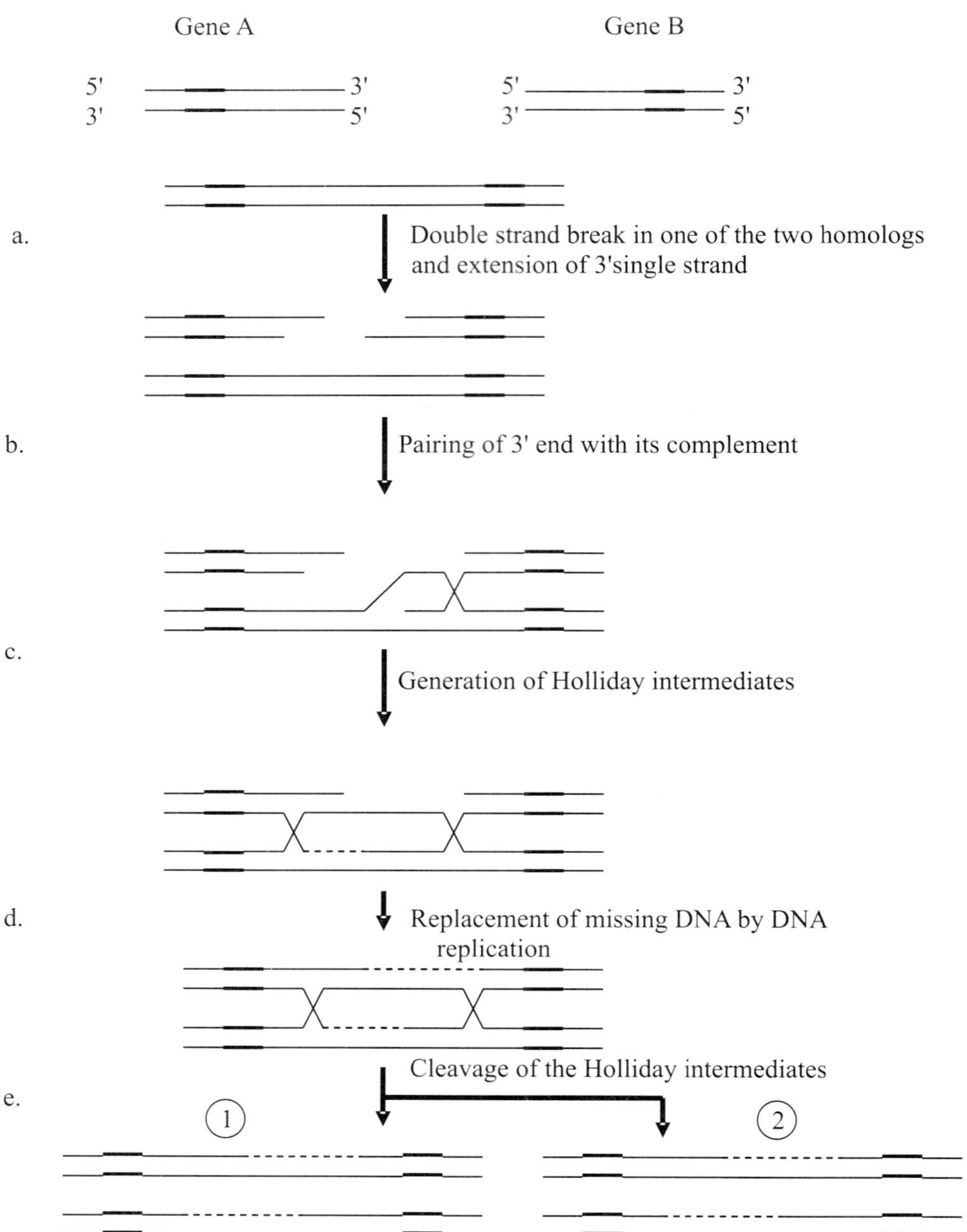

Fig. 4.1 Showing model for double-strand break repair for homologous recombination and formation of Holliday intermediates(a DNA molecule with two crossovers).

4.5.1 Enzymes Involved in Recombination

The enzymes involved in the genetic recombination (*E.coli*) are the Rec B,C,D enzymes coded by genes *Rec* B, C and D, respectively. They unwind and degrade the DNA until it reaches a Chi sequence the recombination signal for a recombinase. At this point the degradation in the 3' end strand stops but it continues in the 5' –terminal strand. Rec A protein promotes all the steps in the recombination whereas Ruv A, B(a product of *Ruv* A and *Ruv* B genes) promote branch migration at higher rate. Topoisomerase, resolvase(Ruv A,B protein), DNA polymerase I and II and DNA ligases are required to complete the recombination. In *E.coli* the Rec A protein promotes both formation of Holliday intermediates and branch migration to extend the heteroduplex DNA.

4.5.2 Hypotheses of Genetic Recombination

There are two mechanisms of genetic recombination: (i) Copy choice(Belling, 1931) (ii) Breakage and reunion (Darlington, 1935) (Figures 4.2A&B). The **copy choice hypothesis** postulates that the paired chromosomes duplicate their genes in the first meiotic prophase prior to the fibers that join them in tandem are developed. Under these conditions, if the chromosomes are twisted around each other, the interconnecting fibers might join some genes produced by different chromosomes at some points and at some other points join adjacent genes produced by the same chromosome. Copy choice implies the pairing of two parental genomes or their subunits. After a part of the genetic information has been copied, the copy process switches to the other of the paired strand and starts to incorporate the existing genetic information in the growing copy. There are two main drawbacks with this theory: First, only the two new strands can become involved in a crossover while the two original strands remain intact though it been proved that 3-strand and 4-strand double crossovers do occur. Secondly, the hypothesis-**breakage and reunion,** requires that duplication of genes must occur during meiosis. However, it is known that gene replication occurs in interphase and thus takes place before the start of meiosis. The gene duplications are the results of various breakage and reunion events that occur more or less randomly in the genetic material. Twisting of the chromosomes during the 4-strand stage of the first meiotic division generates stresses on the chromatids which results in chromatid break. The broken end of one chromatid will join with the broken end of a different chromatid. Breaks and reunion between maternal and paternal chromatids would result in recombinant whereas break and reunion involving sister chromatids will have no genetic consequence. Actually, breaks may occur randomly but misalignment most frequently involves base-pairing of similar as opposed to random base sequences. As a consequence, gene duplication often begets more gene duplication. Among proteins that have been lengthened by contiguous gene duplication are the bacterial and plant ferredoxins, immunoglobulins, calmodulin, plasma albumin, plasminogen, vertebrate fibrinogen chains, transferrinn, various ribosomal proteins and β-galactosidase.

4.6 MEIOTIC VS MITOTIC RECOMBINATION

Meiotic recombination has characteristics which distinguish it from recombination in mitotically dividing cells. In meiosis, two rounds of chromosome segregation follows a single round of DNA replication. After DNA replication, each chromosome consists of a pair of

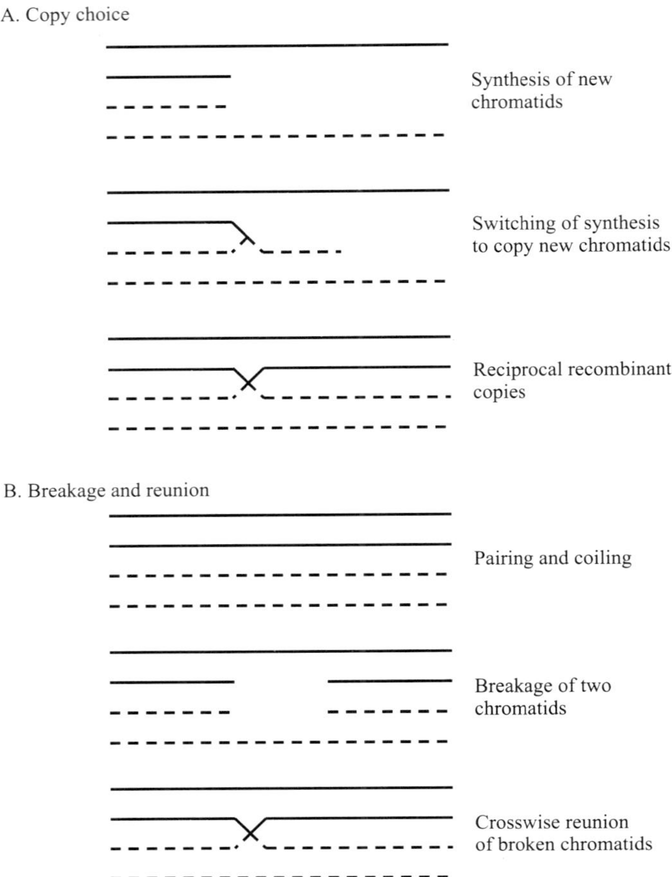

Fig. 4.2 Showing (A) copy choice and (B) Breakage and reunion theories of genetic recombination(adapted from Levine, 1969).

chromatids held together by cohesion. In the first meiotic division there is separation of maternal and paternal chromosomes but for this segregation to occur accurately, the chromosomes first pair with their correct homologous partners and then become connected so that they orient together on the meiotic spindle. Sister chromatids attach as a single unit to microtubule from a spindle pole. Connection is established by the exchange of homologous chromosome arms plus cohesion between sister chromatids. Exchange of chromosome arms between non-sister chromatids yields a chiasma and dissolution of sister chromatid cohesion along the arm allows the homologues to separate at the first meiotic division. The exchange uses a specialized pathway of homologous recombination that repairs DNA double strand breaks(DSBs). Basically the cell damages its own DNA and then uses the repair process to lock homologous chromosomes together. Meiotic recombination is subject to many layers of control: (i) choice of DNA substrate for DSB repair (ii) the distribution and timing of recombination events and (iii) integration of recombination with higher order chromosome structures. Execution of these controls requires meiosis specific factors such as DmC1 in addition to proteins such as Rad51 which function during normal repair of DSBs in other cell types. A central step in recombination involves proteins related to bacterial RecA which catalyzes pairing and exchange of DNA strands between single stranded DNA formed at the break and intact homologous double stranded DNA. Most eukaryotes have two RecA homologues, the ubiquitious Rad51 and its meiosis-specific counterpart Dmc1(disrupted meiotic cDNA1)- meiotic- specific DNA strand exchange protein DnmC1. A hall mark of meiotic recombination is the occurrence of crossovers which are not randomly distributed along chromosomes. Instead the presence of a CO makes it less likely that another will form nearby- a phenomenon called crossover interference. DmC1 may be involved as crossover interference which is lost or diminished when DmC1 is absent.

Mitotic crossing over occurs in Drosophila, numerous fungi as well as in yeast. It also occurs in plants(maize). Mitotic crossing over takes place between homologous chromatids during somatic mitosis. Mitotic recombination differs from meiotic recombination in its frequency. As a rule, mitotic recombination is a rare event and once a crossing over occurs in one chromosome, the chance of another one occurring in the same chromosome or even in different chromosome is reduced considerably. In contrast, meiotic recombination occurs in almost all chromosomes and double crossovers are quite common. Mitotic crossing over is restricted to a small segment of homologous chromosomes in comparison to the entire chromosome in meiotic crossing over. Mitotic recombination can be used for detection and measurement of linkage in fungi, particularly *Aspergillus*. However, in Drosophila the same can not achieved because of few observable differences between marker genes and the inability to isolate recombinant somatic tissues and reproducing them further. Two processes that account for exchange pairing are: (i) the onset of cell division mechanism and (ii) the absence of DNA replication. As cell division processes do not start in mitosis until DNA replication is completed, mitotic recombination is a rare event. On the other hand, cell division processes in meiosis are precocious and occur before DNA replication is complete, thereby accounting for the higher frequency of meiotic recombination (Strickberger, 1970). Both meiotic and mitotic crossing-overs are affected by the same factors such as temperature, irradiation, specific genes, whole chromosome or the entire genotype. Mitotic crossing over will affect the individual both structurally and physiologically. In heterozygous individuals, somatic crossing-over will lead to malfunctioning of parts of organs as they will carry recessive genes in homozygous condition. This could either seriously affect or be fetal to the organism.

4.7 RECOMBINATIONAL HOT-SPOT

It refers to any sequence within a gene or a chromosome at which mutations occur at a significantly higher frequency than usual. In case of T_n 10 mutagenesis insertion occurs at a hot-spot with a symmetrical 6bp consensus sequence(5-GCTNAGC-3) where internal 5-methyl groups at the third position of the pyrimidine is necessary for strong recombination. **Hyper variable regions**- They refer to high polymorphic sequences scattered throughout the human genome which consist of arrays of short usually GC-rich, tandemly repeated units to which no specific functions can yet be attributed. HVRs are thought to be hot spots for recombination. Unequal exchange at meiosis or mitosis or slippage during DNA replication may result in allelic differences in the number of repeated units present at an HVR site and consequently in length polymorphism. HVRs may be used for construction of DNA-finger print.

GC content across a locus also has a weak influence on recombination rates and lower GC ratios generally correspond to lower recombination rates (Yu et al., 2001). Further meiotic crossovers are seen more frequently in GC-rich R and T-bands than in GC-poor G-bands (Holmquist, 1992). This observation directly relates recombination frequency with the cross Giemsa banding of chromosomes.

4.7.1 Site Specific Recombination (SSR)

It refers to any recombination of two DNA molecules which occur only at specific sites or only at particular DNA sequence (which may or may not share homology), for example, the integration of a prophase into the host cells' genome. SSR is mediated by specific proteins-recombinases and are conservative (i.e. occur without any synthesis or degradation of the DNA). It occurs only at specific target sequences and this process can also involve a Holliday intermediate. The enzyme recombinases cleave the DNA at specific points and ligate the strands to new partners. This type of recombination occurs virtually in all cells. The functions of site specific recombination include DNA integration, regulation of expression of certain genes and promotion of programmed DNA rearrangements in embryonic development or in the replication cycles of some viral and plasmid DNAs. In vertebrates a programmed recombination related to transposition joins immunoglobulin gene segments to form immunoglobulin genes during B-lymphocyte differentiation. Small segments of DNA called transposons use recombinase to move within or between chromosomes. Each site-specific recombination system consists of a recombinase and a short (20to 200bp), unique DNA sequence where the recombinase acts. The DNA sequences recognized by site-specific recombinases are partially asymmetric(nonpalindromic) and the two recombining sites align in the same orientation.

Effects of site-specific recombination—The outcome of site-specific recombination depends on the location and orientation of the recombination sites in a double stranded DNA molecule. The recombination sites with opposite orientation in the same DNA molecule will result in an inversion. The recombination sites with the same orientation on one DNA molecule will produce a deletion whereas when on two DNA molecules will generate an insertion. Some genes are regulated by genetic recombination. Gene regulation by DNA rearrangements which move genes and/or promoters is common in pathogens which benefit by changing their host range or by changing their surface proteins and thereby evading immune response. Phase variation in Salmonella is an example of gene regulation by site-specific recombination. In *Salmonella typhimurium* the products of flagellin genes fliC and fljB are different flagellins. The hin gene encodes the recombinase which catalyzes inversion of the DNA segment containing the fljB promoter and the hin gene. The recombination sites consist of inverted repeats and called hix. In one orientation when the promoter is near the fljB, both genes, fljB and fljA are expressed. The product of the fljA gene works as a repressor protein and represses the transcription of the fliC gene. In opposite orientation when the DNA segment containing the promoter for a flagellin gene is inverted by hin recombinase the fljA and fljB genes are not transcribed and only the fliC gene is expressed. The interconversion between these two states, i.e. switching between two distinct flagellin proteins is known as **phase variation**. The other example is of site-specific recombination resulting in alterative expression of two sets of tail fiber genes affecting host range in bacteriophage μ.

Non-reciprocal gene conversion—The rearrangement of genetic information can be through DNA transposition besides homologous genetic recombination and site-specific recombination. It is a specific type of non-reciprocal recombination which converts one allele into another allele within the same genome. For example, if the two strands of a heteroduplex jointly carry different alleles (e.g. one wild-type and another mutant type) then the heteroduplex will have a region of mispaired bases at the mutated site. This mismatch will be recognized by a mismatch repair system which removes nucleotides from one strand and replaces them through repair synthesis using the other strand as template. Consequently the nucleotide sequence of one strand at the mismatch site is converted into that of the other. This process leads to the conversion of one allele(or gene) to another allele (or gene).

DNA transposition—DNA transposition involves a short segment of DNA with the capacity to move from one location in chromosome to another. These 'jumping genes' were first observed in maize by Barbara McClintock in 1940s. It is a class of genetic recombination event in which genetic information is moved from one part of the genome where it is silent to another where it is expressed. Nonreciprocal gene conversion is similar to replicative DNA transposition

4.7 CHIASMA INTERFERENCE

C.I. or chromatid interference refers to the influence of those chromatids involved in one chiasma on the chromatids selected for exchange at an adjacent site. If the involvement of any one strand in the two crossovers is completely at random 2-strand, 3-strand and 4-strand doubles will arise in the ratio of 1:2:1 and this will lead to on the average double recombinant, single recombinant and parental strands in the ratio of 1:2:1 in the double crossovers. But if it is not random then the above ratio will change. Now if the involvement of one strand in one crossover reduces the probability of its participation in a second, there will be a reduction in the frequency of 2- and 3 strand double crossovers and under these situations the proportion of gametes carrying recombinant chromosome will increase at the cost of those carrying parental type chromosomes. **Coefficient of coincidence**-There is another phenomenon in which there is inhibitory effect of one chiasma on the formation of an adjacent chiasma and it is measured by coefficient of coincidence. In regions which are short in terms of map unit the interference is strong but as the distance between the two potential chiasmata increases the interference declines to a point where it no longer operates. Interference usually increases as the distance between loci becomes smaller until a point is reached when no double crossovers occur. **Coefficient of coincidence** (C.C.) is calculated as:

$$\text{Coefficient of coincidence} = \frac{\text{\% of double recombinants observed}}{\text{\% of double recombinants expected}}$$

The C.C can take values from 0.0(complete interference) to 1.0 (no interference) but does not generally exceed 1.0(negative interference, i.e., one crossing over increases the likelihood of another crossing over occurring nearby). In case of positive interference one crossing over interferes with the occurrence of second crossing over nearby. The net effect of double crossing overs is to make the genes appear closer together than they really are. Negative interference is found in the fungus Aspergillus, in bacteriophage, in Neurospora and in yeast, bacteria and other organism. Negative interference leads to a distortion in the genetic analysis when it is necessary to express small chromosomal distances as a proportion of much longer distances over which this type of interference does not operate. Interferences are generally not found to operate across the centromere. C.I. limits the number of chiasmata that may occur along the length of a chromosome arm. Chiasmata distribution is theoretically expected to follow Poisson distribution. Thus there is random distribution of chiasmata in the absence of C.I. and there is random selection of chromatids(random chromatid crossover)(Beadle and Emerson, 1935; Emerson and Beadle, 1937; Rhoades and Dempsey, 1953,1966a) and only two of the four chromatids are involved in the exchange of segments. C.O. per unit of physical length near the centromere is less frequent than in distal regions. The reduction in frequency of C.O. has been attributed to the influence of centromere(Beadle, 1932d; Mather, 1939). Centromere has an inhibitory effect on chiasma formation. The percentage of recombination between any two genes no matter how far apart they are on the linkage map can not exceed 50 with random chromatid crossing over. The recombination frequency varies from near zero to a maximum of 50% depending upon the particular genes involved.

4.8 MAP UNITS AND MAP DISTANCE

The genetic mapping involves the determination of linear order with which the genes are arranged with respect to one another (gene order and the determination of the relative distances between genes (gene distance). T.H. Morgan and his colleagues developed the techniques for gene mapping. The recombination frequencies provide a basis for mapping as the probability of recombination between two linked markers depends on the distance between them. The standard map unit is 1% recombination often called a centimorgan(1 cM) after Morgan. In human 1 cM is ~ 1×10^6 bp but it varies with the locations in the genome and with the distance between genes. In case of whole genome linkage scan in human the researcher is typically faced with several loci of potential involvement in disease process. The limits of each locus are defined by two genetic markers (usually STRs) spanning over several centiMorgans. As 1 cM equals to 1 Mb on average and each Mb contains an estimate average of 15 genes (based on 45, 000 genes in the 300-Mb genome), this may represent several thousands kilobases of DNA and over 100 genes per locus. The genetic distance in Morgans(cM) is the sum of all crossingovers and is therefore equal to half the mean chiasma frequency between the two loci and the map distance is equal to 50 times the mean chiasma frequencies. So recombination frequency, θ = genetic distance in Morgans applies only for intervals where no more than one crossover occurs. If two or more COs occur between two loci then the relationship between θ and cM becomes less clear as the number of COs can rarely be ascertained directly. The relationship between % recombination and map distance shows that the additivity holds up to 20 map units. If the DCOs do not occur then map distances may be treated as completely additive units. Double crossovers usually do not occur between genes less than 5 map units apart. In a two point linkage experiment the greater the unmarked distance between two genes the greater the probability of double crossover occurring without detection and there is thus under estimation of the map distance. Therefore, the most reliable estimates of the amount of crossing over will be obtained from closely linked genes. In Drosophila the DCOs do not occur within a distance of 10 to 12 cM. The minimum DCO distance varies between species. Within this minimum distance (1-10 map units) recombination percentage is equivalent to the map distance but then this relationship becomes non-linear outside this minimum distance. The true map distance will thus be under estimated by the recombination fraction and at large distances (40-50 map units) they virtually become independent of each other (Kosambi, 1944). As the frequency of CO usually varies in different segments of the chromosome therefore the actual physical distances between linked genes bear no direct relationship to the map distance calculated on the basis of CO percentage. Genes that are loosely linked (> 20 map units) can be placed on a map but their location is much more tentative. However, the linear order of genes in the physical map and genetic map should theoretically be identical. Since crossover frequency can vary from one chromosome segment to another, the map distance can not be assumed to be proportional to physical distance along the chromosome. Further as the maximum recombination between linked genes is 50% the genes very far apart on the same chromosome may behave as though they were on different chromosomes. Thus recombination between markers at the opposite ends of the same chromosome can not be distinguished from independent assortment expected from location on different chromosomes. However, in case of former it is possible to connect the distant markers by showing their common linkage to the intermediate loci. Further in case of tetrad analysis in yeast a method is available to rule out the possibility of distant linkage in some instances. In case of distantly linked markers it is impossible to get non-parental ditype asci without a large number of tetratype asci. The recombination frequency in human female meiosis is about 40% higher than in the male. Further, map lengths using DNA markers have come about 30% higher than the estimate from chiasma frequency (Nilsson et al., 1993).

4.8.1 Recombination Analysis and Construction of Linkage Map

It provides an approximate indication of the physical distances between markers but it can place the markers in an unequivocal linear order. The standard procedure for determining the order of genes, i e. making a map of the linkage group is the three or multipoint ***test cross (Sturtevant, 1915)***. A three point test cross data from maize is analyzed here.

In maize(*Zea mays*) the two parents, say A and B were selected. Stock A was recessive homozygous for three traits, ligule-less(*lg*), glossy-leaf(*gl*) and less pigmentated(*b*) and thus had the genotype, *lg lg gl gl b b* whereas the stock, B was dominant homozygous for all these three traits and thus had the genotype as *Lg Lg Gl Gl B B*. The F1 of the cross, A x B had the genotype *lg gl b*/ *Lg Gl B* which is heterozygous at all the three loci. The F1 was crossed to a tester strain which is homozygous recessive for all the three loci i.e. the recessive parent *lg gl b*/ *lg gl b* and the test cross progenies were scored for the three traits and their numbers along with their phenotypes are given in Table below. In the test cross progeny there will be eight phenotypic classes in four reciprocally constituted pairs. One pair will consist of the two parental types, two pairs will be single cross over recombinants (due to single crossing over in one or other of the two intervals defined by the three markers) and the remaining pair will be of double crossover recombinant (due to simultaneous crossing over in both intervals). The double crossovers (two of the eight classes) are the ones which are conspicuously less frequent than the others.

Seedling phenotype	*Number*	*Interpretation*
lg gl b	172	Parental combinations are greater than all recombinant classes
Lg Gl B	162	so all three markers are linked
lg Gl b	6	Double crossovers and the order of the markers must be *lg gl B*
Lg gl B	5	
lg gl B	56	Single crossovers between *gl* and *B*
Lg Gl b	48	
lg Gl B	51	Single crossovers between *lg* and *gl*
Lg gl b	43	
Total	543	

Recombination frequencies and thus the distances between different pairs of loci are as calculated as follows.

1. Between *lg* and *gl* = (94+ 11)/543 = 105/543 = 19.3% and 17.3 (= 94/543) when double crossover is not considered.
2. Between *gl* and *B* = (104 + 11)/543 = 115/543 = 21.2% and 19.1 (= 104/543) when double crossover is not considered.
3. Between *lg* and *B* = (104 + 94) = 198/543 = 36.4%

So the best map is *lg*-19.3-gl-21.2-B and thus the distance between *lg-B* is 19.3 + 21.2 = 40.5 instead of 36.4. The map distance of 36.4(17.3 + 19.1) between *lg-B* is under estimated as the double cross over was not considered. Similarly the values of 19.3 and 21.2 will also be under estimated to the extent that undetectable double crossovers occur within *lg-gl* and *gl*-B intervals. Coefficient of coincidence = $\frac{2.02\ (= 11/543)}{4.09\ (= 1.93 \times .212)}$ = .4938 and interference = .5062.

4.8.2 Tetrad Analysis

A. Segregation of three linked markers in yeast Assuming three linked markers, a, b and c the cross between a b c x + + + will give rise the following possible tetrads. (Figure 4.3). The map order a-b-c with the implications that tetrads of classes 4-7 represent double crossovers is deduced from the observation that classes 1, 2 and 3 are the most frequent. In the single crossover classes 2 and 3 only 2 out four chromatids are recombined. Further, double crossover classes 4-7 are about equally frequent and so all four chromatids are participating in crossovers on equal footing. The result indicates that where there is a crossover in each interval, the involvement of a chromatid in one crossover does not either increase or decrease its probability of being involved in the second. Doubles in which the second crossover involves the same two chromatods as the first (so called two strand doubles), the two chromatids that were not previously involved (four strands doubles) and one of the same and one different(three strand doubles) generally occur in frequencies consistent with the 1: 1:1: 2 ratio expected on the assumption of random choice of chromatid at each crossover.

Tetrad class 1	2	3
a b c	a b c	a b c
a b c	a b c	a b c
+ + +	+ b c	+ + c
+ + +	+ + +	+ + +
Non-crossover	1 crossover a-b	1 crossover b-c

Class 4	5	6	7
a b c	a b +	a + c	a b c
a + c	a + +	a b +	a + +
+ b +	+ + c	+ b c	+ + c
+ + +	+ b c	+ + +	+ b +
2 crossovers 2 chromatids 2 strand double	2 crossovers 4 chromatids 4 strand-double	2crossovers 3 chromatids 3 strand double	2 crossovers 3 chromatids 3 strand double

Fig. 4.3 Showing the possible tetrads.

B. Tetrads types from two marker segregations Distinguishing location far apart on the same chromosome from location on different chromosomes.

Considering 2 : 2 segregation there are just three possible types of tetrad as given leaving aside any other order within tetrads in case of the cross a b x + +.

Parental ditype : a b, a b, + +, + + ;

Non-parental ditype(recombinant type) : a +, a +, + b, + b ;

Tetra type(two parental types and two reciprocally constituted recombinant types): a b, a +, + b, + +.

1. When a and b are linked

If a and b are linked then no crossingover between a and b will give a P tetrad; a single crossover will give a T tetrad and double crossovers in the a-b interval will produce P, N, or T depending upon whether 2, 4 or 3 chromatids are involved in the crossover in the frequency of 1 : 1 : 2. A frequency of crossovers high enough to give as many N as P tetrads(50% recombination) will inevitably give a much higher number of T tetrads. As the mean number of crossovers become large the two as become distributed among the four products at random with respect to the two bs. This means a P : N :T ratio of 1: 1 :4. For short a-b distances N asci are rare and for no a-b distance the ratio T/N will be less than 4.

2. When a and b are not linked

If a and b are not linked then the P : N : T ratio will entirely depend on the second division segregation frequencies of a and b. Segregation of both at the first division can only give ditype tetrads- P and N equally. It both a and b are close to their respective centromeres then the frequency of T asci will be close to zero. Only if at least one is far from its centromere will the T/N ratio become as high as 4.

C. Deducing centromere distances from unordered tetrads - In case of two unlinked markers a and b the frequency of tetratype tetrads will depend entirely on the two frequencies of second division segregation. Assuming p be the second division segregation frequency of a and q the second division segregation frequency of b then the tetratypes can be obtained in the following three ways.

1. When a segregates at the first division and b at the second division.

a b
a +(all tetratypes)
+ b(Frequency(1-p)q)
+ +
T

2. When b segregates at the first division and a at the second.

a b
+ b(all tetratypes)
a +(frequency p(1-q))
T

3. When both a and b segregate at the second division.

a b a b a + a +
+ + + + + b + b 2/4 tetratypes
a b a + a b a +(frequency pq/2)
+ + + b + + + b
P T T N

The total frequency of tetratypes, Tab = (1-q)p + p(1-q) + pq/2 = p+ q -3pq/2

With a third unlinked marker c and with second division segregation frequency r, one can determine Tbc and Tac using the following two additional equations.

Tbc = q + r-3qr/2 and Tac = p + r-3pr/2. Now using the three simultaneous equations one can evaluate p, q and r. This method depends on at least two of the markers showing significantly less than the limiting 67% frequency of second division segregation. It is important to find a marker, say a, which is very close to its centromere and p= 0, q= Tab and q= Tac.

4.9 DETERMINING RATE OF RECOMBINATION

Genetic maps which are based on meiotic recombination,order and estimate distances between DNA sequences that vary between parental homologues. The rate of recombination per nucleotide which profoundly affects the evolution of chromosomal segments, is calculated by comparing genetic and physical maps. In human, physical maps have been constructed using cytogenetics, overlapping DNA clones and radiation hybrids (see

Roy, 2009 for detail) but the ultimate and by far the most accurate physical map is the actual nucleotide sequence.

Sex linkage

After the discovery of sex chromosomes, the genome was partitioned into autosomes and sex chromosomes. The term autosome refers to any chromosome other than the sex chromosome. Sex determination switches come in all shapes and sizes in the animal kingdom from the presence of an X or Y chromosome to environmental and social factors. There are two types of sexes: homogametic sex and heterogametic sex. In case of homogametic sex there is only one kind of chromosome in two copies and thus produces only one kind of gamete. In case of organisms of heterogametic sex an individual carries an unlike pair of sex chromosomes which segregate at meiosis and produces two types of gametes. In case of mammals, insects and certain flowering plants, the heterogametic sex is male and the sex chromosomes are called X and Y. XY is male and XX is female. In birds and reptiles the female is the heterogametic sex and the sex chromosomes are called W and Z and WZ is female and WW is male. In male mammals, the X and Y chromosomes share a region of homology but differ in substantial segments that are entirely non-homologous. The homologous segments pair at meiosis and generally form chiasmata. In most species including mouse and humans the unique X and Y segments are segregated from each other at the first meiotic division. In case of Drosophila, the homology between X and Y chromosomes is very limited and there is no chiasmata or crossing over in male flies in general; nevertheless they pair and follow regular disjunction. In both mammals and Drosophila the Y chromosome carries some of the multiple gene copies encoding ribosomal RNA(rRNA) and in mice and humans it has been shown to contain a gene which switches development of male mode but overall it appears to be poor in genes. The X chromosome, on the other hand is very gene rich in all animals. In human, Y chromosome is essential for maleness but in Drosophila, it is not essential for production of maleness. In *C.elegans*, there is no Y chromosome and XX is a self-fertilizing hermaphrodite and XO is a male.

Genes on the X chromosome are easily distinguishable genetically by their unique mode of sex-linked inheritance called **criss-cross inheritance**. The important features of criss-cross inheritance are summarized as follows.

1. Males inherit their X-linked gene exclusively from their mother whereas females inherit one X-linked set from each parent.
2. Males carry only one allele with respect to the X-linked gene (a condition called **hemizygous**) and thus an allele recessive in female heterozygote will always show its effect in the males.
3. In females the X-linked markers show segregation and recombination in the just the same way as autosomal markers. In males there is no recombination as there is nothing to recombine.
4. As a few mammalian X-linked genes are also located on the Y-chromsome those genes are sometimes called pseudoautosomal. As these genes are present in two copies in both sexes they crossover between X and Y chromosomes with a frequency depending upon their distance from junction between the homologous and nonhomologous segments.

4.10 RECOMBINATION WITHIN GENE

Until 1950s gene was viewed as an undivisible unit of heredity occupying a unique position along the chromosome and determining a specific function. Thus it was thought that recombination could occur only between genes and not within genes but the study of fine structure of the rII locus of phage T4 by Seymour Benzer (1959) showed later to be correct. He isolated a large number of T4 mutants of a class called T4rII which were hypervirulent on *E.coli* strain B but failed to grow on *E.coli* strain K. These mutants were sorted into two complementation groups, A and B by mixed infections; all A+B combinations lysed K cells but A+A or B+B combinations did not. But A+A or B+B mixed infections of strain B usually yielded wild-type phage particles, able to lyse strain K cells with a frequency of between 0.01 and 1%-far higher than the spontaneous mutation rate to wild type of single mutants growing by themselves. The symbols in the matrix indicate the three different results of mixed infections of E. coli K cells by pairs of rII mutants (Figure 4.4a).

(i) 0 shows mutants showing full complementation (one mutant has lost function A and retains function B and the other has lost b but retains A), giving a complete clearing (lysis) of the mixed infected patch on K cell lawn. This implies that the rII locus consists of two subunits, A and B and that the products of both are required to produce bacterial lysis. In other words, the mutant sites are in different cistrons (different functional genes).

(ii) + no complementation but formation of some wild-type phage particles by recombination. In other words, no overall lysis indicating that the mutants have lost the same function but some locally cleared spots due to formation of occasional wild-type phage by recombination between the separable sites of mutation. In other words, mutants are at different sites in the same cistron.

(iii) - neither complementation nor recombination. No lysis at all indicating that the sites of mutation are the same or one is a deletion overlapping the other.

Benzer concluded that different mutations were usually at different sites, separable by recombination. He could position a large number of sites in a linear map by reference to a set of mutants with overlapping deletions (Figure 4.4b). A and B mutants mapped in different through contiguous segments. Mutants 1-10 and 15-21 are apparently point mutants; 5 is at the same site as 6, and 18 and 19 at the same site as 20. Mutants 11,12,13,14,22 and 23 are deletions overlapping sets of point mutations as shown and they partially define the sequence of mutations; the positions within the groups (1,2,3), (8,9) and (18,21) are not defined. Mutants 1-13 form one complementation group(cistron) and mutants 15-23 another. Mutant 14 is a deletion overlapping both groups and complementing nether. The overall picture is one of two adjacent cistrons (genes) with many individually mutable and recombinable sites within each.

On the basis of these results, he proposed the term '**cistron**' to describe the functional gene. It was defined by the comparison of two different ways of combining two linked mutational differences, say a/+ and b/+. The *cis*- arrangement is a b/+ + and the *trans* arrangement is a +/ + b. If a and b fall within independently functional genes(cistrons), *cis* and *trans* will be phenotypically identical- more or less wild-type depending on whether the mutants are completely recessive. But if they fall in the same cistron, *cis* will approximate to

(a)

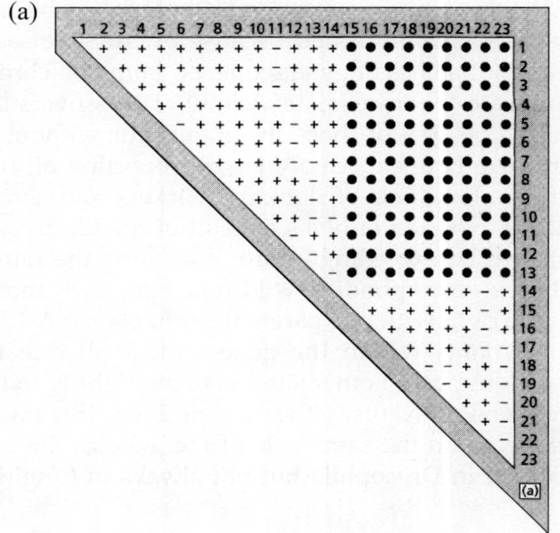

(b)

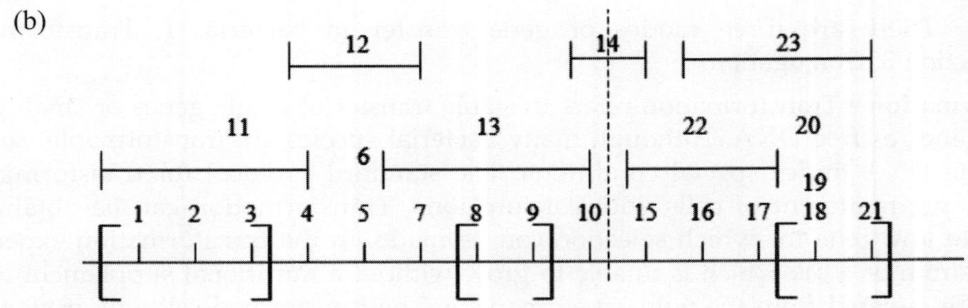

Fig. 4.4 (a) shows the principles of Benzer's(1959) analysis on the relationships between T4 rII mutants. The picture shows one of the two adjacent cistrons(genes) with many individually mutable and recombinable sites within each. Numbering of mutants is arbitrary in the presentation. (b) showing map based on the results of 4.4a.

wild-type and *trans* will give the mutant phenotype. In practice, the study of just the *trans* phenotype- the simple complementation test usually suffices and non-complementing mutants are assigned to the same cistron.

4.10.1 Mapping by Reference to Flanking Markers

The detection of recombination within genes has shown that the mutational sites within a gene are separable by recombination. Now the question is whether such mutational sites fall in a linear sequence and can we map such mutational sites. Assuming that crossing over occurs within genes occur in the same way as between genes then intragenic recombinants would show reciprocal recombination between any markers closely flanking the gene in question. Thus the order of mutational sites within a gene can be established by the use of flanking markers but this type of analysis is complicated because of crossover-associated conversion events-nonreciprocal transfer of patches of gene sequence from one chromatid to its homologue. Conversion tends to occur in the immediate vicinity of crossovers but up to 70-80% of conversions in Drosophila and 50-70% of those in fungi occur without crossing over. As a consequence of conversion there is unexpectedly large proportion of intra-genic recombinants that retain the parental combinations of flanking markers and according to crossover theory such products would be expected only a s a result of double crossing over which with closely placed markers should be exceedingly rare. Assuming the mutant sites within the gene to be as 1 and 2 and their corresponding wild-type sites as + and flanking markers as A, a on the left and B, b on the right with one parental chromosome A 1 B and the other a 2 b, the wild-type(++) recombinants within the gene will, if all due to single crossovers, all be either A b or b B, depending on whether site 1 is to the right or to the left of site 2. If ++ recombinants arise by conversion events, either 1→ or 2→+, this method still works provided the converted segments are on the same side of the gene as the associated crossovers. This assumption seems to hold in Drosophila but not always in fungi(Fincham, 1994).

4.11 GENETIC ANALYSIS OF MICROORGANISMS

Bacteria There are three modes of gene transfer in bacteria. 1. Transformation 2. Transduction 3. Conjugation.

Transformation - Transformation refers to stable transfer of single genes or small group of linked genes as free DNA. Although many bacterial species are transformable, some only assimilate DNA under special conditions. The standard protocol for transforming *E.coli* involves pre-treatment of cells with calcium ions. Transformation can be obtained with respect to any gene for which selection can be made. In the transformation experiment a mutant strain of *E.coli* which is unable to grow without a nutritional supplement is treated with DNA isolated from the wild-type organism. The transformed cell will grow and form colonies on unsupplemented medium. The transformation frequency ranges from one in 100 to one in 1000 depending on the DNA concentration. **Double transformation** refers to transformation of a cell or organism for two traits by two independent sequential transformation steps and thus occurs with a very low frequency. Two genes are acquired in the same event only when they are closely linked. The **co-transformation** refers to simultaneous transfer of two or more physically unlinked genes to animal target cells. It is useful for transfer of a non-selectable gene together with a selectable marker gene. If high DNA concentrations are transfected the procedure will lead to the transformation of the target cells with both genes so that transformants carrying the non-selectable DNA fragments can be selected with co-transformed marker. The incoming DNA is not maintained as additional genetic material but it replaces the equivalent DNA previously present in the genome by recombination.

Transduction - It refers to transfer of gene (or DNA fragment) from one bacterium to another with the help of either temperate or virulent bacteriophage. There are two types of transduction depending on what kind of bacteriophage is being used. 1. Generalized or non-specific transduction and 2. Specialized transduction.

Generalized transduction - In this type of transduction phage can integrate at any position of the the host chromosome and therefore almost any gene can be incorporated into the transducing phage and subsequently be transferred to the recipient cell. The general transducing phages include P22 of *Salmonella typhimurium* and P1 of *E.coli*. These phages replicate their DNA in the host cell and also cause fragmentation of the host DNA. These phage genes direct the host machinery to synthesize proteins for making heads and tails of phage particles. The fragments of bacterial DNA get packaged in the phage heads and thus get transmitted to newly infected cells. As the capacity of the head is limited only small linked group of genes can be transferred together. Like transformation the transduced genes will persist through cycles of cell divisions only if they are integrated into the recipient genome by homologous recombination.

Specialized transduction - In this type of transduction the phage can integrate only at a specific position of the host chromosome and therefore only host genes close to this position can be incorporated into the transducing phage and subsequently be transformed to the recipient bacterium. **Co-transduction** refers to the transfer of more than one gene from one bacterium to another with the help of bacteriophages.

The transducing phages are defective since part of their genome has been substituted by the transduced DNA and so they can only replicate in the recipient cells if complemented by a wild-type helper virus (a virus with one or more functions which the defective virus is unable to perform).

Conjugation - In this method of gene transfer there is unidirectional transfer of a plasmid DNA from a donor to an acceptor bacterium. Random collision between cells of opposite mating types leads to development of direct contact between the donor and the recipient, followed by establishment of a conjugation tube through the transfer of DNA occurs. For the DNA transfer one plasmid strand is nicked at a fixed point, the origin of transfer(ori T) and introduced into the recipient cell beginning with the 5′ –end. The duration of mating determines the amount of plasmid DNA transferred, ie. premature interruption of the conjugation process result in the transfer of only a part of the plasmid. This phenomenon can be exploited in the mapping of plasmid genes relative to the oriT. After the transfer is completed the transferred as well as the retained strands are replicated by DNA polymerase III starting with an mRNA primer.

Conjugation between F⁻ and Hfr cells - The integration of F-plasmid into the *E.coli* genome is not entirely random and certain sites on the *E.coli* map are preferred. The transfer occurs only after the linearization of the circular DNA by breakage. Breakage occurs at one('ori', origin) of the two insertion points of F-plasmids. This portion of bacterial genome is always the first to enter the recipient cell. The 'ori' always marks the head and Hfr the tail of the transferring genome as the position of Hfr plasmid is at the end of the transferring linear genome. As the bacterial chromosome usually transfers only in part the integrated F-plasmid is included only rarely in the transfer. For example, if F is integrated into the chromosome between markers z and a in the sequence x-y-z-a-b-c in an Hfr strain, the F-transfer mechanism results in transfer of a chromosome segment starting with marker a and extending for variable distances in the order a-b-c-x-y-z but very rarely seldom reaching as far as z. The sequence of the genetic markers of each Hfr strain is strain-specific since F - plasmid can integrate at different sites of the chromosome. The transfer usually results only in a **merozygote**(Wollman et al., 1956) which is diploid for only part of the chromosome and haploid for the rest of it. Assuming the integration of F into chromosome to be random then the following sequences of markers would be possible.

b-c-x-y-z-a(integration of F in between a and b)
c-x-y-z-a-b(integration of F in between b and c)
z-a-b-c-x-y (integration of F in between y and z)
or other sequence

Time-of-entry mapping - In this mapping technique suspensions of cells of two strains, one Hfr leu$^+$ lac$^+$gal$^+$ ton^{r}str^s(able to synthesize leucine, use lactose or galactose, phage T1 resistant, streptomycin-sensitive) and F$^-$ leu$^-$ lac$^-$ ton^s str^r were mixed and allowed to conjugate undisturbed for different periods of time after which they were mechanically agitated to separate conjugating cells, diluted and plated on growth medium containing streptomycin and lacking leucine. This selected recombinant cells with leu$^+$ from Hfr and st^r from F$^-$ strain. When the conjugation was interrupted after less than five minutes no cells grew to form colonies. After more than five minutes but less than 10 minutes, str^r leu$^+$ colonies appeared increasingly linearly in number with the time of conjugation and all inhering ton^{s+}, lac$^-$ and gal$^-$ from the F$^-$ strain. After 10 minutes the ton^r marker began to appear in some of the leu$^+$ str^s recombinants and lac$^+$ and gal$^+$ followed after 14 and 21 minutes, respectively. Further the experiment showed that the Hfr markers leu$^+$ ton^r leu$^+$ lac$^+$ gal$^+$ form a series not only in the time they take to enter recombinants but also in their maximum frequency of entry and the longer they delay the more likely they are not to enter at all. Similar results were obtained with other Hfr strains except that the markers transferred vary from one Hfr to another.

Each Hfr strain thus transfers markers from a sector of the map starting at a fixed point and proceeding either clockwise or anticlockwise. Transfer can be broken off at any time and so further a marker is from the leading point the less likely it is to be transferred. Comparison of all the marker time- of- entry sequences from all the different Hfr strains shows that they can be combined to construct a circular time map measuring about 100 minutes in circumference. And after about 100 minutes of conjugation the F-factor is very often found to have been transferred as well.

The markers from the transferred chromosome segment are integrated into the recipient genome by crossing over. Crossovers are usually numerous and so even if markers are transferred together they tend to be integrated independently when separated by more than about 5% of the genome. The time-of-entry mapping predicted that the *E.coli* genome consisted of a single closed –loop DNA molecule which has now been verified by gentle release of cell of a radioactively labeled DNA followed by autoradiography. The problem with the time-of-entry mapping is that the accuracy of the time map over short distances is restricted by the limited precision of the time experiments. To overcome this problem of estimating distances and sequences over short intervals-fractions of a minute on the time map transduction and recently DNA sequencing have been more useful.

Generation of F′ plasmids - The Hfr strain is generated as a result of integration of F-plasmid into the bacterial chromosome. The integration occurs through crossing over between dispersed repetitive sequences present in both the bacterial chromosome(repetitive sequences are present at numerous loci) and the plasmid. As the F-plasmid is a circular DNA so a single crossover will be sufficient to integrate it into the chromosome. Occasionally the integrated F-plasmid gets excised from the bacterial chromosome imprecisely and carries with it one or a few chromosomal genes found adjacent to the point of integration and thus F′ plasmid is produced. The F′ plasmids thus act like natural cloning vector. This event is detected when a marker at the trailing end of the chromosome appears in recombinant cells after only a few minutes of conjugation. Such que-jumping markers are generally associated with F-plasmids carrying chromosomal material. F′-plasmids have the F-plasmid origin of replication(ori) and so they can replicate independently of the chromosome. As they carry chromosomal marker(s) so they become diploid with respect to any genes present in the plasmid. A library of F′ plasmids carrying between them the whole *E.coli* genome in small pieces has been established.

4.11.1 Deletion Mapping

Deletions are often confined within genes but occasionally overlap adjacent genes. Deletion mutants occur rarely and are not able to revert back to wild-type as there is loss of extensive tract of unique genetic material which can not be recovered by random mutation. Deletion mapping refers to the localizing of the positions of deletions in the chromosomes of eu- and prokaryotes. In case of eukaryotes the positions of deletions can be estimated according to whether or not they phenotypically 'disclose' certain recessive mutations(Rieger et al., 1976). All segmental deletions that allow the manifestation of a recessive mutation in the heterozygous state must include the locus of this mutation. Deletion mapping in case of prokaryotes is based on the principle that non-overlapping deletions can generate wild-type whereas overlapping deletions can not. For example, three mutants (a, b and c) each differing from the wild-type by a deletion of the linkage structure can be tested for recombination and the results can be presented as shown below. The two mutants (a and c) recombine and generate wild-type but neither recombines with the mutant b.

	a	b	c
a	0	0	+
b	0	0	0
c	+	0	0

+ = recombination; 0 = No recombination

The recombination results thus allow ordering of the three mutants. In other words, the overlaps of the deletions can be used to define a series of segments in the linear map. In this example, the deletions fall in the serial order a, b, c such that a and b overlap on one side and b and c overlap on the other as shown below.

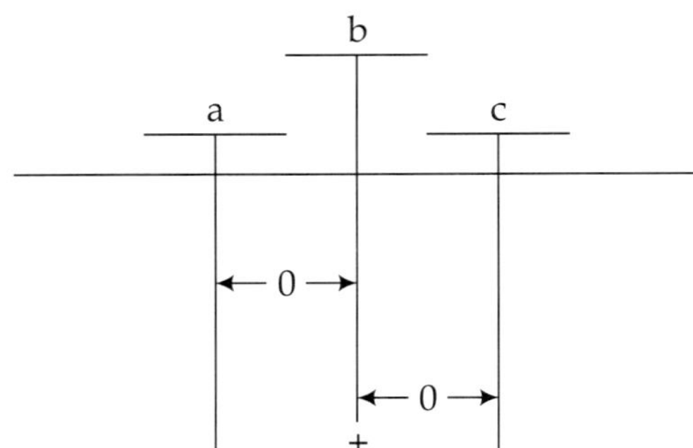

Deletion mapping consists of two steps. In the first step, a set of partly overlapping deletions is mapped as discussed above and in the second step, each of the point mutants is crossed to the set of deletions. Each cross will or will not yield wild-type recombinants depending on whether the site of point mutation falls outside or inside of the segment deleted in the other parent and thus each point mutation can be placed within one of the segments defined by the deletion overlaps(Fincham, 1994).

Study of Chromosome Number Variation

5.1 VARIATION IN CHROMOSOME CAN BE DUE TO VARIATION IN EITHER CHROMOSOME NUMBER OR CHROMOSOME STRUCTURE

5.1.1 Chromosome Number

Variation in chromosome number can be of two types. I. Euploid variation II. Aneuploid variation. The **euploid** variation refers to variation in the multiples of basic chromosome set (the gametic or haploid number of chromosomes), the genome whereas the **aneuploid** variation refers to variation in the chromosome number of particular chromosome(s) of a chromosome set. If X refers to the basic number of chromosomes in a chromosome set then we can have species with 2X, 3X, 4X, 5X or 8X number of chromosomes. When n refers to the gametic (haploid) number of chromosomes, 2n refers to zygotic (diploid) number of chromosomes and thus 2n could be X, 2X, 3X, 4X, 5X or 8X i.e. haploid, diploid, triploid, tetraploid, pentaploid or octoploid, etc., respectively and species with 2n = 3X, 4X, 5X and so on are grouped as polyploids. The term' genome' was coined by Hans Winkler in 1920.

5.2 Classification of Polyploids

Polyploids can be classified as auto or allo polyploids. Kihara and Ono (1926) coined the terms 'autopolyploidy' and allopolyploidy'. Considering two genomes, say A and B which are not identical, **autopolyploidy** refers to the multiple of the same genome either A or B. Thus, AAA or BBB and AAAA or BBBB are called triploids and tetraploids, respectively.

Allopolyploidy refers to the multiple of chromosome sets which are not identical. For example, AABB can be called as allopolyploid. Allopolyploid or amphidiploid can be further classified as 'classical allos' or 'segmental allos' depending upon the degree of similarity between the two genomes, A and B. If A and B genomes are very dissimilar it constitutes '**classical allos** or genomic allos'. When there is some degree of similarity between genomes A and B they constitute '**segmental allos'** and they are characterized by homoeology or partial homology. Homoeology is partial homology during which some chromosome segments may pair and others do not(Huskins, 1932).

5.2.1 Origin of Polyploids

Polyploids can arise because of failure of nuclear division during mitosis or meiosis (Figure 5.1). In mitosis the failure of nuclear division will result in somatic doubling of chromosomes, 2X cells will give rise to 4X cells and thus a polyploidy is formed. During meiosis unreduced gametes(2X) can be produced and this unreduced 2X gametes can fuse

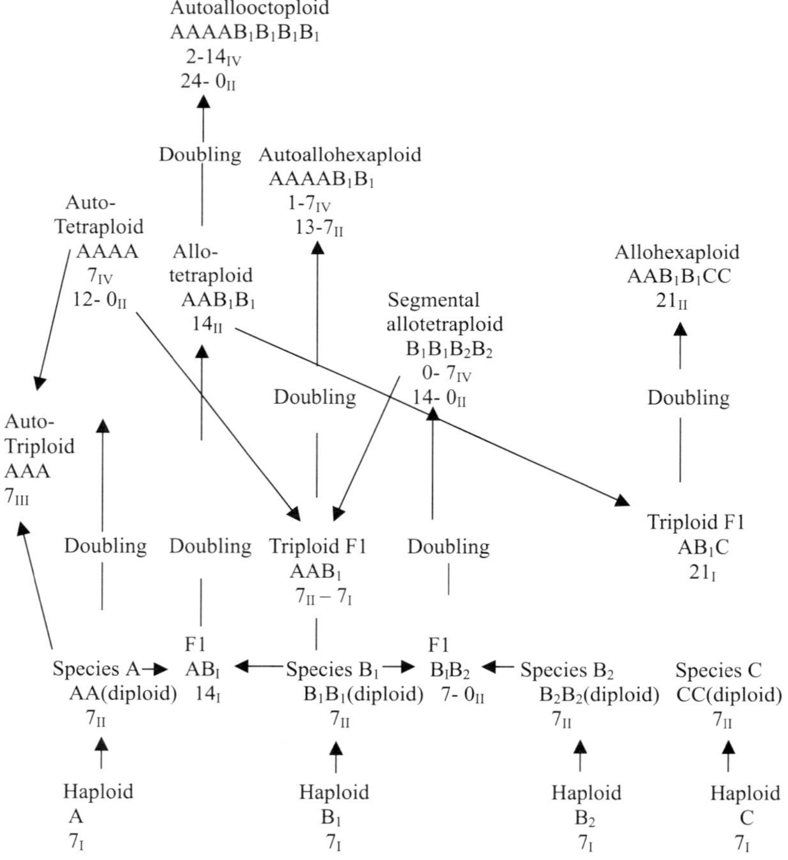

Fig. 5.1 Origin of typical autopolyploids, allopolyploids, segmental allopolyploids and autoallopolyploids (adapted from Swanson, 1972 (Stebbins, 1950)).

with reduced gamete(X) to produce 3X, the triploid, zygote and thus a triploid is formed. The allopolyploids can be formed in two ways. In the conventional approach the F1(AB) of the two species, AA and BB, upon chromosome doubling will result in the production of **amphidiploid,** AABB. Fusion of unreduced gametes will also result in the production of allopolyploids in the following way. AA + B ⟶ AAB, AAB + BB ⟶ AABB, AAB + AAB ⟶ AAAABB. The various diploid, triploid, autotetraploid and allotetraploid crop species and other organisms along with their chromosome numbers are given in the Table 5.1.

Table 5.1 Showing various plant and other species along with their ploidy level, '*C*' values and kilo base pair/haploid genome.

Crop species	*Scientific name*	*Ploidy level*	*'C' value(picograms)*	*Kilo base pair/haploid genome*
Rice(A)	*Oryza sativa*	2n=24		
	O. glaberrima	2n=24		
Maize	*Zea mays*	2n=20	7.8	
Sorghum	*Sorghum bicolor*	20		
Pear millet	*Pennisetum americanum*	14		
Ragi	*Eleusine coracana*	2n=36		
Fox tail millet	Setaria italica	2n=18		
Wheat	*T. monococcum*	14	24.9	
Barnyard millet	*Ehinochloa colona*	2n=36		
Kodo millet	*Paspalum scrobiculatum*			
Little millet(katki)	*Panicum miliare*	2n=36		
Oat	*Avena strigosa*	14		
Rye	*Secale cereale*	14	33.1	
Barley	*Hordeum vulgare*	14	5.5pg	5.3×10^9
Lentil	*Lens culinaris*	14		
Chickpea	*Cicer arietinum*	16		
Pigeonpea	*Cajanus cajan*	22		
Cowpea	*Vigna unguiculata*	22		
Faba bean	*Vicia faba*	14	44.0	
Urd	*Vigna mungo*	22		
Moong	*Vigna radiata*	22		
Pea	*Pisum sativum*	14		
Lathyrus	*Lathyrus sativus*	14		
Cabbage	*B. oleracea*	18		
Turnip	*B. nigra*	16		
Rape	*B. campestris*	20		
Safflower	*Carthemus tinctorius*	24		
Sesame(Til)	*Sesame indicum*	26		
Linseed(flax)	Linum usitatissimum	30		
Castor	Ricinus communis	20		
Niger	Guizotia abyssinica			
Soybean	Glycine max	40		
Sunflower	*Helianthus annuus*	2n=34		
		2n=108		

Lily	Lilium spp.	2n=24		
Berseem	*Trifolium alexandrinum*	2n=16		
Red clover	*Trifolium pratense*	2n=14		
Rye grass	*Lolium multiflorum*	2n=14		
Sudan grass	*Sorghum sudanense*	2n=20		
		Autopolyploids		
Potato	*Solanum tuberosum*		1.9(diploid)	
Banana	*Musa sapientum*	Auto and allo 22 and 33, 44		
Lucerne(alfalfa)	*Medicago sativa*	4x=32		
Sweet potato	*Opomea batatas*	6x=90		
		Allopolyploids		
Peanut	*Arachis hypogeae*	2n=40		
Tobacco	*Nicotiana tabacum*	2n=48		1.6 x 10^6
Coffee	*Coffea arabica*	22,44,88, x=11		
Wheat	*T. aestivum*	42	69.3	5.9 x 10^6
	T. diococcum	28	49.1	
Oat	*Avena sativa*	42		
	A. abyssinica	28		
Cotton	*G. hirsutum*			
Apple	*Malus domestica*	34,51,x=17		
Pear	*Pyrus communis*	34,51,x=17		
Plum	*Prunus spp.*	16,32,48,x=8		
Strawberry	*Fragaria × ananassa*	56, x=7		
Raphanobrassica		2n=36		
Triticale(hexaploid)			84.7	
Tritical (Octoploid)			103.9	
Bacterial virus	Bacteriophage T_4	Haploid(single chromosome)		2 x 10^2
Bacterium	*Escherichia coli*	Haploid(single chromosome)	0.005(0.0036-0.005)	4 x 10^3
Yeast	*Saccharomyces cerevisiae*	2n=4 or 8 and n=2 or 4	0.025, 0.009(haploid)	1.35 x 10^4
Thalecress	*Arabidopsis thaliana*	2n=10	0.2(haploid)	1.25 x 10^5
Nematode worm	*Caenorhabditis elegans*	2n=12	0.088(haploid)	c.1 x 10^5
Fruit fly	*Drosophila melanogaster*	2n=8 and n=4	0.17-018(-haploid)	1.8 x 10^5
Mouse	*Mus musculis*	2n=40	3.0	2.3 x 10^6
Human	*Homo sapiens*	2n=46	2.7-3.5	2.8 x 10^6
Green algae	*Chlamydomonas reinhardi*	Haploid with n=8		
Bread mold	*Neurospora crassa*	Haploid with n=7		
Triturus cristatus(newt)			19(haploid),35(diploid)	
Protopterus aethiopicus	Lung fish		142(haploid)	
Fritillaria assyriaca	monocot	2n=12	127(haploid)	

*A pg of DNA corresponds to approximately 10^9bp.

5.2.2 Chromosome Behaviour During Cell Division

The cell division includes mitosis and meiosis. Mitosis in case of polyploids is quite normal however, meiosis is considerably modified.

5.2.2.1 Meiosis in autopolyploids

Here we will consider the modified pairing behaviour in case of autotriploid(3x) and autotetraploid(4x). In case of autotriploid there will be formation of trivalents but rarely there will also be bivalents and univalents. The pachytene and metaphase I stages in case of triploid will be as shown in Figure 5.2. Triploids in maize have a high % of aborted ovules and pollen grains and are infertile. Aneuploid gametes with chromosome numbers ranging from n+1(11) to n+9(19) are formed as a consequence of the random or near random segregation of each set of three homologous chromosomes. It was observed that pollen grains with chromosome number n, n+1 or n+2 compete successfully than those with higher numbers.

In case of autotetraploid(4x) the orientation of chromosomes at metaphase I depends on the number of chiasmata formed and the position of the chiasmata. There would be a

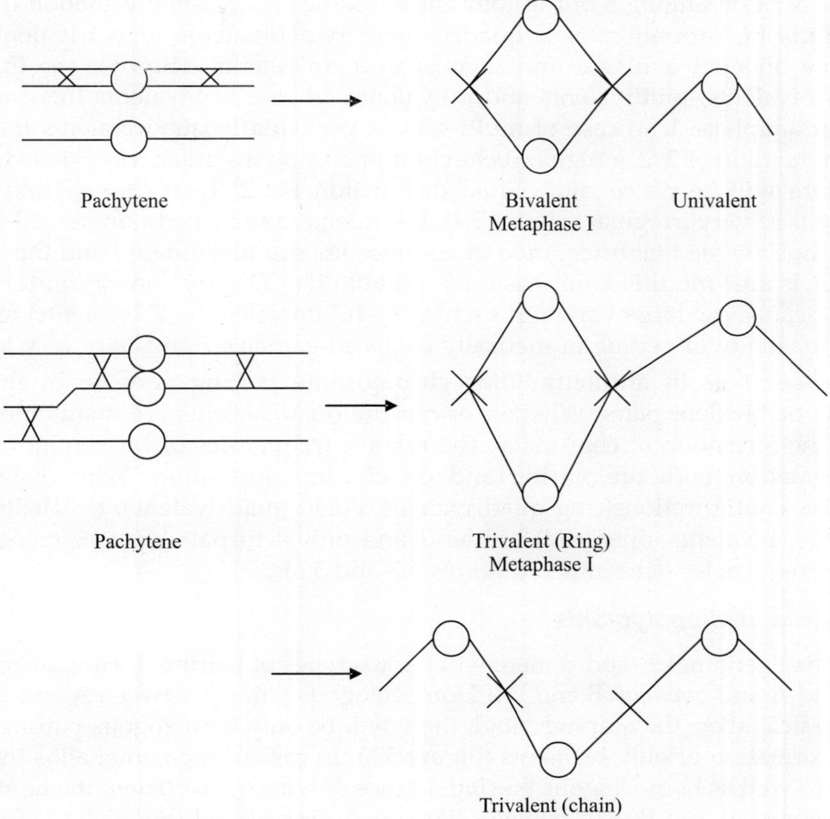

Fig. 5.2 Modified behaviour in autotriploid.

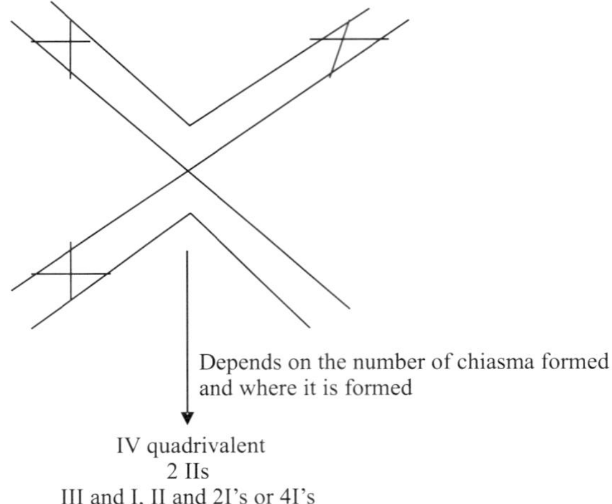

Fig. 5.3 Showing a maximum of three association in four chromosome (auto4x).

maximum association among 3 out of four chromosomes to chiasma formation (Figure 5.3). There will thus be formation of a quadrivalent, two bivalents, one trivalent and one univalent, one bivalent and two univalent or four univalents. Thus we see that there is formation of bivalents, multivalents and univalents. In case of bivalents there will be 1 :1 disjunction at anaphase I. In case of multivalents, particularly quadrivalents there will be separation in the ratio of 2 :2, a regular behaviour or 3 : 1 an irregular way whereas in case of trivalents there will be no regular, equal disjunction i.e. 2 :1. In case of univalents the disjunction will be very irregular (Figure 5.4). Univalents contribute to numerical imbalances to gametes. Thus we see that orientation of chromosomes at metaphase I and the disjunction at anaphase I are modified in case of polyploids. Disjunction in autotriploid(3x) rye(2n=3x=21 can show large variation such as 7-14(7 univalents + 7 bivalents) to 8-13,9-12, 10-11. The probability of getting numerically balanced gametes 7 and 14 is very less.

Thus we see that in autotetraploids chromosome pairing may be in the form of quadrivalents or bivalent pairs. Whether or not the quadrivalents are maintained until MI depends on the formation of chaismata. The relative frequencies of MI configurations thus contain information both on pairing and on chaisma formation. With distal chiasma localization six configurations(ring quadrivalents, chain quadrivalents, trivalents with one univalents, ring bivalents, open(rod) bivalents and univalent pairs) can be recognized and their relative frequencies determined (Figures 5.5 and 5.6).

5.2.2.2 Meiosis in allopolyploids

Considering two genomes A and B there will be two types of pairing: I. Homologous pairing between A and A and between B and B II.Homoeologous pairing between A and B genomes. In case of classical allos, the amphidiploid, there will be only homologous pairing and there will be thus formation of only bivalents (Figure 5.7). In case of segmental allos there will be homologous as well as homoelogous pairing. In case of wheat(*T. aestivum*) the homoeologous pairing is suppressed and thus it behaves like amphidiploid and thus there is formation of bivalents only.

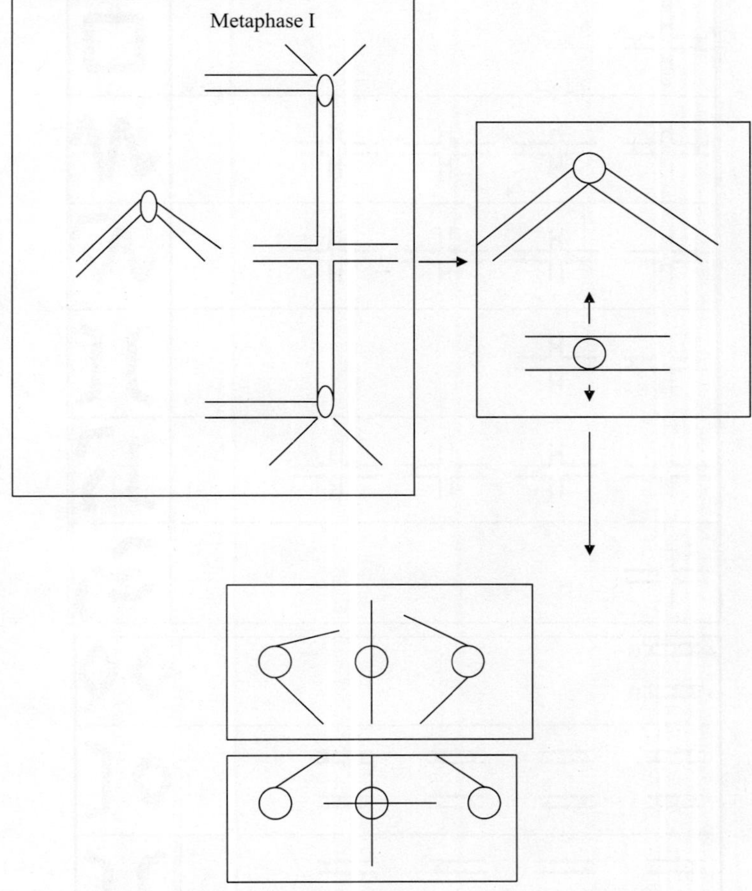

Fig. 5.4 Showing irregular behaviour of univalent.

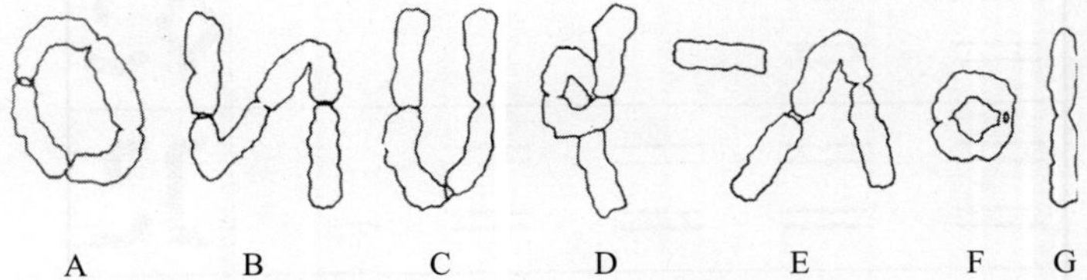

Fig. 5.5 Showing six MI configurations as observed in autotetraploid T. virginiana. (A) Ring quadrivalent(can also have zig-zag shape); (B, C, D) three forms of chain quadrivalent; (E) trivalent with univalent; (F) ring bivalent and (G) open(rod) bivalent. All chromosomes are metacentric or submetacentric but centromeres are not always as conspicuous as shown in the figures (adapted from Sybenga, 1975).

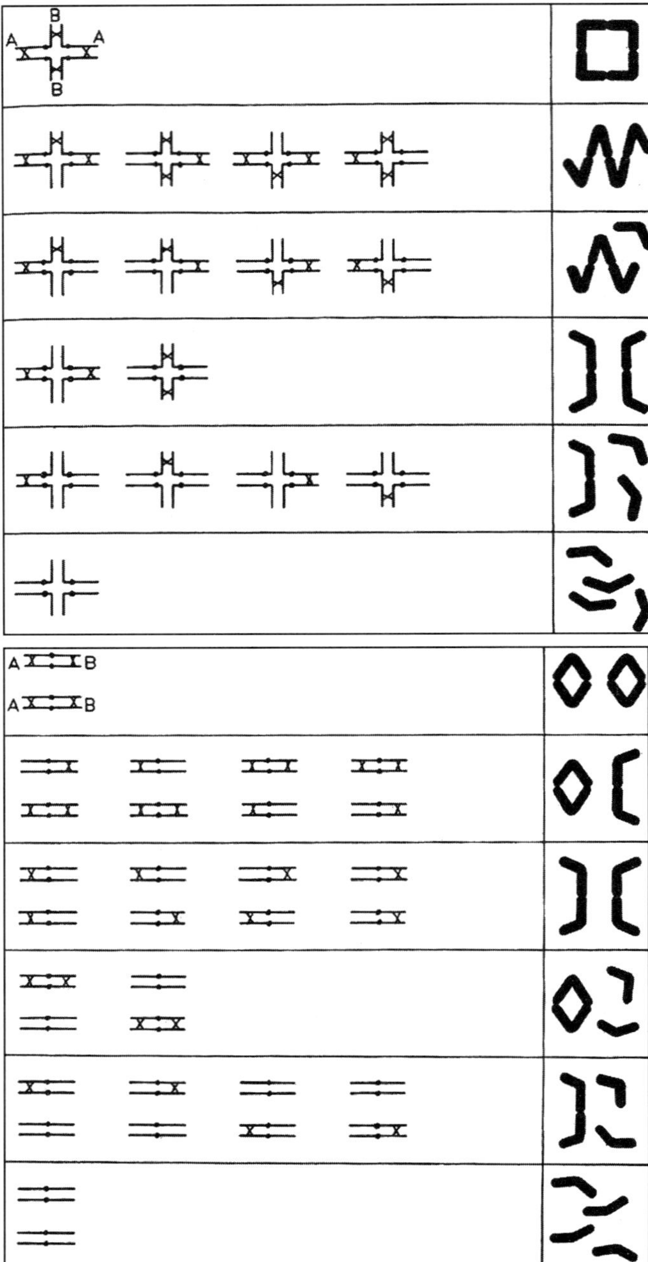

Fig. 5.6 Showing association of four homologous chromosomes in a quadrivalent or in two bivalents and the resulting metaphase 1 configurations with all possible combinations of presence and absence of chiasmata assuming one point of partner exchange, no interstitial chaismata and complete chiasmata terminalization.

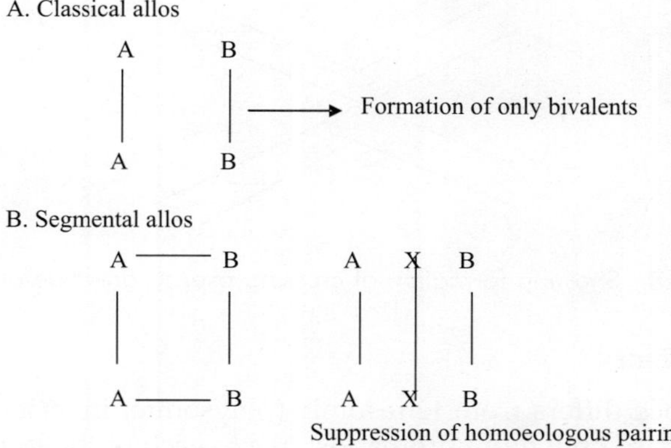

Fig. 5.7 Showing pairing behaviour in classical allos and segmental allos.

5.2.3 Effects of Polyploidy

There will be four types of effects: morphological, physiological, cytological effect and genetic effect.

5.2.3.1 *Morphological and physiological effects*

Polyploidy results in increase in cell and nuclear size. In tomato, the size of guard cells increases, leaves and petals become larger and vigorous. There are changes in the proportion of different parts of a plant. This effect of polyploidy is called '**gigas**' effect. The 'gigas' effect can be modified by selection. The 'gigas' effect is conferred to newly synthesized autopolyploids. Polyploidy can result in less branching, larger fruits, late flowering and fruiting. It may result in less frost resistance in comparison to diploids as in case of tetraploid rye and red clover. Finally, polyploidy can result in higher sugar percentage as has shown in the triploid sugarbeet. Similarly, triploid banana(AAA) is very sweet in comparison to the diploid.

5.2.3.2 *Cytological effect*

Polyploidy will result in sterility. Sterility can be due to irregular chromosome segregation or morphological changes associated with polyploidy. For example, in red clover (*Trifolium pretense*) flowers can not be pollinated by other species because of large size of flowers. The escape from sterility can be due to different mechanisms such as asexual or subsexual reproduction where there is no meiosis such as apomixes in plants and parthenogenesis in animals. Here meiosis is present but modified and there is suppression of first division and no pollination is involved. In another mechanism meiosis is modified, for example, in even number polyploids there is more quadrivalents and bivalents and less trivalents and univalents. In case of autotetraploid(4x) teosinte(maize) at metaphase I there is more bivalents because the crossing over is restricted to one side of the partner exchange (Figure 5.8). The another example is of polyploid wheat where there is bivalents formation because of suppression of homoeologous pairing. Although a segmental allopolyploid there are multivalents at meiosis but due to suppression only bivalents are formed and so it immediately improves fertility.

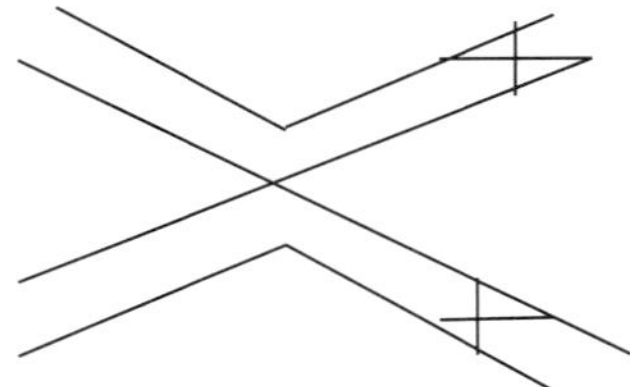

Fig. 5.8 Showing formation of crossing over to one side of parter.

5.2.3.3 Genetical effect

The disomic inheritance differs from tetrasomic (polysomic) inheritance in that the former involves the separation of only four chromatids, one to each of the four gametes whereas the later involves the separation of eight chromatids into four pairs, one pair to each fo the four gametes. Considering one locus with 2 alleles system(A and a) the F1(Aa) between two parents(AA, aa) upon selfing will poduce two types of gametes, A and a in equal frequency and three genotypes, AA, Aa and aa in the ratio of 1:2:1 whereas in case of autotetraploid(4x) the F1(AAaa) from the parents AAAA and aaaa will produce three types of gametes, AA, Aa and aa and its frequencies for a given genotype will depend on the cytological events-formation/no formation of quadrivalents, i.e. the regularity of quadrivalent formation and the crossing over which depends primarily on the distance between centromere and the locus in question. These cytological events will lead to two types of segregation of chromatids: I. random chromosome segregation and II. random chromatid segregation.

5.2.3.3.1 Random chromosome segregation

In this type of segregation sister chromatids (chromatids derived from the same chromosome) never end up in the same gamete. This type of segregation occurs when either there is no quadrivalent formation, i.e. when four chromosomes form two bivalents in the first meiotic phase or there is complete linkage between the centromere and the gene, i.e no crossing over occurs. Mechanisms involved in random chromosome segregation are: 1.Sister chromatids are attached to the same centromere and as a result of which they move to the same pole during anaphase I and 2. The two sister chromatids separate and move to opposite poles during anaphase II. The gametic output assuming random chromosome segregation in case of autotetraploid is be as given in Table 5.2. Assuming random chromosome segregation AAAA will produce only AA type of gametes, aaaa will produce only aa type, AAaa will produce 1/6AA, 4/6Aa and 1/6aa types of gametes, AAAa will produce1/2AA and 1/2Aa and Aaaa will produce 1/2Aa and 1/2aa types gametes. Thus wee can see that AAaa upon selfing will produce aaaa with a frequency of only 1/36 and thus in comparison to diploid inheritance in tetrasomic inheritance the frequency of a recessive homozygote is reduced to a great extent (Table 5.3).

5.2.3.3.2 Random chromatid segregation

Random chromatid segregation occurs when quadrivalents are formed and when there is crossing over between the centromere and the gene in question,i.e. there is no linkage and as a result of which the two sister chromatids end up in the same diploid gamete. There is thus

Table 5.2 Gametic output in an autotetraploid

Parental genotype	Gametes			Divisor
	AA	*Aa*	*aa*	
A_4	1	-	-	1
A_3a	$2 + \alpha$	$2(1 - \alpha)$	α	4
A_2a_2	$1 + \alpha$	$4(1 - \alpha)$	$1 + 2\alpha$	6
Aa_3	α	$2(1 - \alpha)$	$2 + \alpha$	4
a_4	-	-	1	1

Table 5.3 Expected phenotypic ratios in selfed and various cross progenies

Selfing	*$\alpha = 0$ Random chromosome segregation*	*$\alpha = 1/7$ Random chromatid segregation*
AAAA	All A	All A
AAAa	All A	783 A : 1 a
AAaa	35A : 1 a	20.8 A :1 a
Aaaa	3 A : 1 a	2.5 A : 1 a
aaaa	All a	All a
Crosses		
AAAa x AAaa	All A	130 A: 1a
AAAa x Aaaa	All A	51.3A:1a
AAAa x aaaa	All A	27A:1a
AAaa x Aaaa	11A:1a	7.7A:1a
AAaa x aaaa	5A:1a	3.7A:1a
Aaaa x aaaa	1A:1a	0.87A:1a

occurrence of double reduction (Haldane, 1930; Mather, 1936) wherein during meiosis sister chromatids may enter the same gamete which results from the recombination between the centromere and distal genes followed by a particular orientation of the chromosomes on the second metaphase spindle. Mechanisms involved in double reduction are: 1. A single (any odd number of crossing over) occurs between the centromere and the locus and thus sister chromatids become attached to different centromeres. 2. The centromeres with two sister chromatids move to the same pole during anaphase I. 3. Again these two sister chromatids move to the same pole during anaphase II. The coefficient of double reduction(α) refers to the proportion of gametes in which sister genes occur. The double reduction,dr(α) in a tetrasomic polyploidy is calculated as

$$\alpha = q^* e^* 1/6$$

where q is the frequency of quadrivalent formation; e is a function of chromatid exchage between the centromere and the gene in question. When $q = 1$ i.e. a maximum value if quadrivalent always forms and $e = 1$ (a maximum value when there is always a crossover), the maximal frequency of dr(α) will be 1/6. $\alpha = 0.0$ if either $q = 0$ (a minimum value when quadrivalent never forms) or e is zero, again a minimum value when there is complete linkage between the centromere and the locus, i.e. no crossover at all. Considering that random chromosome segregation and random chromatid segregation represent the extreme types of polysomic segregation, a model representing the intermediate between the two types, i.e. quadrivalents are formed sometimes and sometimes they are not and there is partial linkage between the centromere and the gene and thus crossing over sometomes occurs and sometimes does not seem to be the more realistic one and thus the coefficient of double reduction(α) which takes the theoretical maximum value of 1/6 is not the realistic value. Assuming 50% crossing over between the locus and the centromere the probability of two sister chromatids being attached to different centromeres(and thus leading to dr) is 6/7. This probability of 6/7 can be obtained considering a 50% crossingover between the centromere and the gene in a simplex individual (Aaaa) (Figures 5.9 and 5.10). Here one of the eight chromatids has an equal probability of being attached to the same centromere as any other seven chromatids and out of 7 possibilities in 6 possibilities the sister chromatids will be attached to different centromeres and thus not landing in the same gamete and thus only in 6/7 ways the sister chromatids will land in the same gamete. Now assuming random orientation of chromosomes at the metaphase the probability of the two sister chromatids being passed to the same pole in the anaphase I is 1/3 whereas to the same pole in the anaphase II is ½ and thus the realistic maximum value of α = 6/7*1/3*1/2 = 1/7 or 0.1429. Thus if there is random chromatid segregation then the triplex, AAAa genotype produces aa gametes. In this situation there is quadrivalent formation and there are other constraints as explained above which lead to such phenomenon.

Thus we see that in polyploids the segregation ratios are more complex and often distorted. The epistatic gene actions between genes are multiplied and gene dosage effects come into play and all these lead to loss of genetic resolution in polyploidy. The qualitative traits blur into quantative trait. The double reduction, dr(α) and numerical non-disjunction are of little practical importance in distorting the genetic ratios which can not be precisely measued any way and for this reason Mendelian analysis is better applied to related diploid

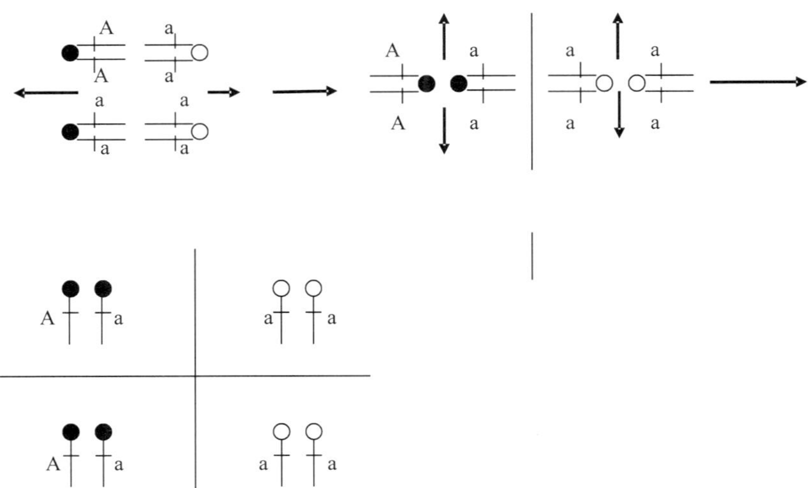

Fig. 5.9 Showing reductional division in simplex quadrivalent with no crossing over between centromere and the locus.

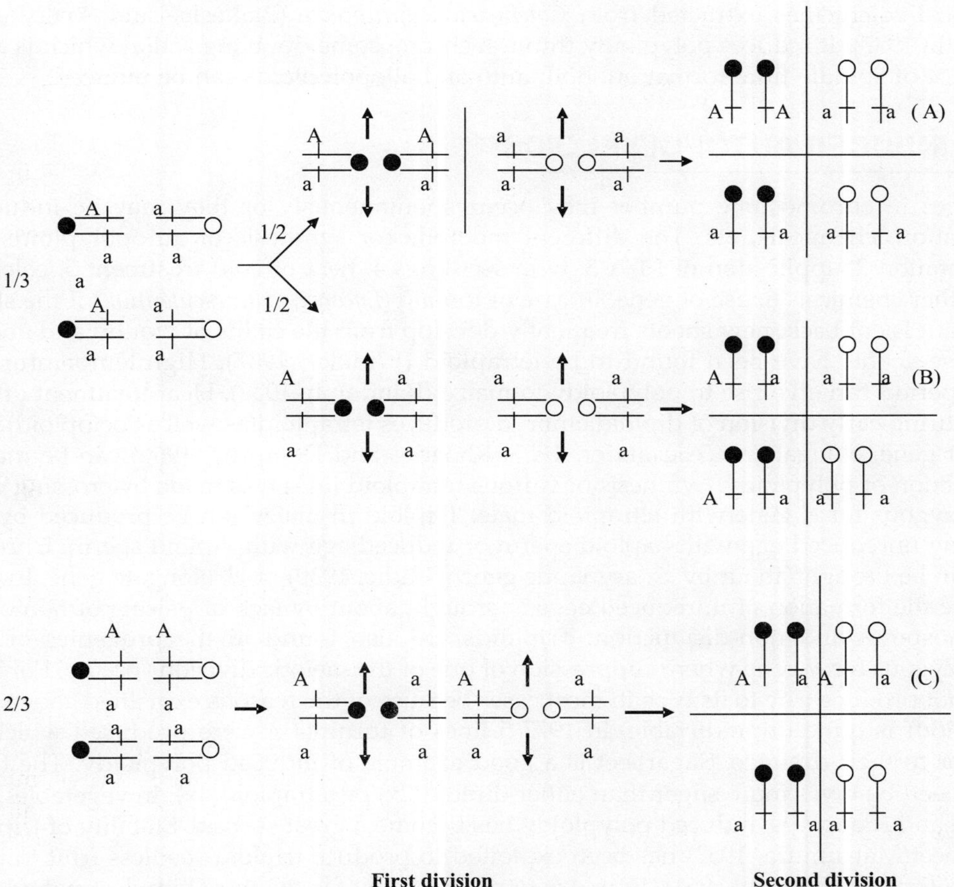

Fig. 5.10 Showing gamete formation in simplex quadrivalent with crossing over bewteen A and a. This results in two Aa and two aa chromosomes. The two Aa chromosomes move to the same pole in the first division one-third of the time and to opposite pole two-thirds of time. When they pass to the same pole in the first as well as second division as shown in A, some homozygotes are formed and the process is called 'double reduction'(adapted from Allard, 1960).

species rather than to such polyploid crops.As Mendelian markers are rarely found in polyploids Mendelian analysis of such polyploids are generally characterized by uncertainty.

Practical aspects of polyploidy

Many plants are natural polyploids such as wheat, tobacco, peanut,etc. Further, polyploidy can be induced and thus use can be made in practical plant breeding. Allopolyploidy and particularly amphidiploid has played an important role in the evolution of genus, Nicotiana, Gossypium,Brassica and Triticum. Polyploids can be induced by the chemical colchicines extracted from *Colchicicum autumnale* (Blakeslee and Avery and by Nobel in 1937). It induces polyploidy through chromosome doubling action which is a result of arrest of spindle fibre formation. Both auto and allopolyploids can be induced.

5.3 INDUCED AUTOPOLYPLOIDS

Changes in chromosome number may occur spontaneously or they may be induced by irradiation, chemicals, etc. The different methods for synthesis of autopolyploids are 1. decapitation 2. application of I I A 3. twin seedlings 4. heat or cold treatment 5. colchicines and other chemicals 6. use of gene. In case of tomato (*Lycopersicum esculentum*) if the shoot of the plant is cut back, new shoots frequently develop from the callus of wound and about 7% of these shoots have been found to be tetraploid (Winkler, 1907). High temperature for a short period can give rise to polyploidy in maize (Randolph, 1932). Heat treatment of young ears during early division of diploid embryo produces tetraploid(as well as octoploid) plants. Use of gene, elongate(*el*) (Alexander, 1957; Rhoades and Dempsey, 1966) can be made for production of polyploidy. Synthesis of various tetraploid lines was made by crossing diploid homozygous for *el* gene with tetraploid male. Triploid in maize can be produced by either crossing unreduced egg with haploid sperm or reduced egg with diploid sperm. Unreduced egg can be brought about by *as*, asynaptic gene (Beadle, 1930) or *el*, elongate gene. In case of *as* gene the formation of unreduced eggs is brought about by lack of pairing of homologous chromosomes and non-disjunction. Triploids are also found in the progenies of plants homozygous for *el* gene where suppression of one of the meiotic divisions occurs. Polyploidy does not produce true to its type in most cases because of the reasons explained above and its behaviour is quite unpredictable. In 1977 5 lines of turnip(4x) were produced which were inferior to their diploids. Sugarbeet is a good example of induced polyploidy. The triploid sugarbeet(3x) gives more sugar than either diploid(2x) or tetraploid(4x). In vegetables, fruits, forage and root crops, induced polyploidy has become a great success. Stability of triploid in sugarbeet(Matsumura, 1953) has been exploited to produce triploid seedless fruit in banana and watermelon(Kihara, 1951). In maize tetraploid is by far the most simplest polyploidy to maintain and utilize experimentally. In horticultural crops particularly the ornamental plants larger and different flowers as a result of polyploidy have got much value and because of the vegetative propagation it is easy to reproduce.

5.4 ALLOPOLYPLOIDIZATION

Allopolyploidy can be produced from autopolyploidy artificially, i.e. from autotetraploid, AAAA to AAA!A!. Differences can be naturally occurring differences or can be induced artificially by mutagenesis. Gaul irradiated 7x barley (*Hordeum vulgare*) for four generations and practiced selection for improved fertility. He got considerable improvement. Doyle(1973) in maize used marker gene- Differential pairing affinity(DPA) for preferential pairing which can be readily induced and thus more quadrivalents means improved fertility in auto 4x rye grass(*Lolium perenne*). Crowley and Rees practiced selection in newly selected auto 4x rye grass and after four generations of selection they obtained plants with more quadrivalents

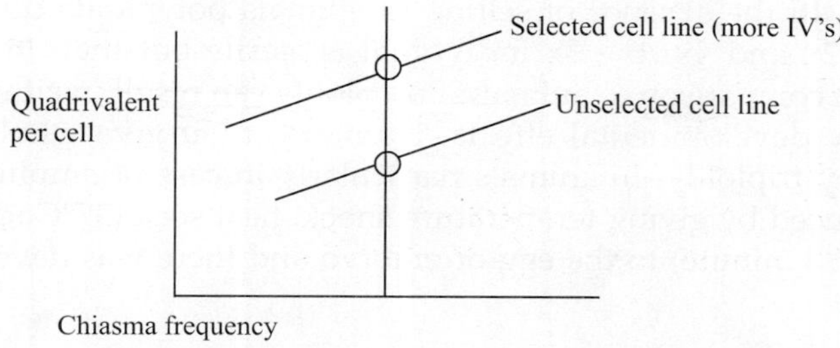

Fig. 5.11 Showing selection for improved chiasma frequency.

per cell which improved the fertility (Figure 5.11). During this process meiosis was modified and there was redistribution of chiasmata with more number of chiasma and so there was more quadrivalent.

5.4.1 Differential Pairing Affinity (DPA)

Two species(A and B) show sufficient differential pairing affinity to prevent the pairing of their homoeologous chromosomes in the allotetraploid(AABB) produced by doubling the chromosome number of the hybrid(AB). DPA is the result of many factors(Doyle, 1979). Chromosome pairing consists of three distinct but sequentially dependent processes. These are congressional pairing, synapsis and chiasmatic pairing. Congressional pairing refers to the movement towards each other of the homologous chromosomes(or segments thereof) across intracellular distances. This probably occurs premeiotically. Synapsis refers to very close alignment of homologous chromosomes which takes place at zygonema. The synaptic forces disappear at the start of diplonema and the chromosomes are held together by chiasmata which were produced by crossing over. The chiasmatic pairing disappears at the start of anaphase. The different DPA factors affect any or all of these processes. DPA factors fall into three groups. Structural non-homology comprises the first type of DNA factors. It results from chromosomal aberrations such as deficiencies, duplications, inversions or reciprocal translocations. Out of these inversions create a large DPA factor. When chromosome aberrations are so small that they can not be visualized using the light microscope, called **cryptic structural changes**, are probably common. The second type of factors is the result of qualitative or quantitative differences in the pairing code. It has been suggested that the pairing code resides in certain palindromic segments of DNA but so far not verified. The third type of DPA factors involves genes that control and modify chromosome pairing. The meiotic process is under genetic control and a large number of genes are known (see chapter 11). One gene, Ph, is known in wheat that prevents the association of homoeologous chromosomes(Sears, 1958; Riley and Chapman, 1958).

5.5 POLYPLOIDY IN ANIMALS

Polyploidy is quite frequent in invertebrates, insects, lower vertebrates like fish. Polyploidy disturbes the sex chromosome mechanism but then in silkworm autoploid the sex chromosome works. In the absence of selfing in animals polyploids do not survive as can happen in a cross, 2x and 4x- the 3x individual is sterile but then in plants isolation of polyploidy is easy in comparison to animals. Polyploidy can result in either one or both sexes sterility. It is purely developmental effects. Human can survive wholly polyploidy. The abortion is caused by triploidy. In animals particularly in case of amphibians polyploidy is most frequently induced by giving temperature shock, heat sock (37°C for a few minutes) or cold shock (1°C for 30 minute) to the egg or embryo and there was development of triploid embryo.

Aneuploidy

6.1 ANEUPLOIDY

Aneuploidy as defined in chapter 5 refers to variation of a single or a few chromosomes. In other words, aneuploidy refers to loss or increased dosages of chromosomes. Aneuploids are produced by loss or duplication of whole chromosomes. Aneuploids can be classified as primary or secondary aneuploids. **Primary aneuploids** involve losses or duplication of complete chromosomes whereas **secondary aneuploids** involve various dosages or combinations of chromosomes that have been derived through misdivision of the centromere such as telocentric and isochromosomes. In the aneuploid there can be **hypoploids**(plants with loss of one or more chromosomes from the normal complement)such as monosomics, 2x-1 and nullisomics, 2x-2 or double monosomics, 2x-1-1 or 2n-1-1-1 which shows the absence of one, two(homologous), two(non-homologous) and 3 non-homologous chromosomes, respectively from the normal disomics, 2x (Table 6.1). Another group is called hyperploids. The plants with one or more chromosomes in addition to the normal diploid complement are called **hyperploids** which include trisomic, 2x+1 and tetrasomic, 2x+1+1, double tetrasomic, 2x+2 where there is addition of one, two different(non-homologous) or a pair of homologous chromosomes, respectively in comparison to disomic, 2x (Table 6.1). In the primary trisomic the extra chromosome is an exact replica of any of the monoploid set(so that three homologous chromosomes are present). **Secondary aneuploids**- Secondary aneuploids are having incomplete chromosomes produced by misdivision of univalents. The telocentric chromosome is formed as a result of misdivision of univalent. The centromere breaks into two half and two telocentric chromosomes are formed as shown in Figure 6.1. In case of hypoploids we can have mono-telocentrics, di-telocentrics or double- ditelocentrics. In case of double-ditelocentric both arms of a chromosome are present as ditelocentric. In case of hyperploids we can have secondary trisomics, tertiary trisomics and telo-trisomics.

Table 6.1 Showing hypo- and hyperploids and their genomic constitution

Aneuploids	
1. **Hypoploids**	
Monosomics	2x-1
Nullisomics	2x-2
Double monosomics	2x-1-1
2. **Hyperploids**	
Trisomics	2x+1
Double trisomic	2x+1+1
Tetrasomic	2x+2
Pentasomic	2x+3
Hexasomic	2x+4
Septasomic	2x+5

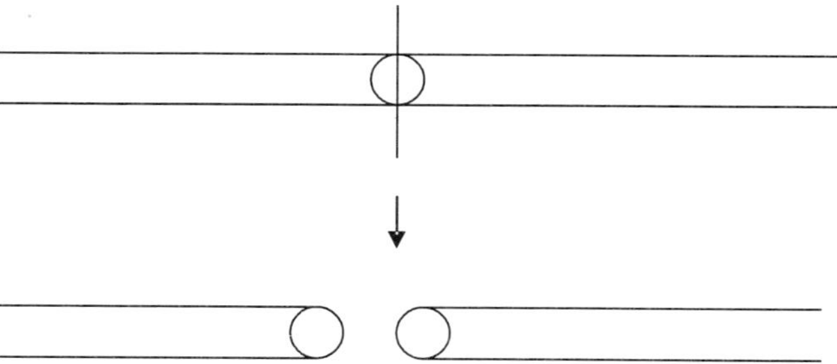

Fig. 6.1 Showing telocentric chromosome.

Blakeslee and Belling (1920) in *Datura stramonium* (2X=14) found secondary and tertiary trisomics. When the extra chromosome is an isochromosome, the trisomic formed is called secondary trisomics (Figure 6.2) and when the extra chromosome is a translocated chromosome it is called tertiary trisomic (Figure 6.3). In case of telo-trisomics the extra telocentric chromosome is derived from one of the two arms of a chromosome.

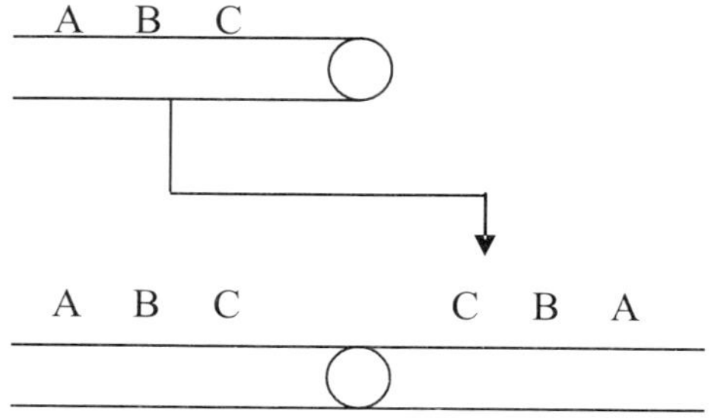

Fig. 6.2 Showing formation of isochromosome.

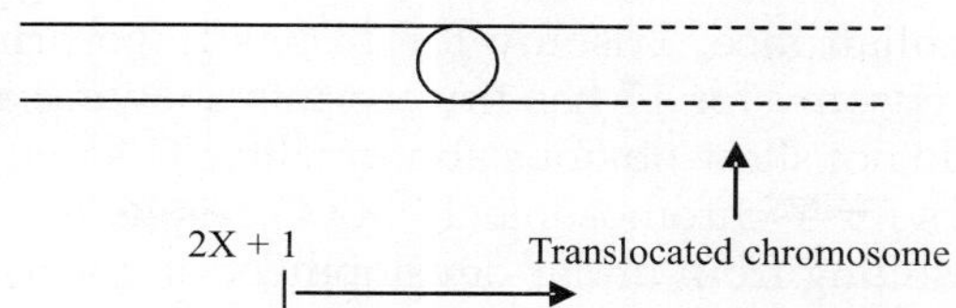

Fig. 6.3 Showing translocated chromosome in tertiary trisomic.

6.2 CHARACTERISTICS OF ANEUPLOIDS

Twelve different trisomics have been constructed in *Datura stramonium* and all differ in phenotypes and fruit size. Aneuploidy affects viability, vigor and fertility of the organism. There are only few species where trisomics are indistinguishable from their diploids, e.g. *Clarkia unguiculata.* As hyperploidy is much more tolerated by an organism than hypyploidy so the trisomics are viable. In 2X plants loss of chromosomes usually results in lethality. In other words monosomics are lethals whereas all trisomics are viable. In Datura four monosomics out of twelve are viable, in 2 out of 12(in tomato) and 4 out of 10(in maize) are viable. In maize all 10 possible trisomics have been isolated. All the primary trisomics are small and less vigorous than their disomic lines and further some can be identified by their unique phenotypes (Carlson, 1976). As most of the nullisomics are sterile either as male or female in wheat they can not be maintained as such and generated afresh from monosomics. Further it is difficult to distinguish monosomics under favourable environmental conditions being from normal when the later are raised under favourable conditions. In comparison to monosomics, trisomics look more similar to normal. Tetrasomics are more vigorous and fertile in comparison to nullisomics. Nullisomics in wheat differ morphologically from one another, have reduced size, vigor and fertility. As many of the nullisomics can not be propogated they have little value in practical plant breeding. Telo-centric trisomics show less sterility than primary trisomics and so they are more useful in genetic studies. In polyploidy plants monosomics and some nullisomics are viable the numbers of which are given in Table 6.2. In animal aneuploidy is uncommon. In Drosophila only monoIV is viable whereas in man none of autosomal monos are viable.In human there are aneploids for both sex chromosomes (Xand Y). The mosomic for sex chromosome, XO condition is called Turners syndrome. In case of Turners syndrome the abnormal female ovaries do not develop normally and so are non-functional. It has got more deleterious effects. In man, there are three autosomal trisomics, 21, 17 and 13. Trisomics involving chromosome number 21 leads to mongolism or Down's syndrome- a condition characterized by mental retardation, a short body with stubby fingers, swollen tongues, monkey like skin, ridges and eyelids fold

Table 6.2 Polyploids showing numbers of viable monons and nullis.

Crops	*Ploidy level*	*No. of viable monos*	*No. of viable nullis*
Wheat (*Triticum aestivum*)	2n=6x=42	21	21
Oat (*Avena sativa*)	2n=6x=42	21	~ 12
Tobacco (*Nicotiana tabacum*)	2n=4x=48	24	0
Cotton (*Gossypium hirsutum*)	2n=4x=52	7	0

resembling that of Mongolian race. Trisomy for 13 has fingerprints showing arches and whorls as well as loops. Trisomy for 17 has fingerprints showing arches and in contrast to those of the other palms do not show obvious abnormality. Trisomics for larger chromosome lead to abortion. Trisomics for Y-chromosome(47, XXY) results in abnormal male known as Klinefelter's syndrome resulting from under development of gonads, XXY is normal, XXYY, the double trisomic shows Klinefelter's syndrome whereas XXXY and XXXY the tetrasomic and pentasomic, respectively show extreme from of Klinefelter's syndrome. Trisomics for X-chromosome(47, XXX), Tetrasomics with X-chromosome(48, XXXX) or pentasomics with X-chromosome(49, XXXXX) are all mentally defective females.

6.3 ORIGIN OF ANEUPLOIDS

Aneuploids are formed as a result of non-disjunction (Figure 6.4). It results into formation of X+1 or X-1 gametes which upon fertilization with X gametes forms either trisomic, 2X+1 or 2X-1, the monosomic. Trisomic may arise as a result of non-disjunction of the ring of four chromosomes at anaphase I in the translocation heterozygotes (Rhodes, 1955). Trisomics are also obtained from segregation of the unbalanced chromosome complement of triploid. In

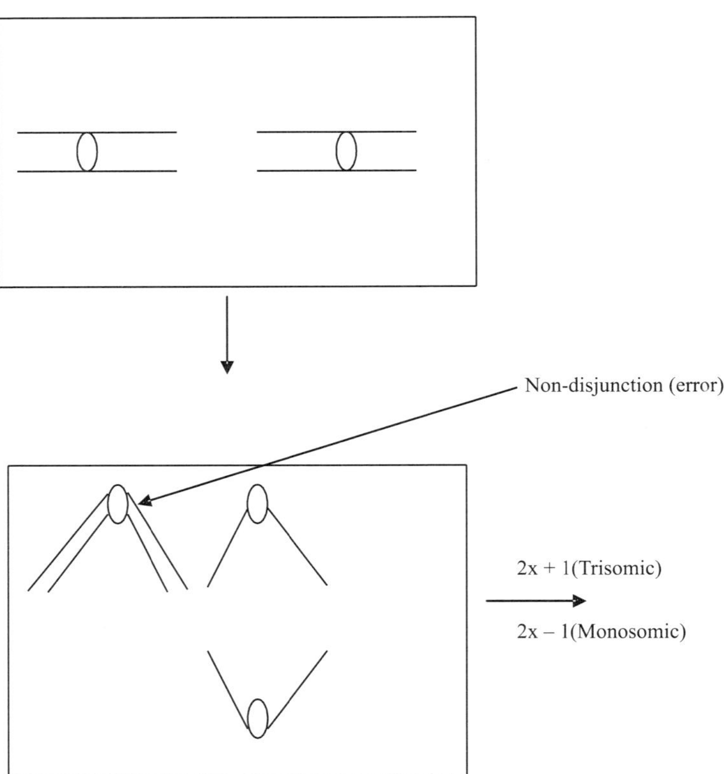

Fig. 6.4 Showing origin of aneuploids by non-disjunction.

Datura in 0.4% there is production of X+1 gametes either because of failure of 1st or 2nd meiotic division. Non-disjunction (N.D.) occurs naturally. It can be increased by temperature shock or other physical or chemical treatments. In man the frequencies in N.D.I and N.D.II are as follows.

	N.D.I	N.D.II
♀	4	2
♂	2	1

Further the frequency of non-disjunction is related to maternal age. It increases with the maternal age. Finally, non-disjunction frequency is higher in female than in males. In wheat most of the monosomics have been obtained from two sources: haploid and asynaptics. There is also high frequency of spontaneous occurrence of monosomics. Ionizing radiation also increases the frequency of occurrence of monosomics. A complete range of aneuploids from nullisomics to nullisomics to tetrasomics for each of the 21 chromosomes are available in the wheat variety, Chinese spring since 1954 (Sears, 1954). A series of monosomic lines of *N. tabacum*, each lacking a single different chromosome of the complement have been produced (Clausen, 1941 and Clausen and Cameron, 1941).

6.3.1 Development of Monosomics in Wheat

Monosomics were derived principally from haploid and asynapsis, from plants nullisomic (2n=40) for 3B which have reduced pairing at meiosis. Backcrosses from or to either of these two types of plants produced a large number of monosomics (Law et al, 1987). And by selecting among the backcross progenies the complete set of 21 monosomics in the Chinese Spring variety was developed. In case of production of monosomic through haploid the haploid wheat is crossed to normal wheat (as male parent) and in the progenies a series of plants with reduced chromosome number are found. Sears built up a monosomic series in varitey 'Chinese Spring' which then served for the development of monosomic series in other varieties. The nullisomics of Chinese Spring were developed by selfing monosomic plants. As monosomics occur spontaneously in varietal population at a frequency of about 1%(Riley and Kimber, (1961) so it is possible to establish a complete set of monosomics in a variety by observation of phenotypes and counts of chromosome. The problem with the development of this type of set of monosomics is that it is a tedious and time –consuming process and it is difficult to select supposedly monosomic plants on the basis of phenotype as most monosomics are not obviously different from euploid. One can also use an existing set of monosomics (monosomics developed in Chinese Spring variety) to develop a further set of monosomics in another variety by means of backcrossing. First the variety in which the set of monosomics is developed, is used as the donor or male parent(or pollen parent) and crossed with monosomic plant used as female. The F1 hybrids would consist of plants with 41 and 42 chromosomes. The F1 hybrids with 41 chromosome (monosomic) is backcrossed with the donor variety and again plant(s) with 41 chromosome, i.e., the monosomic,is selected from the backcross population and backcrossed to the donor variety which is used as the recurrent parent. After a number of backcrosses the selected monosomic becomes homozygous or nearly homozygous for the genes of the donor variety. Similar backcrossing programme is carried out for each of the 21 monosomics to develop a monosomic set in the donor variety. Another way of checking whether or not there is any background effect is to self the monsomic line obtained after a number of backcrosses and isolate the euploid(disomic) line and compare it with the control euploid line. If the derived euploid is indistinguishable from the control euploid then there is no need to further backcross but if both are not phenotypically identical then it requires further backcrossing.

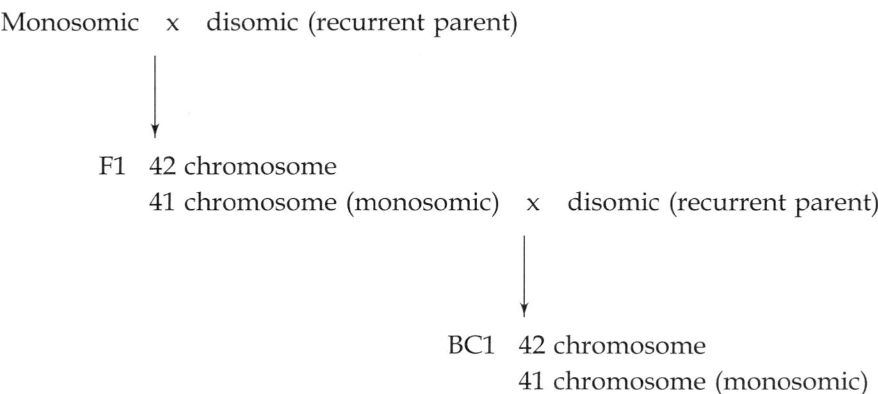

This method of development of monosomics using existing set of monosomics has the advantage that all the 21 monosomics can be developed in less time than other methods. Monosomics also occur spontaneously with high frequency in most varieties. Ionizing radiation increases the frequency of occurrence of monosomic but then it produces other aberrations.

Problems with development of monosomics by backcrossing - There are two problems associated with the development of monosomic lines through backcrossing. 1. Problem of occurrence of **'univalent shift'** and 2. **background variation**. Non-disjunction of chromosomes other than the hemizygous chromosome could occur in the monosomic hybrid during any of the backcross stages which would lead to deficiency for a chromosome other than the one for which selection is being practiced and as a consequence the selected monosomic will be hemizygous for another chromosome. This phenomenon of production by a monosomic plant of offspring monosomic for the wrong chromosome because of the less-than complete regularity of bivalent pairing in hexaploid wheat is called 'univalent shift'(Person, 1956). The 'univalent shift' occurs not only during the development of monosomics but also during maintainance of monosomics. Detection of **'univalent shift'** and selection against its occurrence can be achieved by the use of di-telocentric lines which are available in Chinese Spring variety. The di-telocentric line is used as male parent in cross with appropriate monosomic line used as female parent. The hybrids will have one telocentric chromosome plus either 41 or 40 normal chromosomes. The plants with one telocentric plus 40 chromosomes are critical in identifying 'univalent shift'. Such plants are selected and examined at meiosis and if 'univalent shift' has not taken place then the mono-telocentric plant would show 20 bivalents and one unpaired telocentric chromosome. But if 'univalent shift' has taken place then one can find again 20 bivalents and one univalent but this time one of the bivalents will be heteromorphic and compose of both complete and telocentric chromosomes paired together and the univalent will be complete chromosome and not the telocentric chromosome. The second problem concerns background variation. If some of the genes from the initial monosomic set are still present in the derived monosomics then the result from the genetic analysis involving the newly developed monosomic set will not be accurate. This problem can be solved by developing duplicate lines of each of the monosomics. Each backcrossing programme is carried out twice for each monosomic line and differences between duplicate pairs of monosomic lines would indicate the presence of background effects and need further backcrosses, i.e., increase the number of backcrosses for developing monosomic line. The number of backcrosses required can be determined by comparing the magnitude of the duplicate differences assessed over all 21 chromosomes or within each duplicate pair. The development of duplicate monosomic lines will also provide an insurance against 'univalent shift' as it is unlikely that the 'univalent shift will occur simultaneously in both lines. A further method of overcoming this problem is to extract euploid lines from each of the monosomic lines after a reasonable number of backcrosses have been completed. If the number of backcrosses is sufficient then the euploid derivatives should be indistinguishable from the control euploid or the variety itself.

6.3.2 Production of Monosomic in Tobacco

Initial monosomics in tobacco were developed from interspecific crosses and repeated backcrossing (Olmo, 1935). The pale sterile variety was crossed to Purpurea as male parent. The F1's were having 11 II's plus 26 I's. The F2 generation population produced as a result of selfing consists of a number of simple as well as complex monosomic types of which complex monosomic types were selected and backcrossed to the standard Purpurea variety which resulted in elimination of all but one I. As monosomics often differ slightly from the normal disomic so they must be cytologically identified before using in the backcrossing programme.

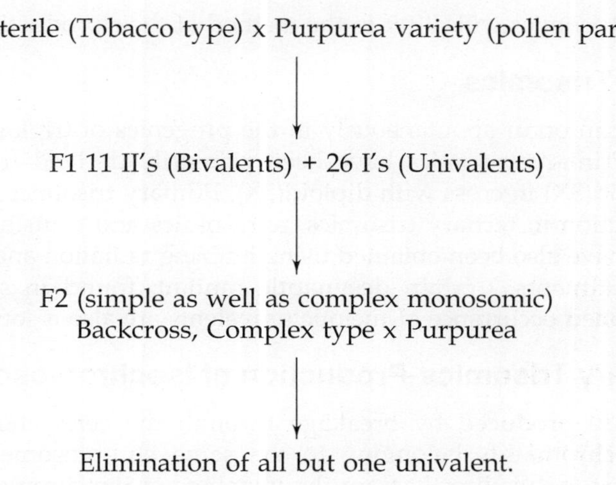

6.4 PRODUCTION OF NULLISOMICS

Nullisomics are commonly obtained afresh each generation from segregating population derived from selfing of monosomic plants as shown below. This is because of the reason that most of the nullisomics are sterile, either as male or female.

Monosomic

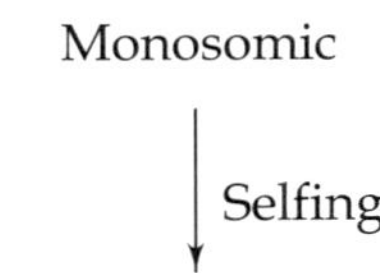

Disomic, monosomic and nullisomic

The three types of offspring are found in varying frequencies for different chromosomes. The frequency of nullisomics may be as high as 10% for monosomic III and as low as 1% for several others. The lower frequency is because of the reason that only a small proportion of male gametes with X-1 chromosome will function in the fertilization (Sears, 1953a). Those nullisomics which can be propagated produce only nullisomic upon selfing.

Nullisomic

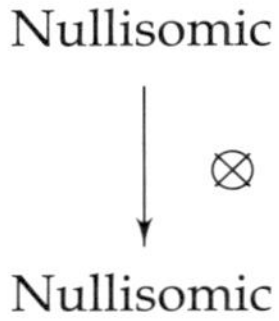

Nullisomic

6.5 PRODUCTION OF TRISOMIC

The three types of trisomics will differ in the method of their production.

6.5.1 Primary Trisomics

Primary trisomics can occur spontaneously in the progenies of triploids and thus it can be isolated from it. Primary trisomics have been primarily derived from the n+1 gametes produced by triploid(3X) in cross with diploid(2X). Primary trisomics have also arisen from the meiotic disjunction in tertiary trisomics, tetrasomics and translocation heterozygotes. Primary trisomics have also been obtained using ionizing radiation and after colchicines and other chemical treatments. Certain desynaptic mutants(found in soybean) that have a genetically conditioned occurrence of meiotic univalents are also a source of trisomics.

6.5.2 Secondary Trisomics-Production of Isochromosome

Isochromosomes are produced by breakage through the centromere (misdivision) and joining across of chromatids belonging to the same chromosome arm. Production of isochromosomes may occur directly from the members of the chromosome complement or indirectly by misdivision of a telocentric chromosome which itself would have arisen by centromeric misdivision. The misdivision occurs when a centromere is incapable of carrying out its function of orienting and moving the chromosomes towards the pole during meiosis or mitosis (Carlson, 1977). Secondary trisomics mostly occur in the offspring of plants with univalent chromosomes (Burnham, 1962). A prime source of misdivision products is the univalent at meiosis.

6.5.3 Tertiary Trisomics

In the tertiary trisomics the extra chromosome is composed of parts of two non-homologous chromosomes. Tertiary trisomics regularly occur in the progeny of translocation heterozygotes. A number of tertiary trisomics have been derived in maize from translocation heterozygotes in which nondisjunction of the ring of four chromosomes at anaphase I of meiosis results in a functional 11-chromosome gamete.

6.6 TELOCENTRIC CHROMOSOMES

Telocentric chromosomes occur spontaneously in the progeny of triploids or monosomics. It can also arise through centromeric misdivision. It can also be induced by radiation and chemical mutagens in primary trisomics or monosomics. Finally it can also arise by breakage of a di-centric bridge in the region of the centromere. So far only 5S and 6L telocentrics have been recovered in maize (Rhoades, 194; Doyle, 1972a).

6.7 MEIOTIC BEHAVIOUR OF ANEUPLOIDS

Normal pairing and disjunction can not occur at meiosis in aneuploids. Meiotic behaviour of trisomic is similar to triploids, i.e. we get trivalent or bivalent plus univalent. In case of monosomics we get only univalents. The gametes formation and the resultant progenies in case of trisomics are as follows.

♀ \ ♂	X	X+1
X	2X	2X+1
X+1	2X+1	2X+2

This shows that there is theoretically 50% trisomics production but actually the recovered frequency is much less. The transmission of trisomic in crosses ♀(2x+ 1) x ♂(2X) about 20% whereas in the reciprocal crosses, ♀(2X) x ♂(2X+1) the transmission is even lower and is in the range of 0.0-20%. The reasons for the reduced transmission of trisomics are: 1. the univalent loss at meiosis in 2X+1 2. the reduced viability of X+1 spores and 3. the reduced viability of 2X+1 zygote. Male gametophyte is more sensitive to abnormality in comparison to female gametophyte. The result of these constraints is that there is production of about 1-10% tetrasomics, about 45% primary trisomics and the rest disomics (Sears, 1954).

The monosomic line under selfing produces the gametes and the progenies in the following manner. There occurs a competition between X gametes and X-1 gametes and male gametes with X are more efficient than gametes with X-1 on the stigma.

♀ \ ♂	X (96-100%) Av. 96%	X-1(1-100%) 4%
X(14-29%) Av.25%	2X(24%)	2X-1(0.1-5%) Av.1%
X-1(61-86%) Av. 75%	2X-1(49-85%) Av. 72%	2X-2(0.6-16%) Av. 3%

The table shows that the monosomics produces on the average 24% disomics, 73% monosomics and 3% nullisomics (Sears, 1954). However the proportion of disomics, monosomics and nullisomics offspring in the self-pollinated progeny of monosomics vary somewhat for different chromosomes. In case of secondary trisomics rings of chromosomes may be seen at meiosis whereas in case of tertiary trisomic chains of five chromosomes are possible at meiosis.

Telocentrics - Most telocentrics are poorly transmitted through pollen in competition with their complete homologue and if there were no transmission of the telocentric a heterozygote having one allele on a telocentric chromosome could be used as male in a cross to the homozygous recessive and the amount of crossing over could be read directly from the percentage recovery of the allele carried by the telocentric. But as there is always some transmission cytological investigation of at least a sample of the critical class is essential to discover how many are carrying the telocentric rather than being crossovers. In case the marker is present on the other arm of the complete chromosome the necessity for cytological examination is eliminated and it would be practical to use the heterozygote as female.

6.8 UTILITY OF ANEUPLOIDS

The utility of aneuploids depends on how easily and how frequently the aneuploids can be synthesized. Transfer of one chromosome is heavily dependent on the tolerability of aneuploid by a particular species. Aneuploids such as nullisomics, monosomics, trisomics and tetrasomics can be used to study the dosage effects of each chromosome and gene. Monosomic can be used 1. for substitution programme and 2. locating genes on specific chromosome. Primary trisomics are used for assigning genetic markers and entire linkage groups to specific chromosomes. Tertiary trisomics and telotrisomics are tools to determine arm location and approximate distance from the centromere(Schulz-Schaeffer, 1980). Secondary trisomics can also be used as efficient tools in linkage mapping(Khus, 1973) and information regarding chromosomal and arm location of a genetic marker, the centromeric position and the proximity of the marker to the centromere can be obtained from the segregating progenies. The aneuploid techniques are the most powerful method of identifying the number of genetic factors segregating in the populations. These techniques partitioned the variation into the effects of individual chromosomes and then into the effects of individual arms or regions of these chromosomes. This method estimates numbers and locations of factors and also the relative size of effects.

6.8.1 Use of Aneuploids in Genetic Analysis

6.8.1. 1 Polyploids-monosomic and nullisomic analysis

a. Locating recessive genes. Monosomics can be used to determine the chromosome carrying a particular gene (Sears, 1953: Kuspira and Unrau, 1959). If all the 21 monosomics of a wheat variety are crossed with disomic variety, the homozygous for the recessive gene to be located and the monosomic hybrids are selected from the resulting progeny, then the hemizygous chromosomes in each of the crosses must have come from the donor variety. Now if a particular monosomic (critical) contains dominant allele then the F1 produced will show segregation and the F2 population derived from the monosomic F1 will be all recessive and the chromosome carrying the recessive gene can be determined which is shown below.

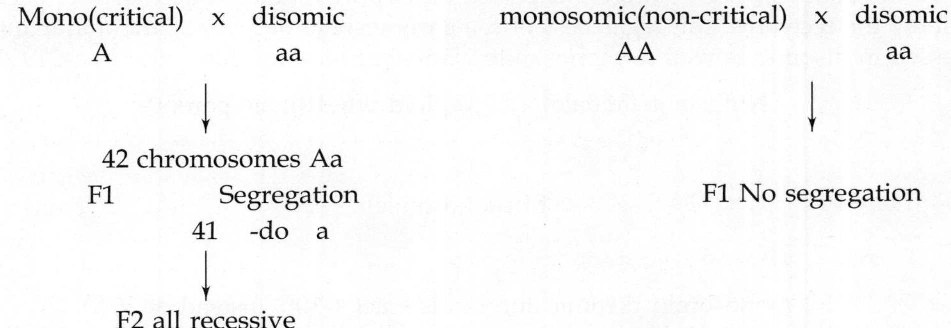

In case of Tobacco the normal recessives are crossed to each of the 24 monosomic dominants and in the critical cross the progeny will all have recessive phenotype.

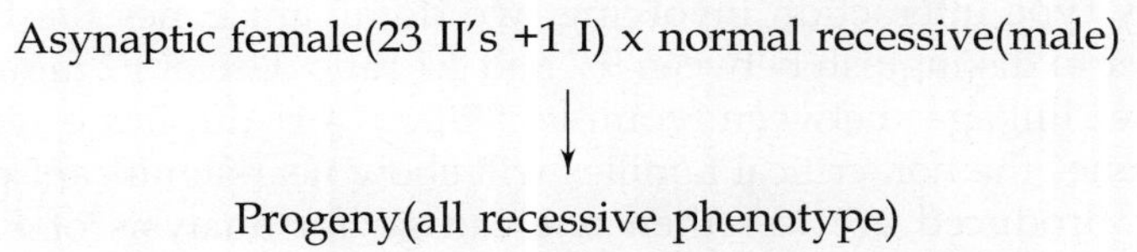

b. Locating dominant genes When the gene is dominant all of the F1 derived from the critical monosomic with disomic with dominant alleles, will have the dominant phenotype and now if the monosomic hybrid is selected and F2 generation population is produced then it would show no segregation. In case of 20 mono(non-critical) x Disomic crosses the F2 population will show 3:1 segregation as shown below.

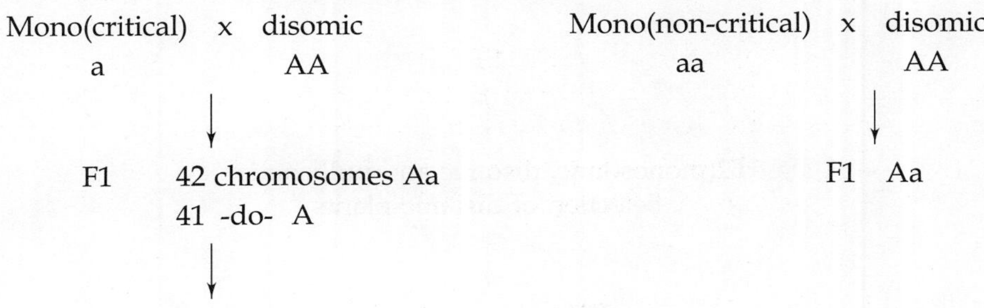

F2	42 chromosomes	AA	No segregation F2	AA segregation in F2
	41 -do-	A	in F2	Aa
	40 -do-	Nullisomic		aa

6.8.1.2 *Nullisomic analysis*

a. Location of dominant gene The hexaploid wheats are crossed to each of the 21 nullisomic dominants and the F1's are obtained. The F1's will all be monosomic which upon selfing will produce nullisomic. In case of critical cross the critical F2 family(the family segregating for disomic, monosomic and nullisomic) will have only a few recessive. Theoretically the frequency of nullisomic will range between 1 to 10% instead of the expected 25% and so in the critical family the segregation will not be in the ratio of 3:1. In non-critical families the usual 3:1 segregation is expected and approximately one fourth of each family will show the recessive trait regardless of the chromosome number as the dominant gene is expressed in disomic as well as monosomic plants but not in nullisomics(Sears, 1953a).

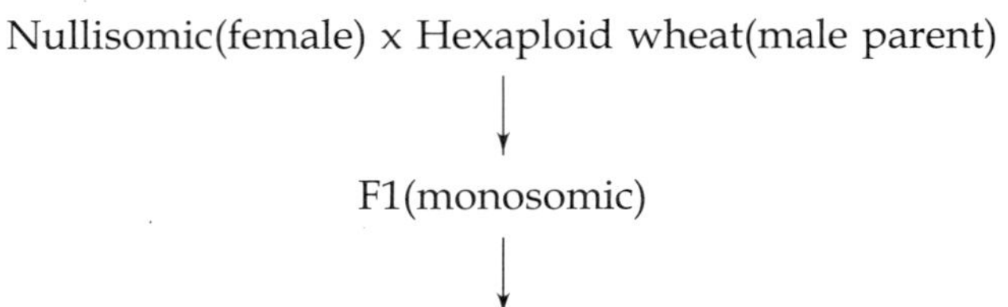

F_2(monosomic, disomic and nullisomic(1-10% instead of 25%)

b. Location of Duplicate/complementary factors Duplicate or complementary factors carried on either the same or different chromosomes can also be located. With duplicate factors on the separate chromosomes two critical families may show 3:1 segregation. In case of complementary type interaction involving two dominant genes the F2 population size must be 95 in order to distinguish between 9:7 and 3:1 ratio. Use of F2 families can also made for detecting the linkage between complementary or duplicate genes. For linked complementary genes the non-critical families will show near-significant departure from 9:7 ratio. If F2 seeds produced are less then one can go for analysis of F3 families. A few cytologically detected disomic plants are selected and selfed to generate F3 families and when a critical chromosome is involved the F3 families will show that the disomic plants were homozygous for the genes present in that chromosome, i.e., these families will not show any segregation.

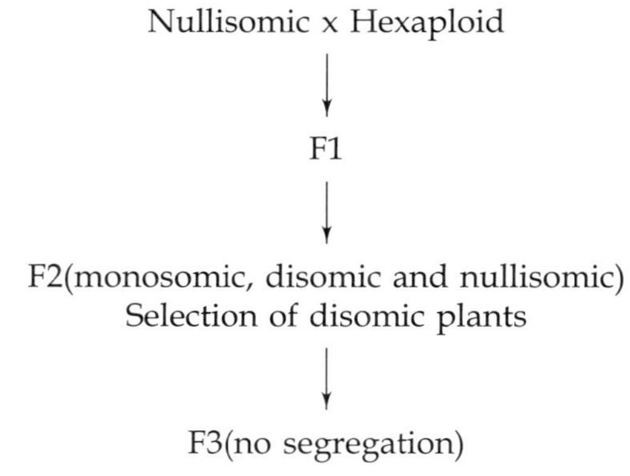

6.8.1.3 *Diploids-trisomic analysis*

In case of diploids monosomics cannot be used as they are lethals and so trisomic is used in the genetic analysis. Plants homozygous for recessive mutant genes in each of the linkage groups are crossed to trisomic plants and the critical trisomic x disomic cross will result in the production of disomic as well as trisomic. The F1 trisomic plants are selected and are either selfed to produce an F2 generation or backcrossed to the recessive parent in the similar test cross. The backcross population will show a phenotypic ration of 5: 1 and the F2 population will show a phenotypic ratio of 35: 1. In practice the loss of univalents during meiosis in trisomic females reduces these ratios but still the ratios will be much higher than 1:1 and 3:1.Thus departures from 3:1 or 1:1 segregation assign the gene in question to the trisomic chromosome (Rhodes, 1955).

	Trisomic (critical) x disomic
	AAA ↓ aa
F1	A_1A_2a select trisomic F1
F1	gametes A_1A_2, A_1a, A_2a, A_1, A_2, a
Backcross	5A : 1a
F2	35A: 1a

Further it can be used in breeding for crop improvement (for further information see Chapter 13).

6.9 TELOCENTRIC ANALYSIS

Telocentric chromosomes are easily identified cytologically and so are of great importance as markers in tester stocks. Telocentric chromosomes have been used in determining the location of genes on the particular arms of chromosomes and in some cases to map the genes with respect to the centromere (Sears, 1966b). Monotelocentrics are very similar to monosomics in behaviour and are fairly easy to maintain. They can be selected from monosomics with a frequency of about 3% in wheat. Ditelocentrics can be selected from progenies of monotelocentrics. They breed true and are generally of high fertility. Ditelocentrics are available for at least one arm of each chromosome and in 8 cases for both arms. Double-ditelocentric lines are present for all chromosomes except 7D in wheat. Mono-isochromosome plants can also be obtained from the segregating progenies of monosomics and they occur at a frequency of ~1%(Law et al, 1987). Like telocentrics they are also cytologically recognizable and in some cases they look more closer to euploid than the monosomics. However they produce more nullisomics than do monosomics when selfed. In Chinese Spring maintainable isochromosome stocks are available for 10 chromosomes.

6.9.1 Location of Dominant Gene on a Particular Arm of a Chromosome- Use of Telocentric Chromosome

Most telocentrics are poorly transmitted through pollen in competition with their complete homologue and if there were no transmission of the telocentric a heterozygote having one allele on a telocentric chromosome could be used as male in a cross to the homozygous recessive and the amount of crossing over could be read directly from the percentage recovery of the allele carried by the telocentric. But as there is always some transmission, cytological investigation of at least a sample of the critical class is essential to discover how many are carrying the telocentric rather than being crossovers. In case the marker is present on the other arm of the complete chromosome the necessity for cytological examination is eliminated and it would be practical to use the heterozygote as female. From a cross to the two ditelocentrics for the chromosome concerned the arm on which a particular gene is located can easily be determined as the recovery of segregant telosomes carrying the gene is only possible for one arm and the frequency with which such telosomes are recovered provides a measure of the distance of the gene from the centromere.

The variety known to be carrying the critical gene on a particular arm is crossed to a variety with di-telocentric for the same arm. The resulting F1 in this critical cross will thus will be heterozygous for the gene but hemizygous for the opposite arm of the chromosome (see Figure 6.5). Thus crossing over which will be restricted to the arm carrying the gene can occur and so segregation of telocentric chromosome will act as a marker for centromere segregation. The F2 or back-cross generation will show such segregation and by scoring the segregation (parental genotypes and recombinants) estimate of recombination frequency between the gene and the centromere can be obtained. Cytological examination will be done to identify the parental and recombinant types. Now if the hemizygous arm also carries a marker then this marker can be used to follow the segregation of the centromere and in this situation there is no need to do the cytological analysis. In the Figure the female parent is a ditelocentric carrying the recessive allele in homozygous condition (aa) and the male is the euploid carrying the dominant allele in homozygous condition(AA). The male also carries another marker, B and is also homozygous, BB. The F1 will be heterozygous, Aa but

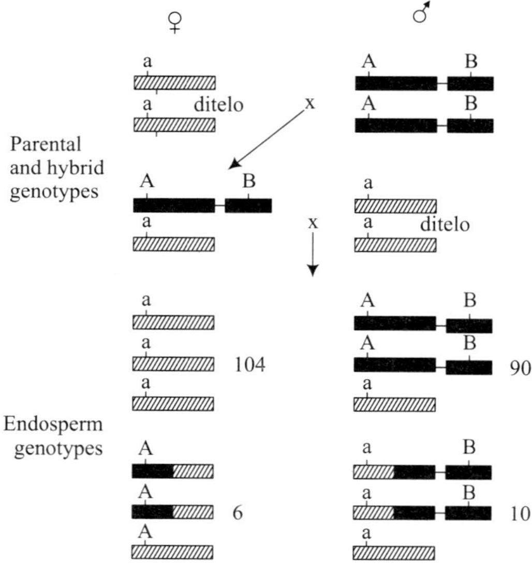

Fig. 6.5 Showing procedure for locating dominant gene using telocentric chromosome (adapted from Law et al., 1987).

hemizygous for B. When F1 is backcrossed to recessive ditelocentric parent, the recombinant and the non-recombinant will appear in the test cross population and from the frequencies of recombinant and parental types, recombination frequency between gene(A-a) and the centromere can be estimated. In the real experiment the Chinese Spring ditelocentric $1A^L$ was crossed to substitution line carrying *T. spelta* 1A. The telocentric chromosome carries a Glu-A1 allele(a) which produces a different protein than that produced by the allele(A) on the long arm of *T.spelta* 1A. The short arm of *T. spelta* 1A carries an allele(B) of Gli-A1 for gliadin protein. Crossing over between gene(A-a) and the centromere will produce two recombinant endosperm genotypes and the recombination frequency between Glu-A1 and the centromere was calculated as 16/210 or 7.6%. Through this telocentric analysis a gene determining high-molecular weight gluten was located on the long arm of chromosome 1A(Payne et al., 1982).

6.10 MONOSOMIC ANALYSIS

6.10.1 Comparison Between Monosomic Series

When two or more monosomic series are available a comparison of the difference between a particular monosomic, say monosomic 4A and its euploid with the difference between the same monosomic, say monosomic 4a and its euploid in another variety will indicate either whether or not the allelic differences occur between the monosomic chromosomes(depending upon whether the magnitudes of differences in the two varieties are the same or not) or that the genes are identical and the differences are due to interactions arising from different genes in their background (Law et al., 1987). In other words, the following comparison is to be made.

(monosomic 4A- euploid R) = (monosomic 4A –euploid S).

6.10.2 Reciprocal Monosomic Analysis

The problems in comparison of different monosomic series are the problems of chromosome dosage and the confounding background interactions. These problems can be overcome through the use of reciprocal monosmic analysis(Mc Kewan and Kaltsikes, 1970) and thus chromosome(s) determining varietal difference can be identified. In this method reciprocal crosses between homologous monosomics developed from two different cultivars, R and S are crossed. In the first cross the monosomic(R) is used as female and the disomic derived from the monosomic(S) is used as male whereas in the reciprocal cross the monosomic(S) is crossed with the disomic derived from the monosomic(R)(Figure 6.6). The F1's in both cases will be monosomic with identical background but will differ in their hemizygous chromosomes and now comparisons between these F1's will show whether or not there exists the genetic differences between the hemizygous chromosomes. These F1's monosomics upon selfing will produce disomics for the critical chromosomes from the varieties, S and R on the segregating backgrounds and again any difference between the disomics populations will show the effects of the different alleles on the two homologous chromosomes.

6.10.3 Backcross Reciprocal Monosomic Analysis

The reciprocal monosomic analysis can still be used even when a monosomic series is unavailable in a particular variety and supposing monosomic series exist only in variety R

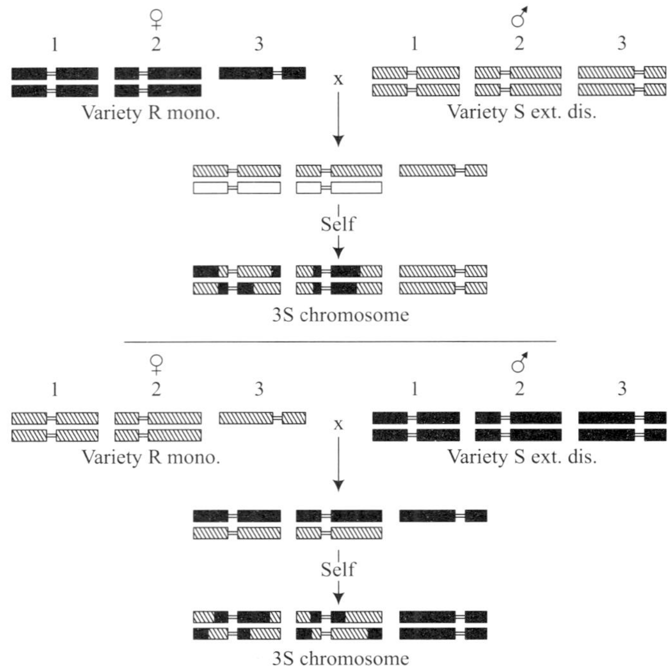

Fig. 6.6 Showing reciprocal crosses between homologous monosomics derived from two different varieties, R and S(adapted from Law et al., 1987).

and not in S then in this condition the monosomic(R) is used as female and crossed to variety S euploid. The F1 is a monosomic hybrid which is then reciprocally backcrossed to the parental monosomic(R) as shown in Figure 6.7. These two backcrosses will give rise to monosomic hybrids having on average identical genetic backgrounds but differing in their hemizygous chromosomes. These monosomic hybrids upon selfing will produce disomic hybrids in which the critical chromosome is homozygous. Comparisons of means of these populations will indicate whether or not there exist genetic differences between the homologues under a variety of environmental conditions.

6.11 INTER-VARIETAL CHROMOSOME SUBSTITUTION

In case of inter-varietal chromosome substitutions only a single chromosome difference occurs in a constant genetic background and the variable products of the cross between a substitution line and its recipient variety can be due only to the gene differences between the substituted chromosome and its homologue. This segregation will provide information regarding linkage, the number of genes determining a trait and their locations. If there is discontinuous variation in the progeny the analysis is simple (for example if the F1 progenies fall into two distinct classes of approximately equal numbers, it indicates that a single gene is responsible) but if the progenies show continuous variation then aneuploids available in the

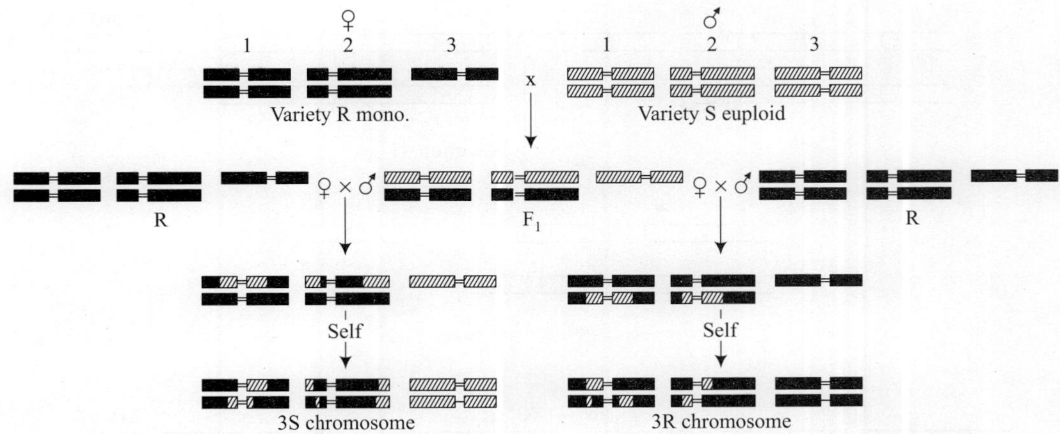

Fig. 6.7 Showing development of backcross reciprocal monosomics hybrids using two varieties, R and S (adapted from Law et al., 1987).

recipient variety can be used to derive homozygous recombinant lines. The F1 between a substitution line and its recipient or between two homologous substitution lines is now crossed as male to the recipient monosomic and from which the monosomic progeny can be obtained which will be hemizygous for either a non-recombinant or a recombinant chromosome. The selfing of these selected monosomic will result in production of euploid lines in which the non-recombinant and recombinant chromosomes are homozygous (Figure 6.8). The Figure shows substituted chromosome carrying the two recessive genes a and b and the chromosome homologous to the substituted homologue carrying the two dominant genes A and B. The selection among the progenies of the cross of F1 with recipient monosomic plant leads to a range of recombinant lines which upon selfing produces true-breeding F1 products or homozygous recombinant lines. These homozygous lines can be replicated and evaluated and comparisons between means and variances of the F1 progenies with those of parental progenies can be made. Further these recombinant homozygous lines can be grown in different environments to study the genotype—environment interaction. Discontinuities among F1 progenies may thus appear which would not be apparent among F2 and backcross generations. Furthermore, the variation observed among the F1 progenies contains no dominance as all the progenies are homozygous. The recognition of discontinuities or recombinant genotypes among the F1 products can also be facilitated by growing along with the progenies of crossing of the substitution line and its recipient with the recipient monosomic. The Figure 6.8 shows substituted chromosome in substitution line carrying the recessive genes a and b being crossed to the recipient variety carrying dominant genes A and B. Selection of monosomics from this cross generates a sample of non-recombinant chromosomes which upon selfing followed by selection gives rise to homozygous parental products. These parental progenies provide a sample of the effects of non-recombinant chromosomes and if any of the F1 products differs from this sample then it can be said that the recombination must have taken place and at least two linked genes are determining the character under investigation. On the other hand if the F1 progenies are

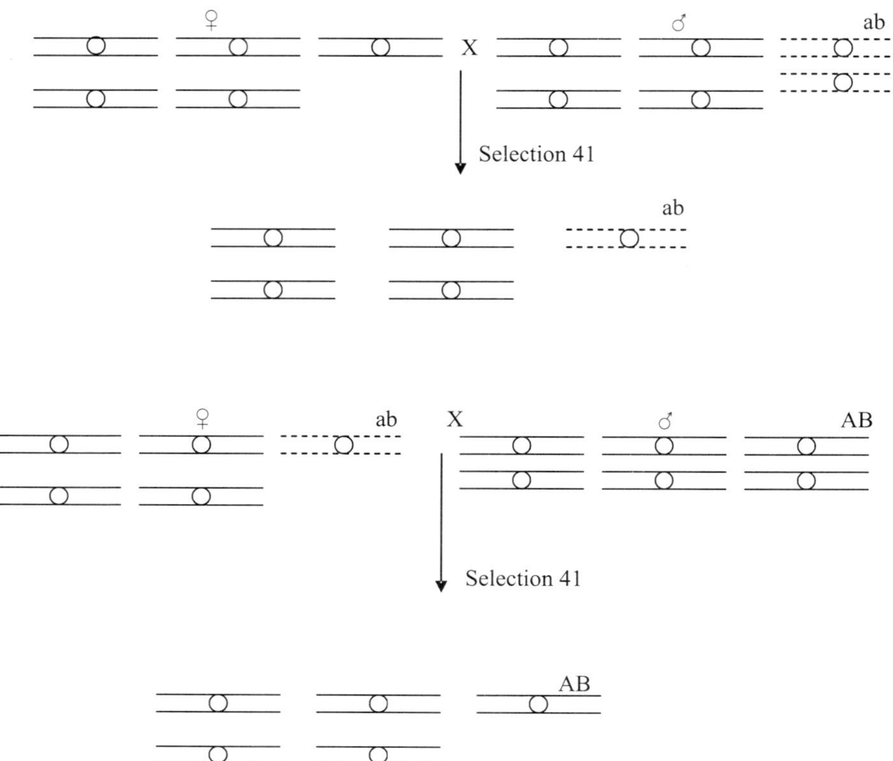

Fig. 6.8 Showing derivation of homozygous parental product by crossing the substitution line carrying the genes a and b, and recipient variety.

indistinguishable from their parental progenies but yet the products of two parents are different then segregation can be said to have occurred at one locus. This method of genetic analysis has been used to identify single genes, often affecting quantitative characters and to determine their locations (Law, 1966, 1967; Worland and Law, 1986). The use of homozygous recombinant lines is the best method available for studying the genetics of quantitative characters.

6.11.1 Substitution Line Analysis

Substitution lines breed true to type and once the substitution lines are established and show negligible background effect it can be used to study the effect of particular differences and further to estimate the genetic effects more precisely than is possible with other methods. The substitution lines are developed through backcross method(see Chapter 13). Variations among substitution lines are not entirely due to the substituted chromosomes but also result from sampling of genes segregating in the background and this could be due to the difference in the number of backcrosses used in their development and the type of marker used. So when using substitution lines it is essential to produce duplicate substitution lines, assess the variation between them and determine the experimental error and thus background segregation must be ruled out while assigning chromosomal differences. The chromosome assays provide information regarding the genes carried by the chromosome and the magnitude of a particular chromosome difference within a particular genetic background. Although the gene x background genotype interaction will be there but assuming such interactions being absent it is possible to predict the phenotype of many genotypes derived from a single set of substitution lines and their donor parents and recipient varieties. The genetic effects can be estimated using methods called triparental crosses 1 and 2 (Law, 1966; Askel, 1967) following the biometrical approaches of Mather and Jinks(1971). The triparental cross 1 involves crossing two substitution lines and their common recipient in all combinations whereas triparental cross 2 involves a single substitution line and its recipient and donor varieties. Scaling tests (Law, 1972) similar to that of Mather and Jinks(1971) can be carried out, each of which tests for the presence or absence of between chromosome interaction, within- chromosome interaction. The expectations of means of the different generations in case of Triparental cross 1 and Triparental cross 2 following notations of Mather and Jinks(1971) are given below. Considering two loci and two alleles at each locus(A,a and B, b) the substitution lines are described in terms of two chromosomes only, AA-aa and BB-bb, the recipient chromosomes being given the lower case(aabb) and the donor chromosomes the upper case(AABB).

Triparental cross 1

The development of different types of progenies and their expectations in terms of m, d, h, i, j and l are given in the table below.

	AAbb	*aaBB*	*Aabb*
Aabb	Aabb m + [ha] - [db] -[jb/a]	aaBb m + [da] + [hb] -[jb/a]	Aabb m + [da] - [db] +[iab]
aaBB	AaBb m + [ha] - [hb] +[lab]	aaBB m - [da] + [db] -[iab]	
AAbb	AAbb m + [da] - [db] -[iab]		

Estimation of genetic effects:

The estimates of additive genetic effects and between- and within- chromosome interaction effects are obtained by different comparisons given below.

Comparisons	*Estimate*
I. Additive effects	
a. ½(AAbb-aabb)	[da]-[iab]
b. ½(aaBB-aabb)	[db]-[iab]
Within chromosome interaction effects	
a. ½(-AAbb+2Aabb-aabb)	[ha]-[jb/a]
b. ½(-aaBB+2aaBb-aabb)	[hb]-[jb/a]
Between chromosome interaction effects	
(AaBb-Aabb-aaBb+aabb)	[iab] + [ja/b] - [jb/a] +[lab]

Triparental cross 2

	AABB	*AAbb*	*Aabb*
aabb	AaBb m + [ha] +[ha] +[lab]	Aabb m + [ha] -[db] -[jb/a]	Aabb m - [da] -[db] -[iab]
AAbb	AABb m + [da] +[hb] +[ja/b]	AAbb m + [da] -[db] -[iab]	
AABB	AABB m + [da] +[db] +[iab]		

The comparisons detecting additive and within-chromosome effects are the same as in case of Triparental cross 1. The between-chromosome interaction effect is estimated by the following comparison.

Comparison	Estimate
(AABb –AaBb- AAbb + Aabb)	[iab] + [ja/b] - [jb/a] - [lab]

In the above model of estimation of genetic effects da and db are the difference between homozygous chromosomes, i.e., AA v aa or BB v bb(additive effect); ha and hb are the within chromosome interaction effect or the difference between the mid parental value of AA and aa and the value of Aa(dominance effect) and similarly for the BB-bb chromosome; iab, ja/b, jb/a and lab are the interactions between chromosomes A-a and B-b, when both are homozygous, both heterozygous or one homozygous and the other heterozygous.

The three hybrids along with the original three parents can be used to produce 21 different generations through intermating and selfing and the different types of genetic interaction effects can be estimated. Further 'scaling tests' can be carried out in order to test the presence or absence of between chromosome interaction. Finally the various parameters describing the various chromosomal effects can be estimated using the methods of weighted and unweighted least squares. For details on model fitting the reader is referred to Roy (2000). Thus through these analyses the major effects of different chromosomes and interactions between them can be determined. Both the triparental analyses provide information regarding the relationships and differences between only two chromosomes at a time and especially in triparental cross 2 the chromosomal effects which are of interest are confounded with the effects of the background chromosomes. This problem of background chromosome effects can be overcome by using a complete analysis of a single substitution set and chromosome(s) contributing to the background effect can be identified and separated. All the 21 substitution lines of wheat variety Hope into Chinese spring(Law and Worland, 1973) have been developed and triparental cross 2 has been applied in which all the 21 substitution lines were crossed to the recipient(Chinese spring) and donor(Hope) parents. Through such analysis additive effects of each chromosome can be estimated and further information regarding complementary and duplicate interactions can be known from the sign of estimates of within- and between- chromosome interaction parameters. The presence of linkage can be detected by studying the means of the selfed and crossed generations resulting from a cross between either two substitution lines or a single substitution line and its recipient or donor variety. However these methods depend on the simultaneous occurrence of epistasis between linked genes before linkage can be detected. Without epistasis linkage can not be detected. Further it is difficult to establish whether more than two linked genes are involved. The complex crossing designs although detects linkage but again it is doubtful if linkages between more than two genes could be recognized.

Haploidy

7.1 DEFINTION OF HAPLOID

Plants having gametic or haploid number of chromosomes are called haploid. Haploid is a saprophyte with gametic chromosome number(one representative of each homologue). The haploids can be monohaploids or polyhaploids. Monohaploids refer to individuals (with x=n) that arise from diploid(2x) species whereas polyhaploids refer to individuals(with 2x, 3x, 4x, etc.) that arise from polyploidy species with 4x(autotetraploid), 6x(autohexaploid), etc. respectively. Haploids are viable. There are haploid organisms such as viruses, bacteria and fungi. In most plants loss of single chromosome results in inviability but tolerate whole set and survive. This is because genetic balance is not disturbed. In comparison to diploids, haploids are not very vigorous. In outbreeding species where the recessive deleterious alleles are not expressed because of presence of dominant alleles, these deleterious alleles are expressed in haploid of outbreeding species and thus there is low viability of recessives. In some species haploids are totally sterile. The haploids look sick, weak and not vigorous. The physiological imbalance results in sterility because meiosis is irregular. For example, in haploid rye 7 univalents or 3 bivalents and one univalent are formed and so not viable gametes are produced and thus there is complete sterility.

7.2 PRODUCTION OF HAPLOIDS

Haploids are produced in different frequencies and not only genera and species are important but strain, seed and pollen parents are important as well. Haploid production varies according to the genotype of the basic material. In other words, haploid production is genotype dependent. Thus genetic factors are involved in the haploid production potential of an individual plant. Haploids can be produced via' normal' embryo and by seed. The former can be accomplished by the following.

7.2.1 Parthenogenesis

In the parthenogenesis haploids arise by the development of an unfertilized egg cells. Here pollination is essential but no fertilization occurs. Haploids also arise spontaneously in many species such as maize (Chase, 1949) at lower frequency (one per 1,000 to 2,000 kernels).

7.2.2 Use of Alien Cytoplasm

The frequency of development of haploid through androgenesis in which there is development of male gametic nucleus in egg cytoplasm is rare. For example, in maize, the average frequency is about 0.11% and it can range from 0.0 to1.8%. The gene marker used for haploid identification is aleurone color marker. In a cross of pp x PP where pp is purple aleurone the parthenogenetic haploid will be having purple aleurone- the haploid seed will be purple in color. In *Brassica napus* the production percentage of haploids ranges from 0.0 to 6.8%.

7.2.3 Other Means of Developing Haploids

Haploids can be produced by other means as well.

1. **Use of foreign pollen or dead pollen** - If a plant is pollinated by foreign pollen (pollen from alien species) or inactive or dead pollen from the different species it stimulates the egg cell to go for haploid production. The source of pollen has great influence on haploid frequency. When *Solanum tuberosum* is crossed with a related wild species, *S. phureja*, differing in ploidy levels, there is production of haploids.
2. **Use of self pollen** - If the pollen of the species is inactivated by irradiation (X-ray, gamma, etc.) or chemical treatment (Toluidine blue(TB), laughing gas and formaldehyde) there is production of parthenogenetic haploids.
3. **Without pollination** - Here if the female parent is given chemical shock or physical methods it gives rise to production of haploids. Here the haploid frequency is low. The physical factors influence meiosis. Low temperature induced haploid production in rye and high temperature induced haploid in maize (Randolph, 1932), rye and *Populus tremula*. Ionizing rays such as X-ray when applied during meiosis or before fertilization induced somatic reduction in water melon. U.V treated pollen in Antirrhinum produced haploids (Knapp, 1939). Kermicle(1969) observed haploid frequency to be associated with a mutation, indeterminate gametophyte(*ig*) in maize. The homozygous *ig* plant used as female produced more than 2% androgentic haploids, a phenomenal increase of more than 1000 fold as compared to the average frequency of androgenesis in maize.

7.2. 4 Haploids Production by Seed

1. Parthenogenesis Parthenogenetic seeds are the haploid seeds, can be used for production of haploids.

2. Chromosome elimination following wide cross hybridization (inter specific and inter generic hybridization) - In this bulbosum technique of haploid production the cross is made between *H. vulgare*, the culrivated rye and *H. bulbosum*, the related wild relative and the haploid is obtained as follows.

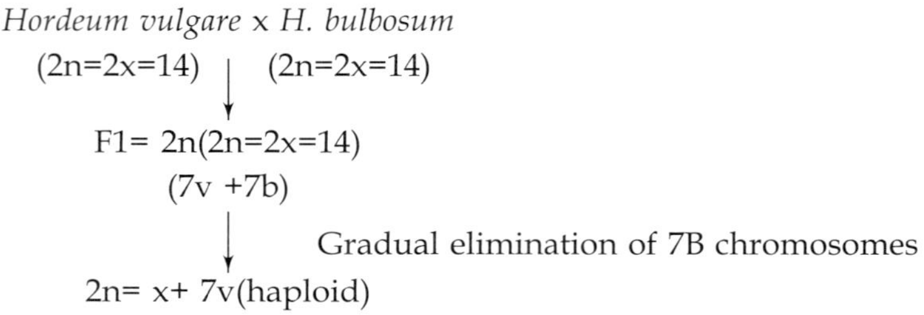

The F1 obtained above is diploid with 7 chromosomes from *H. vulgare* and 7 chromosomes from *H. bulbosum* (Kasha and Kao, 1970). But in the early stage of development there is elimination of all the 7 chromosomes of *H. bulbosum* and the result is the production of haploid with 7 chromosomes from *H. vulgare*. Thus by practicing the selection the frequency of haploid can be increased. By 8th day of pollination only 7 vulgare chromosomes are left. There are other crosses from which haploids can be extracted. One such cross is between *T. aestivum* and *H. bulbosum*.

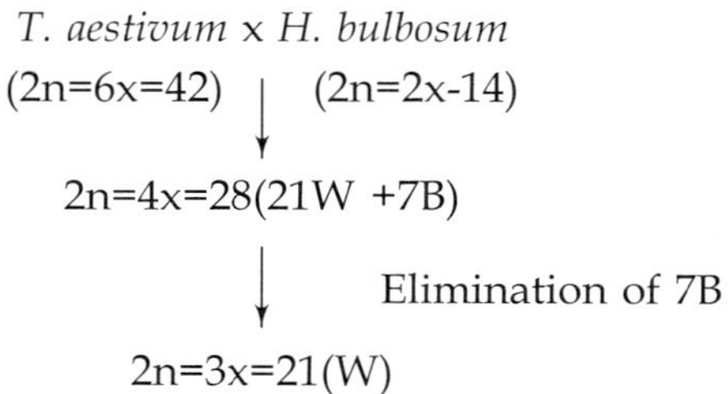

Thus in the similar way as discussed in case of cross between *H. vulgare* and *H. bulbosum* the cross *T. aestivum* x *H. bulbosum* generates haploid (2n=3x=21 W) with elimination of all 7 chromosomes of *H. bulbosum*. 4-17% of the seeds obtained from this cross are haploids.

3. Twin seedlings - Twin seedlings are important source of haploids and it is produced through polyembryony. Polyembryony is the most frequent cause of spontaneous haploid production. Polyembryonic seeds can produce haploid-haploid, diploid-diploid, or haploid-diploid twins. It is thought that in haploid-diploid twins a normal diploid zygote develops together with a haploid synergid into two embryos. The frequency of polyembryony is genetically controlled and applies to androgenesis as well. The frequency of twin seedlings in Capsicum depends on the genotype of the female parent(Campos and Morgan, 1960). As a very small percentage will produce twin seedlings so this method is not important commercially. Twin seedlings result from development of haploid nucleus of embryosac. In rye out of 300,000 seeds 82 twins were obtained out of which one was haploid whereas in Asparagus out of 150,000 seeds 325 were twins of which there were as many as 27 haploids.

4. Development of haploids from pollengrains - Pollengrains can develop into haploids but it does not occur naturally. Development of haploid from pollengrain was first reported in *Datura innoxia* (Guha and Maheshwari, 1964, 1966). There are two ways to produce haploids here: **a. anther culture and b. pollen culture**. Anther culture appears to be the best mechanism for production of haploids. It gives higher yield in terms of number of haploids produced than pollen culture because of the presence of nutritive tissue (Tapetum) in the anther. The mechanism of production of haploid is the same in the two methods. Both anther and pollen are cultured in sterile medium. There are two different pathways in the production of haploids from pollen. In one of the pathways pollen directly gives rise to plantlet as is observed in case of Nicotiana, Petunia and Oryza. But in case of Brassica, Hordeum and Solanum pollen gives rise to callus which upon differentiation produces plantlets. Here the two important things to be kept in mind are I. the stage of development of pollen and II. nutrient medium. The microspore mother cell undergoes meiosis to produce four microspores. Microspores are uninucleate just before 1st mitosis. After 1st mitosis(but no cytokinesis) there is production of vegetative nucleus and generative nucleus in either pollengrain or in the pollen tube. So it is important to choose the right stage of pollen at which haploid induction takes place and the best stage lies between the quartet stage and a stage just past the first pollen mitosis(Sunderland, 1974) and further media requirements are unique in comparison with other types of culture. Pollen is cultured in liquid medium and thus a suspension culture is produced.

5. Somatic reduction - Haploids can be produced by somatic reduction using various chemicals such as colchicines, fluorophenylalanine, chloramphanicol, etc. All these chemicals are mutagenic substances and they induce somatic chromosome reduction.

7.3 IDENTIFICATION OF HAPLOIDS

Haploids can be identified considering the following:

1. Morphology
2. Chromosomal count
3. Presence of nucleoli
4. Marker gene.

Haploids are generally smaller, weak, less vigorous, less fit in comparison to diploids and thus can be recognized. It can further be confirmed by counting the chromosome number. The rapid method for identification of haploid will be to look at nucleoli in interphase nuclei. There are two nucleoli in 2X nucleus whereas there is only one in X, the haploid nucleus. Marker gene can be used for identifying the haploid in the population. We have seen in case of maize that purple aleurone can be used as a marker to detect haploid.

7.4 USE OF HAPLOID

Although haploids are inviable, less fit, less vigorous and small so it can not be used directly but then dihaploids (homodiploids) can be produced using colchicines.

7.4.1 Rapid Production of Pure Lines

These dihaploids are genetically homozygous at every locus irrespective of the genetic constitution earlier and thus pure breeding lines can be produced over night from F1 or F2 segregating generation population which normally requires 6 generations of selfing. Thus there is a great saving of time in the production of homozygous pure lines. Secondly, haploids express all genes in plant regardless of dominance or recessiveness and therefore they can be used in screening new mutations. The disadvantages include the question of cost, requirements of good laboratory facility and good technical staff and also picking of good genotype. In maize breeding dihaploids were produced in 1940 for production of hybrids. Dihaploids in tomato were produced in 1922(Marglobe). Dihaploids have been produced in *N. tabacum*. In rape, *B. napus*' Maris' haploids were produced in 1979.

7.4.2 Dihaploid (2x) Breeding

Breeding methods can be used at the level of dihaploids and improved dihaploids can be further used for the production of improved tetraploids.

7.4.3 Androgenetic Haploids

It is the development of plants from male gametophyte. Androgenetic haploids can be used as a means of transferring cytoplasm between varieties and between species. Androgenetic haploids render it possible to incorporate nuclei into alien cytoplasm(Chase, 1957). Cytoplasm of *Aegilops caudate* and some other Aegilops species induces haploids (11-56%) and twin seedlings(0.5-15%). The plants with the alien Aegilops cytoplasm and the 'Salmon' nucleus were developed by repeated backcrossing of the F1 of (*Aegiplos caudata* x 'Salmon wheat) to 'Salmon' wheat as pollen parent. Transfer of cytoplasm from one species to another will result in the production of male sterility and thus can be used for production of hybrids.

7.4.4 Mutation Breeding

Mutation breeding can be effectively carried out at haploid level as at haploid level all genes will be expressed which is not the case in diploid. So if useful haploids are found as a result of irradiation it can be used for production of diploids. Chemical (EMS, Ethidiumbromide, hydrazine, N-methyyl-N-nitro-N-nitrosoguanidine, N-3-nitrophenyl-N-phenyl urea, N-nitros-N-methyl urethane) or physical treatment (X-ray, γ-ray) can be given in the *in-vitro* culture and mutants can be obtained which can differ in either nutrients requirements or phenotype. In case of higher plants stable mutants must originate from a single mutant cell(otherwise there will be problems of chimeras) but it is very difficult to select and regenerate a single cell mutant from a population of dead or still non-growing wild type cells. Thus screening of haploids for resistance against fungicides, insecticides, bacteriocides, herbicides (2,4-D), pathotoxins (fungal and bacterial toxins) can be conducted and resistant haploids can be selected and further dihaploids can be produced. Similarly haploids can be screened for resistance to Nacl, So2 or Pb salts and if any haploid found resistant then dihaploid can be synthesized. As only cultures from a few haploids can be regenerated to complete plants so most mutation experiments using tissue culture are carried out with diploid somatic sources.

7.4.5 Production of Heterotic Hybrids

Asparagus officinalis with 2n=20 is dioecious and have separate male(XY) and female(XX) plants and males are high yielder than females. From XY male through anther culture haploids (or through polyembryonic seed) with X and Y can be produced and finally homozygous diploids(XX) female and homozygous diploids(YY) supermale can be generated. A cross between XX and YY will produce all male high yielding heterotic and uniform F1plants (Therenin, 1974).

7.4.6 Genetic Study

Haploids can be used to estimate various genetic parameters (see author's book 'Plant Breeding-Analysis and exploitation of variation'). Haploids can be used to transfer genes from polyploids to haploids. Haploids can also be used for production of **allosubstitution hybrids** (Nitzsche, 1972). In case of allopolypoids(AABB) haploids(AB) can be produced which upon chromosome doubling by colchicine will produce homozygous line and hybridization between two alloplyploids will result into production of allosubstitution hybrids. Finally, in ornamental crops haploids can be used directly because of small flowers with prolonged flowering, the latter caused by male sterility (Kostoff, 1941).

7.5 SOURCES OF ANEUPLOIDS

Production of monosomics and trisomics - In wheat haploids are the major source of the trisomics and monosomics (Sears, 1954) and from these complete series of nullisomics, monosomics, trisomics and tetrasomics were developed (see chapter 6). The aneuploids have provided the basic materials for studies related to genetics, cytogenetics, breeding and evolution of wheat. Further haploids of wheat deficient for the chromosome 5B (see chapter 11) provided evidence for location of ph gene preventing homoeologous pairing. Haploids have also provided evidence for the polysomic nature of tetraploids in case of potato and alfalfa.

7.6 HAPLOIDS IN ANIMAL

Haploids also occur routinely among some diploid species of insects (bees, wasps), mites, rotifers and others that, through parthogenesis produce haploid males which produce sperms by mitosis rather than meiosis in addition to the sexually produced diploid females.

Structural Changes in Chromosome

8.1 GENERATION OF STRUCTURAL CHANGES IN CHROMOSOME

Structural change in chromosome involves breaking and joining of chromosome. Ionizing radiations such as X-ray, gamma ray, alpha and beta rays, electrons, neutrons, protons, etc, nonionizing radiations such as UV irradiation and chemicals such as acridine dyes proflavin produces changes in chromosome structure. One of the known effects of ionizing radiations is the breaking of chromosomes and chromatids. Such breaks involve the sugar-phosphate backbone of the polynucleotide strands. When both strands of the double helix are broken one can get a loss of the distal end of the chromosome with the possible generation of a breakage-fusion-bridge cycle after its duplication. Two simultaneous breaks in the same chromosome can lead to a deletion or inversion while simultaneous breaks in nonhomologous chromosomes can lead to translocation. Breaks can also occur in only one of the two strands of the double helix. Two simultaneous breaks in the same strand of the double helix at the DNA replication result in loss of the intermediate segment. The acridine dyes such as proflavin induces the addition or deletion of one or even possible several adjacent nucleotide pairs.

The structural arrangement is stable only in a small number of cases whereas it is mostly unstable. The causes of unstability are: 1. Mechanical inefficiency 2. Genetic loss. For example, as shown in Figure 8.1 below the breakage of a chromosome results in formation of a dicentric bridge and an acentric fragment lying on the metaphase plate. The bridge can break at any point and the acentric chromosme can go to any poles or can be lost. Further, genetic loss could be lethal if that acentric fragments contain some gene very important for the survival.

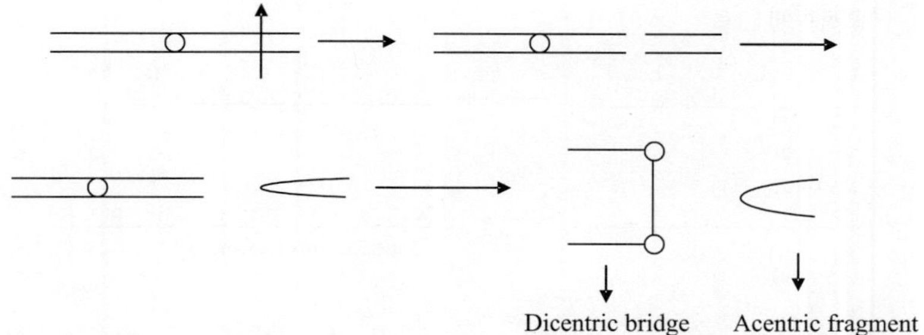

Showing the genetic loss as a result of structural changes because of irradiation.

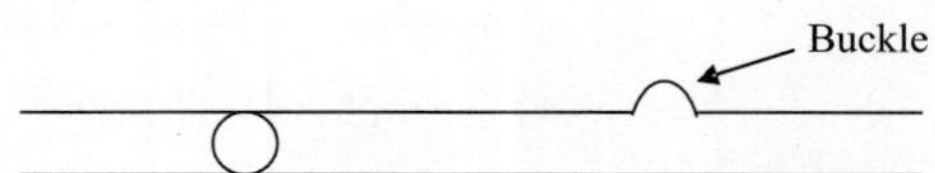

Fig. 8.1 Showing formation of a buckle.

8.2 TYPES OF STRUCTURAL CHANGES

There are three types of structural changes. 1. Deficiency/deletion, duplications 2. Inversions 3. Reciprocal translocations.

8.2.1 Deficiency/deletion

Deficiency/deletion refers to loss of a chromosome segment containing one or a group of genes from the chromosome as shown in Figure 8.2. The deficiency/deletion can be either intercalary or interstitial or terminal. The deficiency can be called interstitial if a non-terminal portion of the chromosome has been lost whereas terminal deficiency is defined as a loss of the tip of the chromosome.

8.2.1.1 Duplications

Duplications or triplications involve doubling or tripling of a particular gene or a group of genes as shown in Figure 8.2. It also includes duplication of base sequences, chromosome segments or whole chromosome sets.

8.2.1.2 Types of duplication

Duplications can be of different types such as tandem, reverse tandem, displaced(displaced, homobrachial, on the same arm), displaced(heterobrachial, on the different arm), transposition(to non-homologous chromologue) and extrachromosome as shown in Figure 8.2 (Swanson, 1957).

8.2.1.3 Origins

Duplications and/or deletions are produced as a result of 1. Breakage and rejoining of chromosomes 2. Error of replication and 3. Secondary consequences of other structural changes such as I. Non-disjunction(produces duplications) II. Segregation from translocation

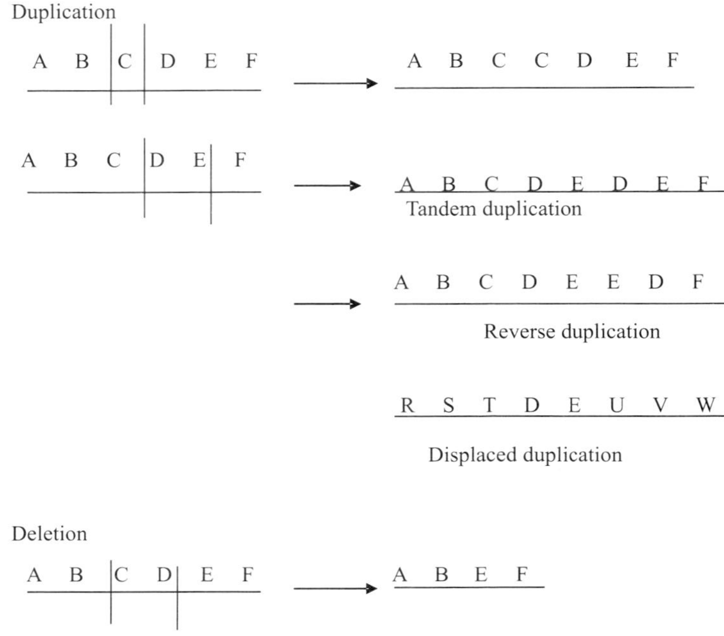

Fig. 8.2 Showing formation of duplication and deletion.

heterozygotes(produces deletions/duplications as a result of CO between non-homologous regions) III. Crossingover between non-homologous regions produces duplication and/ deletions (Doyle, 1964) IV. Crossingover in an inversion heterozygote leads to loss of a chromosome region V. Breakage of dicentric bridge may result in loss of chromatin VI. Duplications are produced as a result of transposition(see chapter 37).VII. Unequal crossing over of chromatids. Duplications occur naturally but it can also be produced through irradiation. Monosomics or monoploids produce gametes deficient for whole chromosome. Chromosome breakage leads to such abnormalities like deficiency, inversion, translocation and ring chromosomes. The cytogenetic stocks such as TB interchange and maize stock with short arm of chromosome 9 duplicated but in reverse order(reverse tandem)(McClintock, 1944) are sources of duplications. The third source is crosses between stocks carrying different inversions, one inversion overlapping the other or one inversion included within the other(Sturtevant and Beadle, 1938) as shown below. Crossing between stocks carrying interchanges/translocations each involving the same two chromosomes also may yield duplications(Gopinath and Burnham, 1956). Treating seeds with radiation(Beard, 1960) should also produce duplication through an exchange of terminal segments.

8.2.1.4 Detection

There are difficulties in detecting duplication and deletion/deficiency. In case of deletion at pachytene stage 'buckle' will appear in the synapsed chromosomes at the point of deficiency (Figure 8.1). Small duplication and deletion has been observed in maize, barley, mouse and human being. The identification of duplications depends mainly on the occurrence of crossingover between supposedly nonhomologous regions. The synapsis and crossingover can occur between alleles of a tandem repeat resulting in production of new alleles. The mutants thus produced are associated with recombination for flanking markers and thus can be identified. Duplication of base sequences can be identified by nucleic acid hybridization.

8.2.1.5 Effect of duplication/deletion

Deletion/deficiency produces two types of effects. 1. Genetic effects 2. Cytological effects. Both these effects are used to detect the presence of deficiency/deletion.

1. **Genetic effects** - The genetic effect of deficiency is called **pseudo-dominance**. If a recessive gene is located in the part of homologous chromosome that corresponds to the deficiency/deletion then it will be expressed in the phenotype of the organism just as if it were dominant.
2. **Cytological effects** - When a normal and a deficient chromosomes are synapsed and as chromosomes tend to pair gene by gene during synapsis a buckling will occur in the synapsed chromosomes at the point of deficiency. Deficiency/deletion has largely mortality effect. The magnitude of effect of deletion depends on the size of segment deleted. Larger deletions are harmful than small deleterious recessive mutation. Larger deletions tend to be recessively lethal. Yellow body in Drosophila and white seedlings in maize are examples of recessive mutations as a result of deletion. The deficiencies such as Beaded, Delta, Notch, Gull, Minute and others act as dominant gene in heterozygous condition. A Notch deficiency involves a loss of about 45 bands on the X-chromosome and it produces a unique indentation of the wing margin in heterozygous females and is lethal in males. Beaded gene has a dominant effect on the shape of the wings and is lethal in homozygous condition. Beaded flies are always heterozygous. *Cri-du-Chat* syndrome in human discovered by Le jeune involves deletion of a part of the short arm of chromosome 5 which leads to besides small head, wide spacing of the eyes and mental retardation. The plaintive cry of the affected infants resembles that of a cat. Deficiencies in corn have been observed where deficiency in male results in pollen sterility. Occasionally small deficiencies in corn that are homozygous act as recessive mutation. In general deficiencies/deletions produce unique phenotypic effect of their own.

Duplication has less bad effect in comparison to deletion on the phenotype. Many copies of a particular gene will result in intensification of phenotype. Sometimes we may find entirely new phenotype as has been observed in Drosophila(as a result of unequal CO) in case of Bar eye where duplicated 16A chromosomal region(16AI to 16A 6) of 4X-chromosome results in Bar eye whereas the triplicated 16A chromosomal region results in ultra-Bar and theses differ from wild type. The triplicated region(16A I to 16A6) is formed as a result of unequal crossing over between two homologous chromosomes having a duplicated Bar region. Ultra Bar is dominant to both Bar and wild type alleles and reduces the number of facets in the compound eye still further. Compound loci consisting of tandem repeats at the R and A locus have been observed in corn. Duplication and evolution of certain genes such as hemoglobin and myoglobin have been traced and the cause is the occurrence of crossingover between non-homologous regions. Synapsis and crossingover between alleles of a tandem repeat result in production of new alleles. The *Adh* locus(alcohol dehydrogenase) also in corn shows presence of an adjacent duplication.

Transmission through gametes - In general, deficient gametes that are transmissible survive only through female and are not competitive in the efficient pollen screen, although exceptions are known(McClintock, 1944). The duplication should be egg transmissible and although it will not be competitive in pollen it should give viable pollen grains.

8.2.1.6 Uses of deletion/duplication

The deleterious mutant phenotype resulting from deletion can not be used but can be useful in locating genes. It is used in haploid micro-organisms. It can also be used for mapping gene in higher organism. For example, if AA parent in cross with aa parent is irradiated, the F1 can be screened for expression of recessive gene and if that gene is expressed it is probably due to deletion which can be cytologically verified (Figure 8.1). In practice, plants homozygous(aa) or heterozygous(Aa) for the recessive are pollinated by irradiated pollen from a homozygous(AA) dominant parent. Fertilization of a recessive female gamete by a male gamete deficient for the corresponding dominant allele will produce a recessive zygote, deficient(hemizygous) for the locus. In other words, loss of the chromosome segment carrying A due to irradiation will result in progeny having the recessive phenotype. The cytological examination of the plant at the pachytene will reveal a chromosome pair with the normal member of pair having a 'fold out' or loop at or near the position of the recessive gene. The deficiency can be either intercalary or terminal but in either case, if long enough, it may produce a heteromorphic bivalent at the diakinesis or metaphase. Although the length of deficiency will vary in different plants hemizygous for the same locus but the one with the shortest recognizable deficiency will provide a reasonable accurate determination of the position of the gene. This will also ensure that a second deficiency in the same plant and too short to be detected cytologically is not at the same locus. The position and extent of the loop in the spore mother cells of the same plants may even vary due to variable pairing- portions of which are nonhomologous. The presence of cytological markers such as knobs or chromomeres is very helpful in identifying the chromosome involved and the position of the loop in the chromosome. The deficiency method of locating genes has been used in maize and tomato and can be readily applied in species that are easily crossed. In a species with poor pachytene cytology such as barley the mutant trait might be introduced into a stock segregating for male sterility. Male sterile plants homozygous for a viable recessive mutants or plants from a stock segregating for a lethal is pollinated by irradiated pollen from a normal stock. Cytological examination of the somatic divisons of the root tips of the hemizygous seedlings will reveal lethals or viable mutants if the length of deficiency is sufficient enough to be recognised. The presence of a heteromorphic bivalent at diakinesis of meiosis will indicate viable mutants. The lethals have potential use in producing true breeding heterozygotes. The deficiency in conjunction with an interchange as a marker can be applied for locating genes in species not having analyzable pachytene chromosomes. As a result of radiation or other mutagenic treatments one can obtain a large number of deficiencies in a particular chromosome and thus one can construct a cytological map and a comparison between genetic map and cytological map can be made.

Duplications are of more use in comparison to deletion. In barley improved production of α-amylase by duplication of this locus is one very useful example of duplication. Further, duplicated genes may mutate to new genes and thus is a step in evolution. Experimentally produced duplications provide a method for producing stocks homozygous for genes which determine two desired but allelic characters previously found in only separate lines. The duplications must be generated in individuals heterozygous for the characters.

8.2.2 Inversion

It is an intra-chromosomal rearrangement of a gene or a group of genes. In inversion there is no addition or loss of chromosome segment and only the gene order on the chromosome is changed. Inversion requires two breaks in the chromosome.

There are two classes of inversion: 1. Paracentric inversion 2. Pericentric inversion.

In the **paracentric inversion** the inverted segment does not include centromere whereas **pericentric inversion** includes centromere within the inverted segment and thus the position of centromere is changed as shown in Figure 8.3b.

8.2.2.1 Effect of inversion

Inversion results in change of gene order and since contiguous genes are often involved in the same biochemical pathway it may result in alteration of gene function. Such change of gene function due to change in location of the gene is called **position effect**. Position effect will actually appear as mutation and may be dominant, recessive, etc. The effect will disappear once the gene is put back to its original location.

8.2.2.2 Detection of inversion

We can have both genetic and cytological evidence of occurrence of an inversion. **Genetic evidence**- There will be suppression of crossingover and possibly the appearance of a mutation like position effect. **Cytological evidence**- There will be formation of a loop when ever synapsis occurs in an inversion heterozygote. Pericentric inversion will usually result in a morphological change of the chromosome due to change in the centromere position and arm ratio and this type of chromosomal aberration will be easily detected in the karyotype.

8.2.2.3 Complex inversion

If more than one inversion is found in a chromosome it is called complex inversion. There are three types of complex inversions (Rieger et al., 1976). 1. independent inversiont 2. direct tandem inversion 3.reversed tandem inversion 4. included inversion 6. overlapping inversion (Figure 8.2).

8.2.2.4 Meiosis in inversion heterozygote

Paracentric inversion will not lead to karyotypic differences whereas pericentric inversion will lead to karyotypic differences because of change in the position of centromere. In both types of inversion a loop shaped configuration is producted at the pachytene by homologous synapsis in normal/inversion heterozygotes but the subsequent cytological consequences of the two types of inversion is strikingly different. There are formation of the dicentric bridges and acentric fragments produced by CO in the heterozygous paracentric inversions which are absent in pericentric inversions. The anaphase bridges and acentric fragments can be formed because of other reasons such as due to spontaneous breakage and fusion of chromosomes during meiosis as well as chromosome breakage and sister chromatid reunion.

Consequences of crossingover in a paracentric inversion heterozygote is shown in Figures 8.3(a, c, d, e). Figure 1a shows that two breaks are required for production of

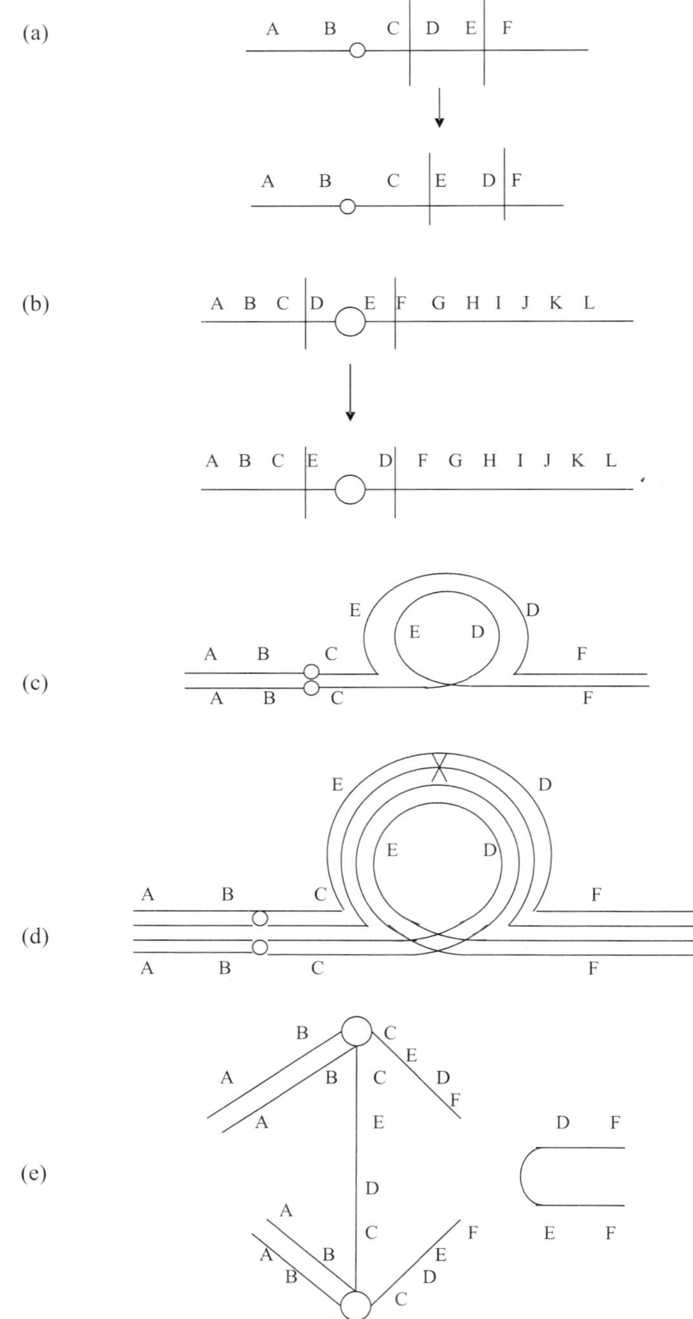

Fig. 8.3 (a, c, d, e) Paracentric inversion and consequences of crossingover in a paracentric inversion heterozygote (b) Pericentric inversion.

inversion. The chromosomal segment containing genes D and E is reversed and thus the gene order which was originally A,B,C,D.E,F now becomes A,B,C,E,D,F. In both para- and pericentric inversion a loop shaped configuration is produced at pachytene stage by homologous pairing in the normal/inversion heterozygote. The reverse pairing loop formed in Figure 8.3c shows chiasmata occurring in the inverted loop between E and D is seen in the diplotene stage 8.3d. Figure 8.3e shows the chromatid bridge at anaphase I and an acentric chromosome. As the chromatid can break at any point gametes will be formed with duplication and deficiencies and thus fertility will be affected. Table 8.1 shows the anaphase configurations as a consequence of various types of exchanges and the 2 strand, three strand and four strand doubles are shown in Figures 8.3a-e.

Table 8.1 Showing cytological and genetical products of crossing over in case of paracentric inversion

		Microspores produced		
Configuration	*Type of tetrads*	*Viable non-crossover*	*Viable double crossover*	*Abortive*
Anaphase I				
1. No bridge, no fragment	a. No exchange b. 2-strand double in loop	4 2	0 2	0 0
2. 1 bridge, 1 fragment	a. Single exchange in loop b. 3-strand double in loop	2 1	0 1	2 2
3. No bridge, 1 fragment	a. 3-strand double with one exchange in loop and one in proximal region	2	0	2
4. 2 bridges, 2 fragments	a. 4-strand double in loop	0	0	4
Anaphase II				
1. No bridge	a. From all above types except #3			
2.1 bridge	a. From #3 above			
3. 2 bridges (one in each sister cell)	a. Certain triple exchanges where two are in loop			

The table (adapted from Carlson, 1977) shows that the dicentric bridge along with an acentric fragment is characteristics of a paracentric inversion. It results either from a single chiasma in the inversion loop or a three strand double in the loop. A single exchange in the loop may also produce a no bridge, one fragment configuration if a second exchange occurs between the centromere and the inversion loop. A four strand double exchange produces two bridges and two fragments. Assuming double chiasmata be occurring in a ratio of 1:2:1 for two strand/three strand/four strand events all the doubles can be estimated from the two bridges with two fragments data. The table further shows that the anaphase II of meiosis has relatively few bridges in comparison to the anaphase I. Finally, the bridges are formed

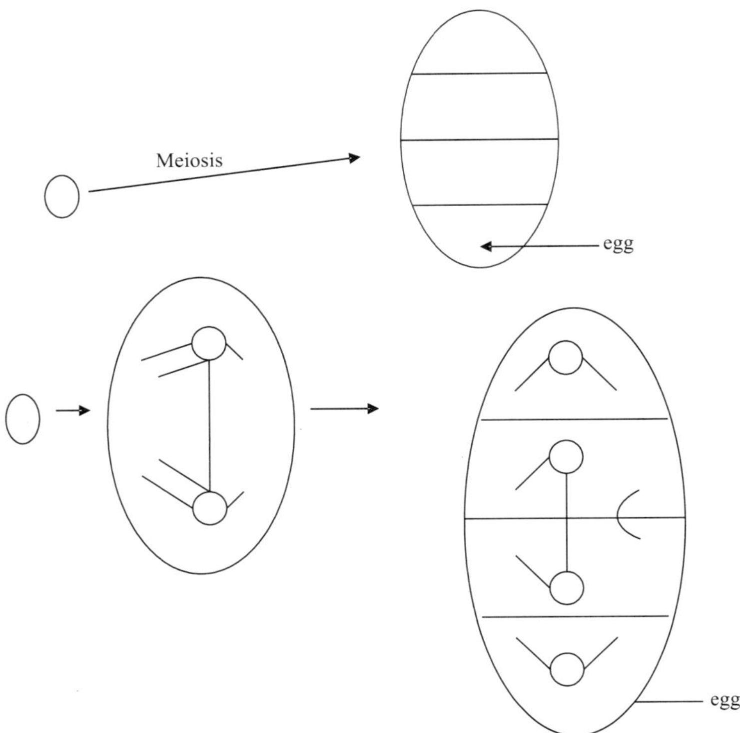

Fig. 8.4 Showing formation of fertile female gametes in case of paracentric inversion in both plants and female Drosophila.

primarily from three strand doubles where one exchange is proximal to the loop and one exchange is within the loop. Theoretically there should not be any difference in expected frequencies of non-crossover, double crossover and abortive microspores and megaspores but as following meiosis in the female three of the four products die and only the basal megaspore survives, and gives rise to egg (Figure 8.4), and further if there is preferential inclusion of certain chromatids in the basal megaspore then the products of crossing over in an inversion heterozygote will differ in the male and the female. Beadle and Sturtevant(1935) suggested that chromatids involved in a dicentric bridge are confined to the central nuclei of Drosophila tetrad due to failure of bridge breakage and only non-crossover or double crossover balanced chromatids are included in the egg. This hypothesis was supported by selective orientation observed in corn where delayed breakage of the anaphase I bridge could orient noncrossover chromatids to the outer poles of the linear tetrad and the basal megaspore develops from one of the outer poles- a case of selective orientation and this would result in higher female fertility than male. But no evidence of such difference in fertility between male and female in case some inversions was attributed to presence or absence of selective orientation because of varying genetic backgrounds (Rhoades and

Dempsey, 1953). Certain genetic backgrounds produce a significantly lower rate of crossing over in female resulting in low frequency of dicentric bridge and there by higher fertility. Alternatively difference in rates of crossing over in the male and female may control differences in male and female fertility. In general, female fertility was found to be markedly higher than fertility in the male.

The paracentric inversion suppresses crossing over through production of dicentrics which results in production of gametes with deficiencies and duplication of chromosome material and thus its loss through inviability or selective orientation. Since only chromosomes in which no crossing over occurred survive, inversion appears as dominant that results in the *suppression of crossing over*. Crossing over may occur frequently within an inversion loop but the chromatids involved in the chiasma are lost and only the relatively rare double crossover chromatids(produced as a result of two strand or three strand double chiasma) are balanced and survive. Pericentric inversion also suppresses crossing over since deficient-duplicate chromatids are produced as a result of crossing over inside the loop. As no dicentric bridge is formed and selective orientation is not possible, so egg and pollen sterility occurs at similar rate with pericentric inversion. The percentage of inviable gametes in both paracentric and pericentric inversions varies with the amount of crossing over within the inverted loop and with long inversion a maximum of 50% pollen sterility corresponds with the maximum rate of recombination. Thus the production of inviable gametes varies with the amount of crossing over within the inverted loop.

8.2.2.4.1 Bridge fusion bridge cycle (BFBC)

It is a process which may arise from the formation of dicentric chromatid or chromosome and thus there are two types of BFBC.1. Chromatid type BFBC and 2. Chromosome type BFBC.

1. Chromatid type BFBC When dicentrics are produced by crossing over in inversion heterozygotes, bridges are formed in Anaphase I and in the subsequent male and female gametophytic mitosis. If a chromosome forms (replicates) daughter chromatids after the break has occurred then the homologous portions of the two daughter chromatids might fuse, i.e the daughter chromatids join together at the broken ends and this will lead to the formation of a dicentric fragment having two centromeres and an acentric fragment. In anaphase both centromeres are attached to the spindles and pulled in opposite directions and the chromosome between them forms a bridge. As the anaphase proceeds, tension mounts on the chromosome and as a consequence of which results in breakage of the bridge. The acentric fragment is not incorporated into the nucleus and thus is lost. The broken chromosome each in its own cell can repeat the process of duplication, fusion(formation of a new dicentric bridge) and breakage at the next cell cycle and thus establishing a BFBC which continues through successive generations. BFBC is referred to as chromatide type since fusions occur between chromatids rather than whole chromosome. The breakage of the bridge is a random process and thus one would expect the BFBC to produce very complex deficient-duplicate chromosomes as a result of random bridge breakage but this is not always true. McClintock(1941b) observed that the bridges tend to break at the same place in successive divisions. The interpretation given was that the fusion of sister chromatids following breakage may be weak or incomplete. Miles (1971) suggested that breakage at certain chromosomal sites leads only to weak fusion of sister chromatids in successive generations such that all dicentric bridges break at the same position and a stable phenotype results in the endosperm.

The chromatid type BFBC may be formed by action of controlling elements as well as through recombination between non-homologous chromosomes. The excision of a controlling element from a chromosome may cause breakage of the chromosome and subsequent fusion at the broken sticky chromosome ends starts the BFBC.

2. Chromosome type BFBC If the broken chromosomes are introduced into the zygote from both the male and female gametes, the broken ends may join together to form a dicentric. The dicentric bridge now spans two centromeres rather than connecting one centromere. In anaphase depending upon the centromeric orientation two bridges or no bridges may be produced. If two bridges appear then breakage of the bridges in anaphase will pass two chromosomes with sticky ends to each pole. The two broken chromosomes of each daughter cell will fuse with each other to generate a new dicentric and thus the BFBC is initiated and continues in successive generations. The chromosome type BFBC occurs in both sporophyte and the endosperm and the variegation for sporophytic markers can be observed (Carlson,1976). A special example of chromosome cycle is found in ring chromosome. Occasionally sister strand crossing over(sister exchange) in a ring chromosome produces double dicentric ring chromosome seen as a double bridge in anaphase of mitotic division(McClintock, 1938b). Breakage of the double ring in anaphase gives rise to a rod chromosome with two broken ends and the fusion of the two chromosome arms restores the ring chromosome structure.

The chromatid and chromosome types BFBC are although distinct phenomena but a chromatid type BFBC can be converted to a chromosome type through nondisjunction (Schwartz and Murray, 1957). If the bridge produced by a chromatid cycle fails to break and the dicentric moves to one pole then a chromosome type BFBC is established in successive divisions.

8.2.2.5 Consequences of inversion

The genetic consequences of inversion are thus 1. the production of duplication or deficiencies and 2. the effect on fertility and are shown in Table 8.2 in case of plants and Drosophila.

Table 8.2 Showing types of inversion and their effects.

Types of inversion	*Meiotic effect*		*Fertility*	
			Female	*Male*
Pericentric	Duplication/deficiency	Plants	SS	SS
		Drosophila	F	SS
Paracentric	Bridge + fragment	Plants	F	SS
		Drosophila	F	F

SS = semi-sterile; F = fertile

Half of the gametes are carrying duplication or deficiency in pericentric or paracentric inversion and thus 50% gametes are unviable and will give rise to 50% unviable offspring. In case of paracentric inversion in female Drosophila and plants gametes are fertile because upon meiosis four haploid daughter cells are formed. The lower cell contains full set of genes(i.e. it contains chromosome which are not involved in crossing over) which forms egg and so it is fertile (Figure 8.4). In male Drosophila there is no crossing over. The suppression of crossing over in the inverted region will lead to the preservation of combination of genes of the inverted region and thus this unique combination of genes which is called **super gene** and which during evolution may form new gene due to mutation. While inversion may reduce crossing over within the inverted segment it increases the rate of recombination elsewhere as is shown in Drosophila and corn(Bellini and Bianchi, 1963). Thus inversion would not only protect certain blocks of inverted segment but also promote exchange of variability in other regions, thereby still getting heterosis in cross between the two races(Wilkes, 1972). Inversion could be large or small. Some of the large inversions are homozygous in various wild populations of Teosinte and this is expected as inversion homozygotes(structural homozygotes) will be fertile whereas the inversion heterozygote will induce a high rate of sterility. Further small, paracentric inversion reduces crossing over without sterility and thus small inversions would survive as heterozygotes and be able to spread in a population if it is linked to genes determining fitness traits. The small inversions are usually noticed as loose pairing in the heterozygote rather than in loops usually characteristic of heterozygous large inversions.

8.2.2.6 Uses of inversions

Long inversions have harmful effects and that is why short inversions or inversions in regions with low recombination frequencies have been found in corn. The harmful effect of inversion is directly proportional to the chiasma frequency within the inverted loop. There are several applications of inversion. The inversion can be used as suppressors of crossing over. Suppression of crossingover in inversion heterozygotes can theoretically have beneficial effects. Suppression of crossingover prevents the separation of those genes in the inverted section of the chromosome except with genes in a similarly inverted section of a homologous chromosome. The net effect of inversion is to bind the genes of the inversion to form a 'super gene'. This permits beneficial gene complexes to form through mutation within the inverted section without periodic disruption due to crossing over. On the other hand suppression of crossing over limits the number of new combinations of genes and thus it is an evolutionary drawback. Thus inversion has got both deleterious as well as beneficial effects. Inversion demonstrated the control of centromeres over chromosomal disjunction. Cytological and genetical data on crossingover in a paracentric inversion provided evidence for the absence of chromatid interference in corn (Rhoades and Dempsey, 1953). Paracentric inversion can be used for locating (cytological placement) genes affecting agronomic traits (Dobzhansky and Rhoades, 1938; Sprague, 1941). The steps employed in using inversion stocks for locating gene are the same as those for use of all-arms marker series of interchange stocks and are as follows.

1. Inversion stock is crossed with the stock carrying the contrasting trait to be investigated
2. Backcross the F1 to the parental stock that is recessive for the character
3. The backcross progeny is grown and each plant is classified for sterility and for the trait
4. Compare the grades of expression for the trait and their frequencies among the sterile and fertile plants

It can be used if the sterility of the heterozygotes is sufficient to distinguish them from normal plants. The paracentric inversions show very little ovule sterility and so pollen classification must be used whereas in case of barley the heterozygotes show about 30% male and female sterility and so either ovule or pollen sterility could be used for classification. A more satisfactory analysis of complex traits would be possible if inversion is used in conjunction with interchange stocks. Finally pericentric inversion with very asymmetrical breakpoints as shown below could be used in another way. If the shift in centromere position produces a chromosome which is morphologically distinguishable from all others then the two types of homozygotes can be separated from the heterozygotes in the segregating F2 population or normals and heterozygotes in backcrosses.

8.2.2.7 Inversion in natural population

Inversion occurs naturally in population (in various pulse crops) and they may be fixed in homozygous condition. Differences between species may be due to one or more inversions. Tomato and potato differ by five major paracentric-like inversions and thus may have played a major role in species differentiation between tomato and potato. Inversion may also occur as what is called floating condition where both homozygotes and heterozygotes exist and thus there is existence of polymorphism.

Inversions can have a significant role in the genetic isolation of sympatric species when they are tightly linked to the definitive alleles that separate these species. When the balance within combination of alleles essential for survival of one species is disturbed by hybridization with other species, the associated inversion(s) would become heterozygous and as a result of which the hybrid type will be partially sterile/or recombination within the inverted segment will be restricted. Thus inversion provides one of the mechanisms for maintaining intact a cluster of alleles involved in the determination of a particular complex trait in spite of repeated backcrossing to the sympatric species and the introgression of such inversion(s) from one species to another would be difficult. Four known large inversions and indication of various small inversions are there in certain races of Teosinte. Chromosome structural change which is a significant component of the isolating mechanisms between *C. arietinum* and *C. echinospermum*, has also occurred within the cultivars, one accession (no. 58) differeing from *C. reticulatum* and other accessions of *C. arietinum* by a paracentric inversion and an interchange and this is commonly observed within pulse species.

8.2.3 Reciprocal Translocation

Reciprocal translocation is discussed in a separate chapter 9.

8.3 G1 AND G2 ABBRRATIONS

Treatment of cells to mutagenes in G1stage of cell cycle results in chromosome aberrations whereas exposure to mutagenes of cells in G2 leads to chromatid aberrations (Figure 8.5A). Figure 8.5B shows different types of G2 aberrations and their corresponding metaphase and anaphase configurations.

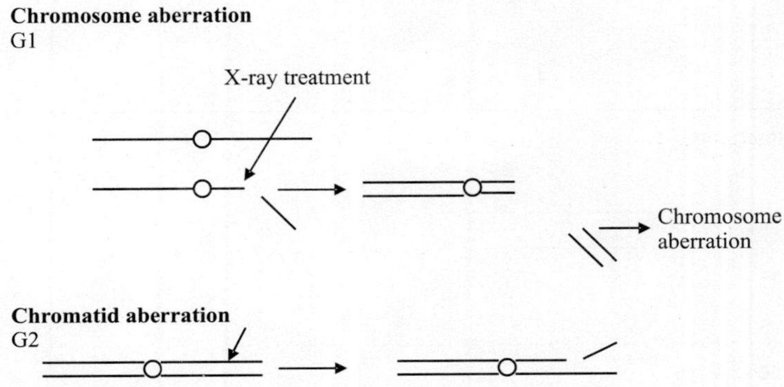

Fig. 8.5A Showing chromosome and chromatid aberrations.

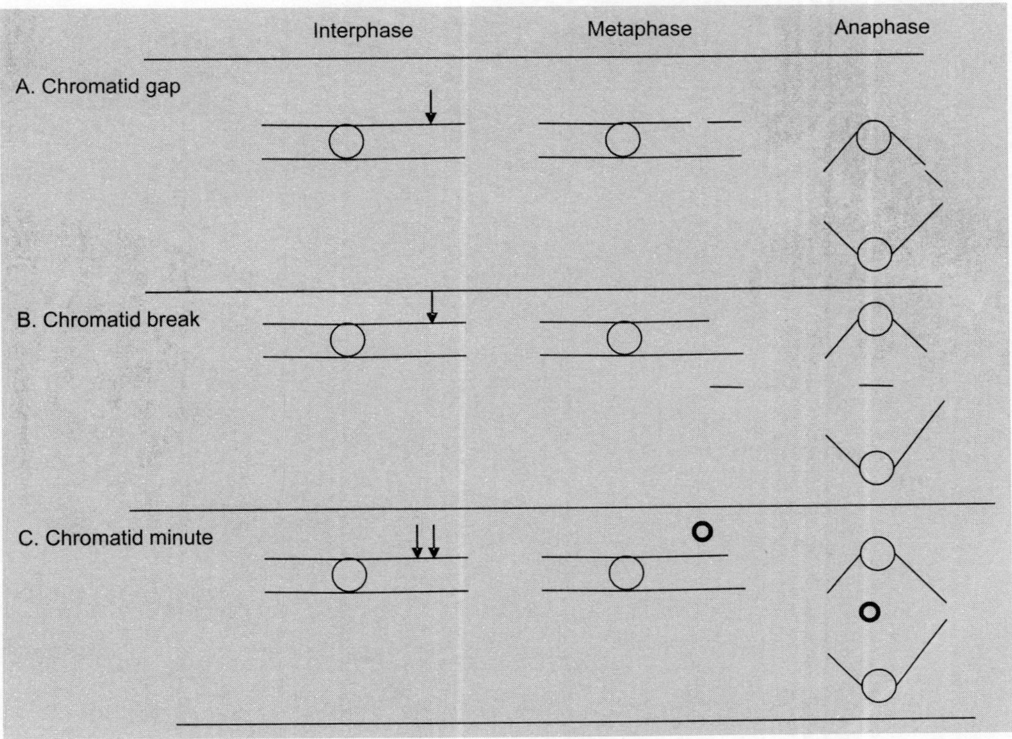

Fig. 8.5B contd.

Fig. 8.5B contd.

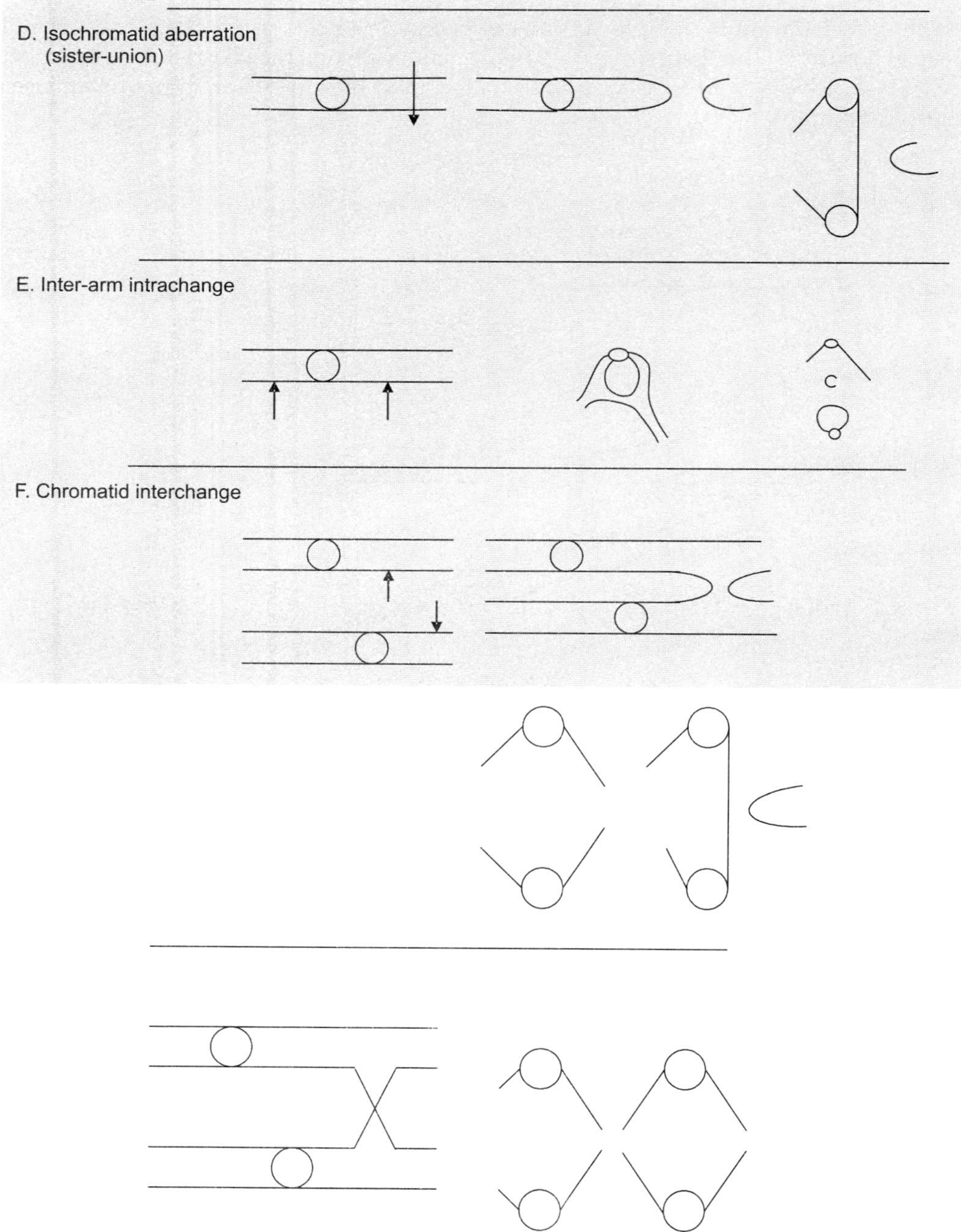

Fig. 8.5B Showing various chromatid (G2) aberrations.

Reciprocal Translocation

9.1 DEFINITION

Translocation refers to **interchange** of segments between non-homologous chromosomes. **Reciprocal translocations** are called **interchanges** and involve inter-chromosomal rearrangement. The rearrangement may be either within a chromosome, between members of a pair of homologues or between non-homologues. They may result in inversions, insertions, duplications, deficiencies or interchanges. Individuals heterozygous for a rearrangement of segment of chromosome are called '**structural hybrids**'. It requires two breaks. In reciprocal translocation as shown in Figure 9.1, there is reciprocal translocation of end portion of chromosome between two non-homologous chromosomes. There is another type of reciprocal translocation- called **insertional** or **shift translocation** wherein there is reciprocal translocation of middle portion of chromosome between two non-homologous chromosomes.

Insertional or shift translocation requires three breaks as shown in Figure 9.1. When the interchanged end segments belong to opposite arms of the two members of homologous pair, pseudoisochromosomes result (Caldecott and Smith, 1952) observed in the progeny of X-rayed barley seeds, also observed in maize. Normal chromosomes are designated as 1 and 2 and interchanges are as 1^2 and 2^1, the superscripts indicating the pieces received in the exchange. Break points in the chromosomes are much more frequent in the heterochromatic region, have been found at specific sites on the chromosome arms and detection of exchange point can be made in pachynema. The '**simple translocation**' involves the attachment of a piece of one chromosome to the end of an unbroken chromosome but there is no evidence of occurrence this type of translocation. It is because of the fact that there is non-fusibility of the intact chromosome ends (Stadler, 1932). Translocation can also involve interchange of

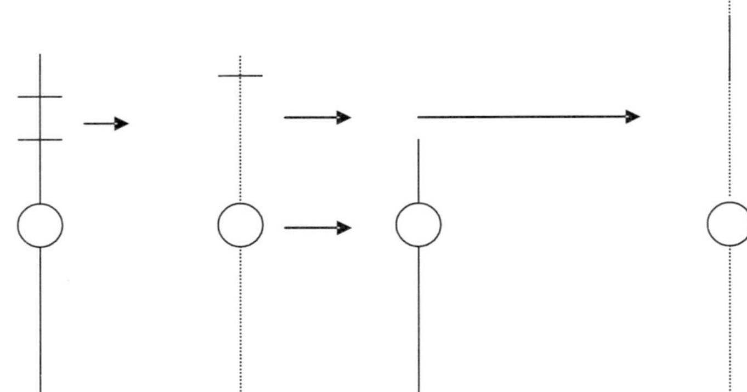

Fig. 9.1 Showing insertional translocation or shift.

segments between the B-chromosome and members of the normal chromosome complement (A chromosomes).

9.2 ORIGIN OF TRANSLOCATION

Translocation can be of spontaneous origin or can be induced by ionizing radiations such as X-ray, γ-rays, fast and thermal neutrons and other radiations (Anderson et al., 1949) and chemical mutagen treatment such as sulphur and nitrogen mustards (Auerbach, 1951). Various studies have indicated overall higher mutagenic efficiency of neutrons over X-rays and even more, chemical substances as well as the relative higher incidence of chromosomal aberrations of both structural (mostly translocations but also isochromosomes and dicentrics) and numerical (aneuploidy) type. Translocations can also arise by crossing over between homologous duplication regions in partially homologous chromosomes (Kihara and Nishiyama, 1937). Besides these above mentioned two causes spontaneous association between heterochromatic regions (Kostoff, 1938), interlocking of bivalents at meiosis(Sax and Anderson, 1933), accidental entanglement of chromosomes (Darlington, 1931), breakage of chromosome spontaneously (as misdivision of centromere in wheat (Morrison, 1954) or in response to external factors) and stick gene in homozygous condition in maize (Beadle, 1937) can give rise to translocation. Heterochromatic regions of the chromosomes are especially susceptible to chromosome breakage (Longley, 1961). There is a non-random distribution of breakpoints with exchanges near the centromere being preferred (Jancey and Walden, 1972). Further, there is non-random joining of broken chromosomal regions which indicates that this event is being regulated.

9.3 METHODS FOR IDENTIFYING THE CHROMOSOMES INVOLVED

The different interchanges are crossed and the F1's are examined cytologically in meiosis. A ring of 6 chromosomes indicates that one chromosome involved in two interchanges is the same. A ring of 4 chromosomes indicates that the two chromosomes involved in one are different from those involved in other. If the same chromosomes are involved, one pair or a quadrivalent will be observed depending upon whether the breaks are at similar or at very different positions.

9.4 EFFECT OF RECIPROCAL TRANSLOCATION

Reciprocal translocations can have genetic as well as cytological effects.

9.4.1 Genetic Effects

Reciprocal translocation can show two types of genetic effect. 1. Lethal effect 2. Positional effect. If vital genes are involved in translocation then it could produce lethality. As in inversion the genes involved in reciprocal translocation have new position on the different chromosomes and different positions with respect to centromere, they might show ***position effect***. Another effect of translocation is a change of linkage group for those genes located in the translocated segment. The individuals show characteristics and usually predictable changes both in chromosomal behaviour and in the heredity. Translocations are common varietal characteristics in a number of plant species(Burnham, 1956) and they are also involved in species differentiation.

9.4.2 Cytological Effect

On the cytological side we find the formation of a cross-shaped configuration at the pachytene stage or wherever chromosome are paired. Individuals heterozygous for one interchange, two pairs of chromosomes are usually associated in a ring or a string or chain at meiosis.

9.4.3 Meiosis in Translocation Heterozygote

Reciprocal translocation can be transmitted to gametes. Two pairs of chromosomes are usually associated in a ring or a string or chain at meiosis. The factors to be considered here will be 1. pairing multiple 2. chiasmata in pairing segment 3. orientation of multiple and 4. separation at anaphase I. If chiasmata is formed in A,B,C and D then at metaphase I there wilı be formation of a quadrivalent ring and if chiasmata is formed in A,B and C then there will be a quadrivalent chain formed at metaphase I. Where there is chiasmata in A and C, two bivalents will be formed and when it is in A and B, one trivalent and one univalent will be formed. Thus in individuals heterozygous for one interchange two pairs of chromosomes are usually associated in a ring or a string or chain at meiosis. The group of 4 chromosomes includes a normal and an interchanged member of each of the two pairs. The pairing of homologous portions results in a cross-shaped configuration observable at pachytene. At diakinesis and metaphase I of meiosis it opens up into a complex of four chromosomes associated mainly at the ends (Figure 9.2).

9.4.3.1 Orientation of chromosomes

Considering an open ring of four chromosomes the multivalents may assume either a zigzag arrangement or remain as an open ring. The zigzag or twisted orientation leads to alternate chromosomes of the ring passing to the same pole at anaphase I whereas in case of open ring

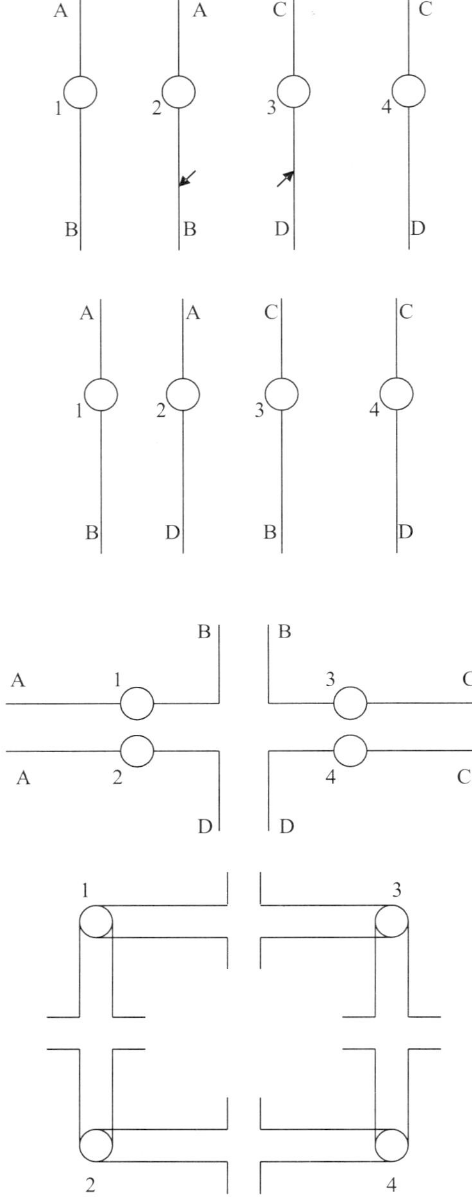

Fig. 9.2 Meiosis in interchange (translocation) heterozygote.

orientation it will result in adjacent members of the ring passing to the opposite poles. Assuming chiasmata in A, B, C and D, mostly the separation will be in the ratio of 2:2, i.e., a type of segregation in which 2 chromosomes pass to each pole but occasionally there will be 3:1 separation resulting in (n-1) and (n+1) gametes. In higher plants n-1combinations abort, n+1 pollen rarely function but n+1 ovules do function and produce 2n+1 or trisomic individuals which have one or both of the translocated chromosomes.2-2 or 3-1 is a type of orientation in which two chromosomes on opposite sides of the ring are directed to opposite poles but ones adjacent to them are not co-oriented towards either pole and this may disjoin to give 2-2 or 3-1 segregation. In case of no crossingover and assuming 2:2 segregation the separation will be of the following three types (McClintock, 1945).

1. **Adjacent and homologous(adj-2)** - Adjacent and homologous chromosomes will move to the same pole. Thus 1 and 2 will go to one pole whereas 3 and 4 will go to opposite pole (**Figure 9.3a**). Here homologous centromeres pass to the same pole. Gametes produced by such orientation will be unbalanced (deficiency and duplication) and will be unviable.
2. **Adjacent and non-homologous(adj-1)** - Here 2 and 4 will go one pole and 1 and 3 will go to the other (**Figure 9.3b**). Here non-homologous centromeres pass to the same pole. Like above here as well gametes produced will be unbalanced in terms of gene content(deficient and duplicated) and thus will be unviable.
3. **Alternate orientation** - In this type of orientation alternate chromosome of the ring will go to the same pole and thus 1 and 4 will go to one pole and 2 and 3 will go to the other pole. Here alternate centromeres pass to the same anaphase pole. Here gametes formed are balanced and they are viable. In this type of orientation the parental chromosomes go to one pole and both translocated chromosomes go to the other (**Figure 9.3c**).

9.4.4 Effect of Crossing-over

The crossingover can take place in either interstitial segment or in any of the other regions of the chromosomes in the interchange complex. Considering the different types of orientation there is on an average 50% alternate and 50% adjacent orientation taking place. In case of **completely directed segregation** (alternate chromosomes in the interchange pass to the same pole 70 to 90%) there will be normal pollen seed set only if there is no crossing over in the interstitial segments. Crossing over in interstitial segment is followed by alternate or adjacent -1 disjunction which results in half of the resulting products of meiosis to be normal and half carry duplication plus deficiency which is a source of spore abortion in plants. Chromosomes that have crossed-over in an interstitial segment pass to opposite poles and hence adjacent-2 segregation would not be expected. When CO is followed by alternate segregation the CO chromatids are in the spores which are deficient and abort but when followed by adjacent-1 segregation the CO chromatids are in the spores that do not abort, are either normal or interchange combinations. Hence in a species with all or an excess of alternate segregation CO is reduced in the regions included in the interstitial segments due to abortion of CO chromatids. The shapes of the multiple are affected by a number of factors such as size, morphology of chromosome, size of interchanging segments, etc. If the interchanged pieces are long the configuration at pachytene of meiosis is a 4-armed or cross-shaped, the two translocated arms alternating with the centromere bearing remainders of the two chromosomes. If CO. has occurred in each of the four arms then the configuration at diakinesis and metaphase I will be ring. If one long piece has exchanged with a very short one the metaphase I configuration will be chains (rings will be rare or non-existent) and the

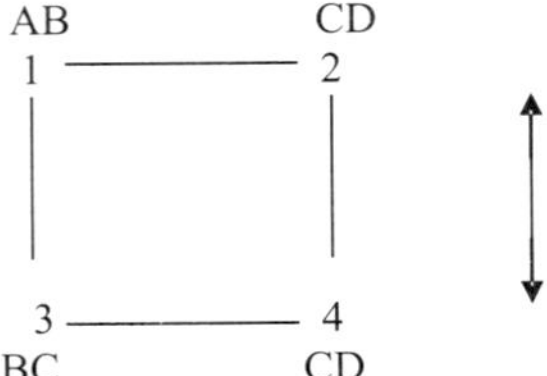

a. Adjacent homologous orientation generating unbalanced (deficient and duplicate) gametes

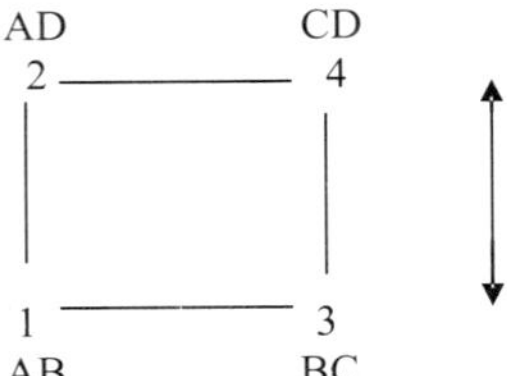

b. Adjacent nonhomologous producing unbalanced gametes.

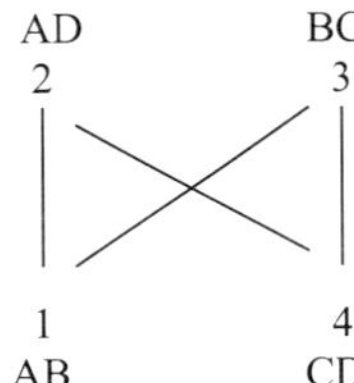

c. Alternate orientation (1 & 4) or (3 & 2) generating balanced and viable gametes.

Fig. 9.3 Showing orientation of multiple and its effect.

configuration at pachytene will be T-shaped. If both pieces are short rings may not form or may be rare but chains may be most frequent but two pairs will be common also. In species with short chromosomes or with low crossingover frequency, a translocation may be present but not observed by any observable association of more than two chromosomes. If the exchanged parts are of similar size and there is similar position of centromere then we will see more alternate orientation. In case of species with short chromosomes we will see less than 50% alternate orientation. In maize, pea, sorghum versicolor, petunia, there is approximately 50% alternate orientation but in barley it is high (greater than or equal to 67%). In rye it is higher because natural selection favours alternate orientation. Also in Datura, Oenothera, *T. monococcum, T. durum* and thus there is directed segregation. Regularity of alternate segregation in Oenothera can be attributed to the fact that all of the chromosomes, despite many translocations, have median centromeres a feature that permits mobility of chromosomes on the metaphase plate. This means that translocations are approximately equal in length. Chiasmata at the end of the chromosomes and few in numbers show alternate orientation. In pea there is evidence of structural differences between northern and southern variety *pumilio* populations with reduced fertility in the translocation heterozygotes produced by crossing between them. Further the IV-VI translocation has been recorded in *Pisum sativum* cultivars and thus could have introgressed from var. *elatius* and/or southern var. *pumilio* (Ben-Zelev and Zohary, 1975). There has been karyotypic as well as cytoplasmic differentiation in pea.

Interstitial translocation - In case of interstitial translocation there would be production of 50% unbalanced gametes irrespective of the types of orientation discussed above. Further, intersertional segment could be other pairing segment than what is shown in Figure and thus the consequences of crossing over are different in different types of segments.

9.5 CONSEQUENCES OF INTERCHANGE HETEROZYGOSITY

The effects of interchange heterozygosity are 1. semi-sterility 2. effect on recombination. The factors affecting the sterility are the morphology of the synaptic complex, the relative sizes of the exchanged segments of chromosomes, the position of centromeres, the flexibility of chromosomes and the frequency and the distribution of chiasmata (Lewis and John, 1963; Rickards, 1964; Sybenga, 1972).

Semi-sterility - The condition in which 50% of the zygotes abort is termed 'semi-sterility (Belling, 1914) but when average frequency of viable combinations is about 75% and hence the sterility is about 25% (termed partial sterility). Semi-sterility is due to 1. adjacent orientation which results in production of gametes which are unbalanced(duplication/ deficiency) and 2. 3:1 separation and thereby production of aneuploidy. Plants differ from animals in that the former show gametes plus gametophyte inviability whereas the later

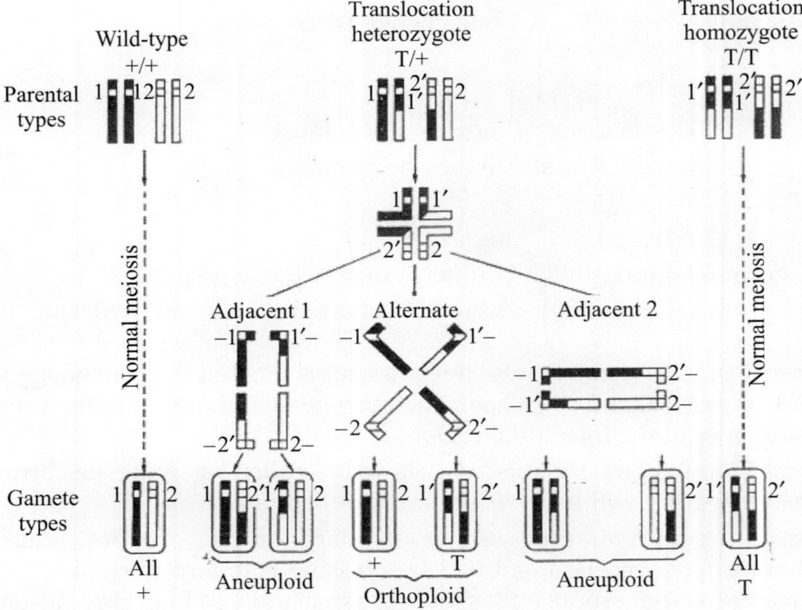

Fig. 9.4 Showing production of gametes by wild types, translocation heterozygotes and translocation homo-Zygotes.

show zygote inviability. For example, in man there is translocation mongolism or Down's syndrome. There is translocation between chromosome 21 and other autosomes, eg. 14/21 translocation. When the following two genotypes are crossed there is production of triplicate chromosome 21 and so mongloid phenotype is produced as shown below.

14, 14, 21,21 x 14, 14/21, 21/14, 21

Gametes 14, 21 ↓ 14/21, 21

14, 14/21, 21,21

9.6 BREEDING BEHAVIOR OF INTERCHANGE HETEROZYGOTE

When interchange heterozygote is either selfed or sibbed it produces 1 normal homozygote : 1 interchange heterozygote as shown below. Upon selfing/sibbing a ratio of 1 standard normal homozygote: 2 interchange heterozygotes: 1 interchange homozygote is expected among the offspring. Since the two types of homozygotes are fertile the ratio is 1 fertile: 1 semisterile or partial sterile.

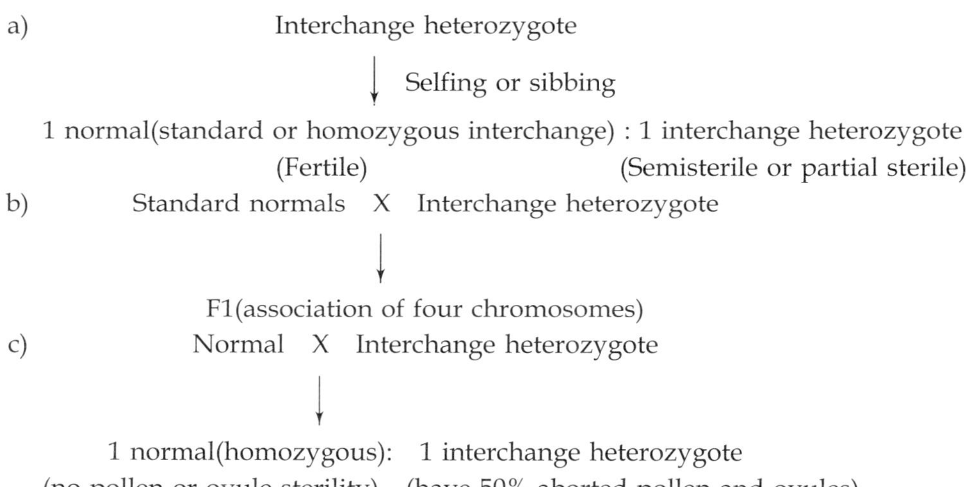

When normal (homozygous for the interchange) is crossed to interchange heterozygote there is 1: 1 ratio of normal and interchange heterozygous individuals in the progeny(with or without crossing over in the interstitial region).

The normal (homozygous) plants will show no pollen or ovule sterility whereas the interchange heterozygotes will have 50% aborted pollen or ovules.

The normals are of two types either standard or homozygous interchange. The interchange homozygote can be identified by crossing standard normal with interchange homozygote which would produce F1 showing association of four chromosomes partially sterile in species without completely directed segregation. This test has been the standard procedure for isolating lines homozygous for interchanges in maize, also in barley, *Datura* and *Nicotiana* and *T. monococcum*.

Considering two non-homologous chromosomes 1 and 2, Figure 9.4 shows the gametes produced by wild type, translocation heterozygote and translocation homozygote and Figure 9.5 shows the progenies produced as a result of mating between translocation heterozygotes.

9.6.1 Effects on Recombination

Interchange heterozygosity will affect the recombination. We will consider its effect in the following two cases. 1. **Recombination of unlinked genes** - Considering heterozygote(AaBb) for two loci(A,B) with independent assortment we get equal frequencies of gametes, namely, AB, Ab, ab, aB. But we do not get this in case of interchange heterozygote as there is no recombination taking place between A and B. AB,CD (parental chromosomes) and AD, CB (interchanged chromosomes) are transmitted together. There is no independent assortment for these two chromosomes.

Recombination of linked genes - The localized chiasmata reduces the crossing over in both the non-homologous chromosomes.

9.7 SIMPLE TRANSLOCATIONS

A simple translocation is one in which a segment of one chromosome is transferred to a nonhomologous chromosome. There will be formation of some gametes with deficiency and other gametes with a duplication and again the nature and nature and extent of the abnormality will determine the fate of the offspring that are produced by such gametes.

9.8 INTERCHANGE IN NATURAL POPULATION

The behavior of interchanges in natural populations depends on whether the species is self-fertilizing or cross-fertilizing, size of the interchange and the degree of sterility present in the interchange heterozygote in relation to the propagating potentialities of the species. If the interchange heterozygotes show considerable sterility and have a small 'margin' they are not likely to survive in cross-fertilizing species and they may become established as homozygotes in a self-pollinating species. Thus interchanges occur as fixed homozygous II, NN as in *Oenothera strigosa*(2n=2x=14) and here the alpha and beta Renner complexes are contributed by egg and sperm, respectively and are quite similar in phenotype. If they are in a species in which the heterozygotes show high fertility(directed segregation) the interchange may float in a cross-pollinating species or again may be established as homozygotes in a self-pollinating species. Thus **interchanges can occur as fixed heterozygous**. Again in case of *Oenothera strigosa, biennis*, NN and II are lethal and only NI survives. The absence of the homozygous combinations can be explained through **balanced lethals,** each genome carries a recessive lethal which is non-allelic to that of the opposite genome and so only the heterozygous combinations in which each lethal is masked by its normal allele, survives. The *strigosa* group possesses a permanent hybrid structure. The *biennis* group is cytologically similar to *strigosa* but morphologically distinct. Structural heterozygosity with the formation of translocations is also found in other plants. In the plant, *Rhoeo discolor*, all the 12 chromosomes form giant multiple (ring of 12). There is no independent assortment. The segregation is alternate and no crossing over occurs in interstitial segments. Similarly races of Oenothera have been established in nature which are self-pollinated and in which all the 14 chromosomes are associated in a large ring. The segregation is largely alternate, resulting in the production of usually two gametes types. They are nearly true breeding because of presence of balanced lethals or lethals of various types and they are able to perpetuate any heterosis if present. There is selective advantage of the big ring in maintaining the hybrid vigour shown by heterozygote and the bigger the ring the greater the likelihood of permanence of the heterozygosity in subsequent generation (Stinson and Steiner, 1955). Further, lethals or deleterious mutations would occur from time to time giving a further selective advantage to the heterozygote.

9.10 FLOATING INTERCHANGES

When lethals are present structural heterozygosity becomes more or less enforced. But when lethals are absent the ring structure is impermanent and translocations are floating rather than fixed in the population(Swanson, 1972) and NN, NI, II will co-exist in the same population and shows polymorphism.

9.11 PRODUCTION OF LARGER RINGS

A single translocation would give rise to a ring of four chromosomes in meiosis. A translocation between a member of the ring and another chromosome would generate a ring of six and thus by successive translocations rings of all the chromosomes are built. The ring thus produced will be unstable because structural homozygosity would result and individuals with variable sized rings and numbers of bivalents would be expected. Structural heterozygosity is enforced only when lethal genes are included in the ring of chromosomes and the end point is a ring involving all the chromosomes with different lethals in each of the two haploid sets.

9.11.1 Methods of Producing Larger Rings

Rings which include all or most of the chromosomes can be produced by either intercrossing translocation stocks or by successive X-ray treatments. In the first method intercrossing different lines is followed by selection of plants with the larger rings for further intercrosses (Gairdner and Darlington, 1931). They obtained higher rings by successive backcrosses to plants having interchanges of opposite parents so as to extract the interchanges from both sides of the ring. To build a large ring by a series of crosses between interchanges selected for their long differential segments, it is required first to produce permanent O6 combinations and then by intercrossing them to produce a still larger ring. Here crossingover in the differential segments (the region between two translocation breaks in the same chromosome) is necessary to add the additional interchanges. Crossingover in the differential segments may occur rather freely and they can be recovered in maize and so the objective was to build a translocation stock which will produce a ring of 20 chromosomes when crossed with a standard normal stock. A ring of 6 built by intercrossing should have one differential segment(crossing between two interchange stocks, for example, 1-2 x 2-3 or 2-3 x 3-4 or 3-4 x 4-5 in which breaks in chromosome common to two interchanges, i.e., chromosomes 2, 3 and 4 constitute a differential segment and a crossing over in this segment will combine the two interchanges and thus a ring of 6 will be formed in each hybrid); a ring of 8 should have two and a ring of 14 should have five such segments or possibly more. Thus a ring of 20 in maize may have 6 differential segments with one chromosome made up of parts of four different chromosomes. The different steps involved can be summarized as follows(Burnham, 1962).

1. Cross between four single interchanges stocks such as 1-2x2-3 or 2-3x3-4 or 3-4x4-5.
2. Crossing will lead to the production of a ring of 6 chromosomes.
3. Screening for the crossover in the differential segment in each cross(Chromosome2,3 and 4, respectively). There will be production of three 2-interchange stocks such as 1-2-3, 2-3-4 and 3-4-5
4. Intercrossing these three stocks such as 1-2-3x2-3-4 and 2-3-4x3-4-5. Intercrossing will lead to the production of two rings of four in each hybrid
5. Selfing of these hybrids will produce four types of offspring and one of the four combinations has the additional interchange. Test crosses are used to select the 3-interchange stocks, 1-2-3-4 and 2-3-4-5 from the crosses, 1-2-3x 2-3-4 and 2=3=4x 3=4=5, respectively
6. Intercrossing 1-2-3-4 with 2-3-4-5 will produce two rings of four in the hybrid
7. Selfing of this hybrid will again produce four types of fertile offspring and from which the 4-interchange stock, 1-2-3-4-5 is selected by test crossing. This 4-interchange stock upon crossing with normal one will produce plants with all 10 chromosome of maize in one ring

The steps involved in the second method of production of multiple interchange stock were : 1. First a ring of 8 was produced by X-raying a stock homozygous for one interchange. 2. Then a line homozygous for the interchanges in the O8 was X-rayed to produce lines which differed from it by one translocation 3. One of these homozygous lines produced a O10 when crossed with a standard normal stock. Thus largest ring with ten chromosomes was constructed in maize. Yamastuta(1951) produced a ring including all the 14 chromosomes of *T. monococcum* but parents of the cross contributed interchanges. In barley a stock which will give O8 + O6 or a O10 + O4 in crosses may be more usuable then one with a complete O14. The larger ring in barley was produced by successive X-ray treatments(Nishimura, Niizeki and Sato, 1952; Nishimura and Kurakami, 1953) in the following manner.

1. From the several O4 produced by the first X-ray treatment homozygous translocations are established.
2. These are then X-rayed to produce further translocations.
3. Whether the new and the previous translocations have one or more chromosomes in common is determined by crossing with standard normal types.
4. Those lines which give a O6 on crossing with the normal are selected and then pairs of lines which give O6 + O6 when crossed with each other are selected. The homozygous interchange stock from each O6 line is to be established.

5. One of each pair of independent O6 lines would then be given the third X-ray treatment to produce a O8 which when crossed with other line member of the pair would give O8+O6.They recognized that (i) in such a stock(O8+O6) there would be two or more arms without translocations than in one which would give a complete O14 and (ii) those with short translocated pieces would be more efficient in avoiding crossingover between the translocation stocks and the stock with which it is crossed.

9.11.2 Application of Interchanges

There are different uses of interchanges.

1. Interchanges can be used as 'Segmental substitution' in case of interspecific gene transfer which explained in later is Chapter 13.
2. It can be used as 'Intra-genome rearrangement'. Genes from autosomes can be transferred to sex chromosome and thus can be made sex linked trait.
3. It can be used for locating unknown gene(s) and mapping.

9.11.2.1 Methods for locating gene(s)

Here interchange heterozygote is used as marker. Using interchanges as markers has the advantage that they do not affect the expression of the character and in the backcrosses an interchange behaves as dominant marker for partial sterility located in the two interchanged chromosomes at the point where the original breakage and exchange occurred. The method for locating gene is as follows. The pairing relations at the pachytene of an individuals, heterozygous for a reciprocal chromosomal interchange and for a gene locus on one of the chromosome involved in the interchange can be represented as shown in Figure 9.2. As we have seen that with respect to the centromeres there may be three types of disjunction (alternate, adjacent-1 and adjacent-2) from the configuration interchange. We now also know that adjacent-1 or adjacent-2 disjunction will give rise to a deficiency for part of one chromosome but a duplicate for part of the other. Such gametes do not function either in pollen or in ovules and results in partial sterility observed in interchange heterozygote. Alternate disjunction on the other hand will lead to normal gametes(N) or to gametes(T) in which the chromosome segments are rearranged but which still contain the complete chromosomal complements are therefore functional. If now an occasional crossingover occurs between the centromere and the gene locus Aa, the resulting recombination will find A linked with the T gametes and a with the N gametes in certain proportions of cases. If p is the recombination frequency then the different types of gametes and their frequencies will be as follows.

Type of gamete	AT	AN	aT	aN
Frequency	p	(1-p)/2	(1-p)/2	p/2

The frequencies of these gametes can be determined directly by crossing the heterozygote to a stock with the normal chromosome arrangement and homozygous aa. In the progeny of such crosses the interchange point behaves as a dominant character, giving partial sterility, so that both partial sterility and the A vs a classification will give 1:1 ratios. In an F2 population, however, union of TXN gametes results in partial sterile individuals(S) while those of NXN and TXT gametes will result in fully fertile(F) individuals. These will occur in the ratios of 2:1:1, respectively, the later two classes being phenotypically indistinguishable.

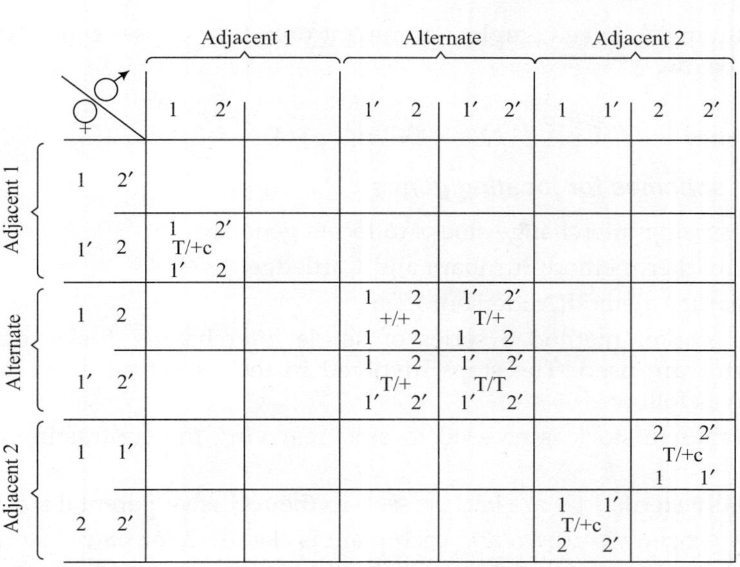

	♀ \ ♂	Adjacent 1		Alternate		Adjacent 2	
		1 2′		1′ 2	1′ 2′	1 1′	2 2′
Adjacent 1	1 2′						
	1′ 2	1 2′ T/+c 1′ 2					
Alternate	1 2			1 2 +/+ 1 2	1′ 2′ T/+ 1 2		
	1′ 2′			1 2 T/+ 1′ 2′	1′ 2′ T/T 1′ 2′		
Adjacent 2	1 1′						2 2′ T/+c 1 1′
	2 2′					1 1′ T/+c 2 2′	

Fig. 9.5 Showing zygote types resulting from matings between translocation heterozygotes. The blank squares represents inviable combinations. The translocation heterozygotes (T/+) followed by c are derived from complementation.

Assuming similar recombination frequencies in male and female gametes the F2 genotypes and their frequencies will be as given in Table(Kramer, 1964) below.

Female gametes	*Male gametes*			
	ATp/2	*AN(1-p)/2*	*aT(1-p)/2*	*aNp/2*
AT p/2	F	S	F	S
AN(1-p)/2	S	F	S	F
aT(1-p)/2	F	S	F	S
aNp/2	S	F	S	F

As some of the sixteen genotypes can not be distinguished either phenotypically or by segregation in their F3 progenies this will lead to differential classes as shown in Table below.

Sterility	*Genotypes*			*Sum*
	AA	A-	aa	
S				1/2
F				1/2
Sum	1/4	1/2	1/4	1

Further assuming A to be complete dominant only four classes can be differentiated in the F2 as given below.

F2 phenotype	AS	AF	aS	aF
Expected frequency	1-p(1-p)/2	1+ 2p(1-p)/4	p(1-p)/2	1-2p(1-p)/4

9.11.2.2 Other schemes for locating genes

The two schemes using interchange stocks to locate gene are:

1. All arms marker method(Burnham and Cartledge, 1939)
2. Linked marker method(Anderson, 1956).

In all arms marker method a series of single interchange stocks that mark all the chromosome arms are used. The steps involved in the use of all arms marker series of interchanges are as follows.

1. The interchange stock is crossed to stock carrying the contrasting character to be studied.
2. The partially sterile F1's are backcrossed to the recessive parental stock.
3. Backcross progeny is grown and each plant is classified for partial sterility and for the character. The parents and the F1's should also be included in the test. In case of quantitative traits backcrosses to both parents may be required.
4. The grades of expression of the trait and their frequencies among the partially sterile and fertile plants are compared.

Linkage is detected if the frequencies of the trait within these two classes differ significantly. The parental classes should be in excess but exceptions may occur in studies of quantitative traits(Lim, 1964). As an observed linkage can be with either or both of the two break points in that interchange, tests with additional interchanges are required to distinguish the three possibilities. Each break point in the interchange showing linkage is checked by a test with an interchange stock involving only that chromosome in common, the two breaks in the common chromosome being close to each other.

Linked marker method - This method uses a series of single interchange stocks in which a gene for an endosperm trait is close to the break point of one of the interchanged chromosome. Classification of the seed for the linked endosperm marker is substituted for pollen fertility classification in the segregating backcross or F2 progenies. One series uses interchanges with one break close to one endosperm marker gene, waxy locus(*wx*) in chromosome 9 whereas another series uses interchanges with one break close to another endosperm marker gene, sugary(*su*) in chromosome 4 in maize and these two series can complement each other. Interchanges which have distal breakpoints(at 0.5 to 0.6 of the arm) and form mostly ring configuration in the heterozygote are superior for linkage test and if properly selected they are very efficient as markers. The steps involved in gene-linked marker method using the 1-9 interchange stock are as follows(Burnham, 1964). The interchange stocks can be dominant or recessive for the character to be studied and so the steps will differ in these two cases.

A. **When the interchange stocks are recessive.**

1. Cross the recessive *wx wx* T1-9 interchange stock with dominant stock *WX WX*
2. The F1 is partial sterile which is backcrossed to the recessive parental stock, *wx wx*

3. Separate the *WX* and *wx* seeds and plant them in separate but adjacent rows
4. Compare the grades of expression of character and their frequencies in the two rows (row from *wx* seed and row from *WX* seed). As plants from *WX* seeds have received the non-interchanged parental combination of chromosomes, the normal chromosomes, they would be fertile whereas plants from *wx* seeds have the interchanged chromosomes and so they will be partially sterile. If the trait under study is linked to the endosperm marker, the *wx* locus then the entire series of interchanges involving chromosome 9 will show the same degree of linkage. But result showing linkage only with T1-9 would indicate one of the genes determining that trait to be located on chromosome 1.

B. **When the interchanged stocks are dominant**

1. Cross the dominant *wx wx* T1-9 interchange stock with stock, *Wx Wx* carrying the recessive character
2. The F1's are partially sterile and are backcrossed to the *Wx Wx* stock
3. Backcrossed progenies are evaluated for the character (either waxy/no waxy or sugary/no sugary) and for the presence or absence of the interchange. The backcross progenies will have plants of two genotypes, *Wx wx* and *Wx Wx* which can be distinguished by the presence or absence of *wx* seeds on mature ear. The backcross progenies are classified for sterility by examining pollen or the open pollinated ears.
4. Compare the grades of expression and their frequencies in the normal and interchange carrying classes.

Although exact localization of gene loci along the chromosome can be made from the study of translocation and deletion which occur frequently and detected easily but the problem is the difficulty of determining exactly, genetically the point of breakage. Possible confusion concerning homology when one or both breaks fall within bands and the possible occurrence of an extreme mutation (amorph) having the same phenotypic effect as a true deletion of this region near the points of breakage can arise.

9.12 METHOD FOR USING INTERCHANGES IN LINKAGE TEST IN SELF-POLLINATED CROPS

The various steps involved in this modified method are as follows.

1. Cross each interchange stock with a stock carrying the contrasting trait.
2. Harvest seed separately from each individual partially sterile F1
3. Raise the F2 progenies from each cross and select the fertile individuals for increase
4. Raise the F3 progenies from the selected fertile F2 plants
5. Identify the two types of homozygotes, i.e., interchange and normal through test crosses based on standard normal. Half of F3 lines are expected to be homozygous for the interchanged and half for the normal chromosomes. Further, the progeny from the test crosses of the homozygous interchange lines will be partially sterile and those from test crosses of the lines with normal chromosomes will be fertile.

6. Comparison is made between two groups of homozygotes for frequencies and degree of expression of the trait being inverstigated.

Johnson (1964) used this modified method for studying malting quality and other traits in barley.

9.13 MINOR USES OF INTERCHANGES

Interchanges can be used for production of duplications and finally interchanges can be used for biological control of insect-pest (Robinson, 1976) using male sterile technique or semisterile using interchange stocks. During meiosis in the translocation heterozygote the synaptic forces between homologous loci results in the formation of a cross-shaped Figure 9.2 (Belling and Blakeslee, 1924). Figure 9.4 shows three gametes production by wild types, translocation heterozygote and translocation homozygote. The translocation heterozygote (T/+) produces six possible types of gametes whereas the both the wild type(+/+) and translocation homozygote (T/T) produce only one type of gamete. The subsequent pattern of segregation lead to reduced fertility of translocation heterozygotes and which may be utilized for insect control. As we have seen there are three patterns of segregation possible and only alternate segregation leads to the formation of orthoploid gametes. Both adjacent types produce aneuploid gametes having an excess of one parental type of chromatin. The term '**aneuploid**' refers to describe a gamete in which any part of chromatin whether more or less than one chromosome extent is present in an increased or decreased amount relative to the remainder of the chromatin(Muller, 1954) and its converse is **orthoploid**. In animals the aneuploid gametes themselves are functional and not lethal and any lethality is only expressed after fusion. There are a number of factors affecting the segregation of translocation heterozygotes and hence their sterility. The crucial property of the inheritance of semi-sterility by translocation follows from the segregation of translocation gametes and wild- type gametes from translocation heterozygotes and consequently mating between translocation heterozygotes and wild-types will produce progenies in which half are translocation heterozygotes and half are wild-types. All the gametes produced from translocation homozygous individuals are orthoploid. Translocation homozygotes can be found amongst the progeny from mating between translocation heterozygotes. In such matings certain combinations of aneuploid gametes may be complementary, i.e. in a zygote the duplication and deficiency of one gamete is complemented by the deficiency and duplication of the other thus producing a viable embryo (see Figure 9.5). Complementation always produces translocation heterozygotes and in this way the proportion of translocation heterozygotes from such matings is increased above the Mendelian expectation. Further the survival of such complementary zygotes increases the fertility of matings between translocation heterozygotes above that which would be expected from simply taking the product of the female and male heterozygotes fertilities. For example, if male and female translocation heterozygotes have fertilities of 50% when test crossed to wild-type then assuming no complementation the fertility of matings between translocation heterozygotes would be 25%. The occurrence of complementation raises this fertility by an amount depending upon the frequency of the gametic types segregating from the translocation heterozygotes.

9.14 USE OF TRANSLOCATION BETWEEN B CHROMOSOME AND A CHROMOSOME

Gene mapping can also be done by utilizing translocations to make chromosomal segments hemizygous. If a recessive mutant is combined with a chromosome lacking the dominant allele, the mutant will be expressed and thereby located. The most useful and promising method of locating recessive gene involves translocations between the B chromosome and members of the normal chromosome complement(A chromosomes). The A-B translocations are capable of generating hemizygosity for regions of genome without limitations imposed by gametophytic inviability (Carlson, 1979).

9.14.1 Behavior of A-B Translocation During Transmission

Roman (1947) showed the behaviour of B chromosome involved in translocation with A chromosome during pollen mitosis using the recessive genetic marker present on A chromosome (see chapter 2 for more detail). As B chromosome has no clear or specific effect on the phenotype, it can be eliminated. Therefore, the heterozygous A-B translocation needs to contain only three rather than four chromosomes. Crosses between a homozygous A-B translocation plant used as female and a normal plant lacking B chromosome will produce heterozygotes, A,A^B,B^A which will form trivalent rather than quadrivalent during meiosis. The A and B chromosome will disjoin regularly from the trivalent but the translocated chromosome, B^A chromosome does not regularly disjoin from the pairing A chromosome. Instead, the B^A chromosome moves to the pole randomly with respect to the A chromosome, producing A, AB^A, A^B, A^BB^A gametes in equal amounts, of which the balanced gametes, A^BB^A and A are transmitted whereas A^B which is deficient, aborts. The gametes with AB^A which is duplicated for a segment of A is transmitted readily through the egg but less well through the pollen and with large duplications the transmission through pollen is greatly inhibited. Thus A-B translocations have the unique ability to transmit sperm with a duplication or a deficiency for specific A chromosomal regions. The pollen is not affected by the genic imbalance and thus various types of individuals are produced with A-B translocation. This transmission of deficient sperm in crosses of A-B translocations provides a hemizygous technique for locating genes on the chromosome. In fact, because of non-disjunction of B^A chromosomes in the microspores second mitosis, hypoploid endosperms and embryos are frequently produced in the F_1's in which any mutation occurred in the corresponding A chromosome segment, being hemizygous, is expressed. In the TB -4a translocation, the point of interchange in chromosome 4 is in the short arm approximately 0.25 of the length of arm away from the centromere. The breakage point in the B chromosome is at or near the junction of the proximal euchromatic region with the distal heterochromatin. The translocation between these two chromosomes leads to the formation of two new chromosomes, A^B and B^A designated as 4^B and B^4 . The 4^B chromosome contains 4L, part of 4S and the noncentric portion of the B distal to the break point whereas the B^4 contains the B centromere, proximal euchromatic segment of the B and the distal region of 4S. The Su locus determining of chromosome 4 is in the region of 4S that was transferred to the B^4 .When the plants homozygous for the interchanged chromosome with the *Su* alleles in both B^4 chromosomes were used as female and crossed with recessive su pollen parent the kernels in the F1 were *phenotypically Su*. However the kernels in the reciprocal cross(RF1) showed *su* phenotype. These kernels lacked the B^4 chromosome in the endosperm but the embryos of such kernels had two B^4 chromosomes. Similarly many of the Su kernels had embryo lacking the B^4 chromosome which presumably was present in double dose in the endosperm. The difference in the constitution of the embryo and endosperm showed occurrence of non-disjunction of the B^4 chromosome in the second pollen mitosis. The 4^B chromosome on the contrary behaved normally in pollen transmission and it is the B4 and not the 4^B chromosome which undergoes mitotic non-disjunction. At least 20 different A-B translocations are required for this purpose, two for each chromosome with points of interchange close to the centromere and in different arms but so far only a series of translocations covering two-thirds of the known genetic linkage map of corn has been assembled(Neuffer and Beckett, 1971). Rakha and Robertson(1970) devised a scheme for producing new A-B translocations through crossingover. Here the aim is to combine an A-B translocation with an A-A translocation in one parent. The two translocations thus involve a common A chromosome which encourages crossingover to occur between them. If the B^A chromosome is chosen so that it can pair with the homologous A^A chromosome proximal to the break point of A-A interchange, crossingover will transfer an interchange segment from a distal region of A^A to the B^A chromosome. Thus the new B^A chromosome will carry parts of the two A chromosomes and the region newly joined to the B^A chromosome can be analysed for mutant genes and thus location of gene can be assigned.

Burnham (1956) cited 24 examples of general and specific uses of interchanges ranging from genetic application to correlation between genetical and cytological information to breeding application for development of inbred lines, hybrids, inducing male sterility and transfer of gene for resistance and other trait besides locating gene for resistance and linkage determination if the marker gene is close to the point of interchange. The other minor uses of interchanges are the production of duplications and biological control of insect-pest(discussed above). Using interchanges stocks insects can be made male sterile or semi-male sterile which can be released in the population and the progenies thus developed would be unviable and male sterile (Robinson, 1976).

9.15 ROBERTSONIAN TRANSLOCATION

In this type of translocation centromeres of the two chromosomes break and centromeres containing long arm fuse whereas centromeres containing short do not fuse as shown in Figure 9.6 which is ultimately lost. Thus centric fusion occurs in the Robertsonian translocations which occur naturally in humans. This results in the reduction of the chromosome number (2n = 45). The breakage occurs more often in the centromere because of the reason that the heterochromatin is located close to the centromeres. The human chromosomes most often involved in the Robertsonian translocation are the acrocentrics of D group (13 and 15) and of G group (21 and 22). This is because of the reason that these chromosomes are the nucleolus organizers and are found associated with each other in the cell. The most common D/G centric fusion seems to be the one between chromosomes 14 and 21(Hamerton, 1971b). A trivalent consisting of three chromosomes, 14^{21}, 14 and 21, is formed in the metaphase I. In case of regular co-orientation the trivalent will separate in such a fashion that the translocated chromosome 14^{21} will go to one pole and the other two chromosomes will go to another pole and this type of disjunction will result in balanced

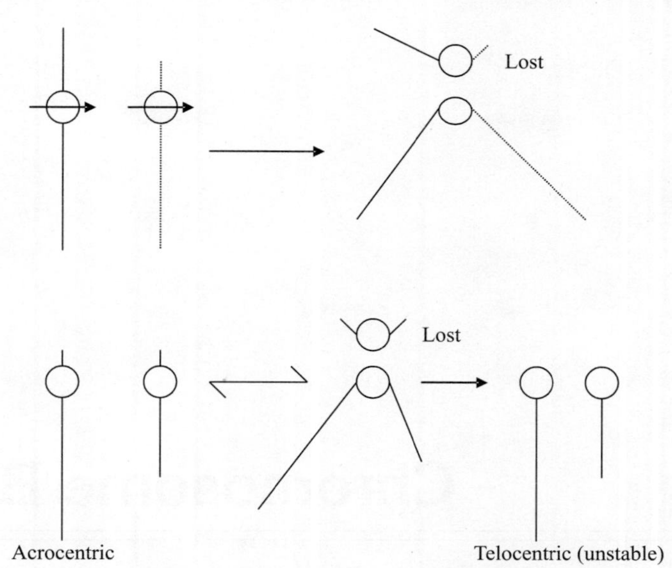

Fig. 9.6 Showing formation of Robertsonian translocation.

gametes. However, in case of non-disjunction, the two adjacent chromosomes (14 plus 14^{21} or 14^{21} plus 21) will go to the same pole. Now if a gamete with such a combination gets fertilized the individual will have 2n=46 but it will carry the equivalent of an extra chromosome and Down's syndrome results from such chromosome combination, 14^{21} plus 21. Thus Robertsonian translocation is associated with an increase in the frequency of children with Down's syndrome and this can be easily identified. Another type of Robertsonian translocation involves the short arm of Y chromosome and the long arm of chromosome 15 (Subrt and Belhova, 1974). It widely occurs in animals and thus induces polymorphism. When centric fusion does not occur then it results in production of telocentric chromosomes as shown in Figure 9.6 which are unstable. In wheat telocentric chromosomes can be generated.

Chromosome Behavior

Variations in chromosome behaviour during nuclear cell division can be due to two reasons: 1. environmental and 2. genetic reason. The genetically induced variation can be either sporadic although there is no experimental evidence for it or constitutive. Genetic control could be simple or polygenic. **Experimental evidence for genetic control of chromosome behaviour**- The evidence has come from the mutants affecting different chromosome behaviour.

10.1 CHROMOSOME CONDENSATION

There is one gene *el*(elongate) (Rhoades and Dempester, 1966a), the recessive gene which is involved in chromosome condensation in maize.

10.2 SPINDLE STRUCTURE

Another recessive gene, divergent spindle(*dv*) in maize which determines spindle structure (Clark, 1940) produced by U.V. irradiation and causing abnormal spindle formation. During cell division chromosomes are distributed to daughter cells by the mitotic spindle. This system requires spatial cues to reproducibly self-organize. Such clues are provided by chromosome-mediated interaction gradient between the small guanosine triphosphate (GTP ase) Ran and importin-B. This produces activity gradients which determine the spatial distribution of microtubule nucleation and stabilization around chromosomes and which are essential for the self-organization of microtubules into a bipolar spindle (Caudron et al., 2005).

10.3 CENTROMERE ORIENTATION OR ANAPHASE DISJUNCTION

There are *clarent* in Drosophila (Sturtevant, 1929) and sticky gene(*st*), a recessive gene in maize (Beadle, 1932a, 1937; Schwartz, 1958 and Philips, 1973) which prevent regular disjunction. It causes stickness, clumping and sometimes breakage of chromosomes and is located on chromosome 4 of maize. **Co-ordination of events**- The gene, *p*, a recessive gene, controls polymitotic(Beadle, 1931) division in maize in which there is repeated division of haploid pollengrain indefinitely. There is another gene *m* which causes mitosis with unreduced chromosome and there is failure of cytokinesis(Beadle, 1932b).

10.4 CHROMOSOME PAIRING AND CHIASMATA FORMATION

Pairing (alignment of homologous chromosomes), synapsis (zipping of the paired homologs by synaptonemal complex proteins) and recombination (exchange of chromosome arms) are pre-requisites for accurate chromosome segregation in meiosis.

10.4.1 Pairing

Synaptic mutants - Two types of mutants were observed. 1. Asynaptic(as) mutants (Beadle, 1930) and 2. Desynaptic mutants(Miller, 1963; Baker and Morgan, 1969) have been found. In case of former pairing is defective, pairing is partial or wholly asynaptic while in case of later the pairing is normal but chiasmata formation is defective. Pairing is controlled by a number of genes (Riley and Law, 1965). *Ph* gene (Sears and Okamto, 1958; Riley and Chapman, 1958) controls homoelogous pairing in wheat. It suppresses homoeologous pairing and promotes homologous pairing. The poor homologous synapsis(*phs*1) gene in maize is required for pairing to occur between homologous chromosomes(Pawlowski et al., 2004). In the phs1 mutant homologous chromosome synapsis is completely replaced by synapsis between nonhomologous chromosomes. In other words, homologs do not pair and the synaptonemal complex, a protenaceous structure, forms between nonhomologous chromosomes. The *phs*1 gene is also thus required for installation of the meiotic recombination machinery on chromosomes as the mutant almost completely lacks chromosomal foci for the recombination protein RAD51.Thus in the *phs*1 mutant, synapsis is uncoupled from recombination and pairing. The protein encoded by the *phs*1 gene likely acts in a multiple process or co-ordinate pairing, recombination and synapsis.

10.4.2 Chiasmata(X)- formation

The desynaptic mutants can have two types of effects. a. Partial or complete but general effect on all chromosome pairs of genome. b. Specific chromosome effect in which there is desynapsis of a particular chromosome. In case of *Hypochaeris radicata*, 2x=8 one pair shows desynapsis in 90% cells whereas in case of *Crepis capillaries*(2n=6), there are three genes, A,C,D desynaptic for each of three pairs and thus there is separate gene control. In case of micro-organisms such as Bacteriophage and Neurospora genes have much more local effects and region specific effects on recombination (Catcheside, 1968). In case of Rye(2x-14) the recombination depends on the localization of chiasmata. With respect to centromere X-ta can be distal, interstitial or proximal. It has been observed that in case of X-ta in the distal region there is more recombination. The difference in such behaviour is simply genetic.

10.5 POLYGENIC CONTROL OF CHROMOSOME BEHAVIOUR

Rees(1961) while experimenting with inbred lines practiced selection for two traits, X-ta frequency and X-ta distribution and found selection to be effective for these two traits, i.e the mean and variance of these two traits changed significantly. Chinnicci practiced selection in both directions,ie. selection for higher recombination frequency and selection for lower recombination frequency and after 33 generation of selection he obtained low line in which recombination frequency reduced from 15.4 to 8.5 and in the high line the recombination frequency increased from 15.4 to 22.1%. These results suggest that the recombination frequency is under control of polygens. When two parents differing in chiasma frequency were crossed and F1, F2, F3, F4 and F5 generations were studied for this trait then it was concluded that there existed variation between genotypes with respect to mean chisma frequency (Rees, 1956b). Variation was also observed within genotypes. Variation in chiasma frequency was observed both within and between individuals. F1 showed more chiasma frequency and was related to heterosis. One family from F3 to F5 has a mean frequency lower than either parent showing transgressive segregation for this particular trait. Variation in X-ta frequency induced by the same environmental variation is less between individuals of the same heterozygous genotype than those of the same homozygous genotype. Average chiasma frequency can vary between genotypes so also can the distribution of chiasmata between and within cells. Thus variation at all three levels affect the gametic output. Finally segregation revealed variation affecting all aspects of chromosome structure and behaviour. It includes neo-centric activity, differences in chromosome coiling, in the terminalization and localization of chiasmata, in disjunction properties at anaphase and in the spindle behaviour.

10.5.1 Effects of Heterochromatin

Heterochromatin (B-chromosomes, telomeres and knobs) may be responsible for a variety of meiotic alterations. Use of heterochromatin can be made for increasing crossingover, changing the position of crossingover, reducing pairing between homoeologous chromosomes and increasing pairing between homologous chromosomes.

B chromosomes - The well known genetic effect of B chromosomes in maize is non-disjunction of the B centromere at the second microspore division (Roman, 1947). This behaviour in combination with A-B translocations in maize has been used effectively to attack a number of genetic problems (see Chapter 9 for detail). Modest increases in crossing over in particular regions of chromosomes were also found in the presence of B chromosomes in maize (Hanson, 1968; Nel, 1973). Significant changes in chromosome pairing determined by B chromosomes has been reported in Lolium species hybrids(Evans and Macefiels, 1972). The bivalent frequency in the diploid hybrids between *L. temulentum* and *L. perenne*, in the presence of a pair of B chromosomes from *L. perenne* reduced from 6.8 to 3.6. Further induced tetraploids of hybrids without B chromosomes had several multivalents in comparison to when B chromosomes were present. The B chromosomes, therefore, suppress pairing and chiasma formation between homoeologous chromosomes and thus B chromosomes could be used to bring in meiotic stability in synthetic disomic polyploids. Similarly, in wheat the presence of B chromosomes from *Aegilops mutica* or *Ae. spletoides* prevents pairing of homoeologous chromosomes in *T. aestivum* x *Ae. speltoides* hybrids that lack chromosome 5B. The presence of B chromosomes compensated for the absence of 5B in preventing homoeologous pairing (Dover and Riley, 1972).

Telomeres - Telomeric heterochromatin of rye(*Secale cereale*) may affect homologous chromosome pairing in hexaploid triticale. With isogenic lines differing only in the presence or absence of telomeric heterochromatin of a rye chromosome(probably 7) Merker(1976) showed that in the absence of this heterochromatin the number of bivalents at metaphase I was reduced by one-third. Roupakias and Kaltsikes(1977) reported that overall chromosome pairing was significantly higher in plants without this heterochromatin(telomeric heterochromatin of short arm of chromosome 6 of rye) as measured by reduction in open and increase in closed bivalents. It seems that elimination of the large blocks of telomeric heterochromatin of rye chromosomes might increase the meiotic stability of triticale as this heterochromatin may be responsible for some pairing failures in triticale.

10.6 DISTINCTION BETWEEN MEIOTIC AND MITOTIC CELL CYCLE

Meiosis can be distinguished from mitotic cell cycle beginning with the initiation of pre-meiotic DNA synthesis (Baker et al., 1976). Entry into meiosis in higher plants is genetically controlled and there are mutants which are incapable of premeiotic DNA synthesis. Further there are mutants affecting transition from meiosis I to meiosis II. Meiotic mutants affect the normal pattern of chromosome bahvior starting with the initiation of premeiotic DNA synthesis and before the reinitiation of mitosis after completion of the second meiotic division. The following generalization has emerged from the study of meiotic mutants.

1. Some of the metabolic processes involved in meiotic development and recombination are also involved in somatic cells in processes involved in DNA metabolism necessary to ensure chromosome stability both spontaneously and after irradiation.
2. Exchange is essential for normal disjunction in organisms in which recombination normally occurs. The genetic control of exchange occurs at two levels: one that is region-specific and one that acts on the chromosome as a whole. In case of later, one type of mutant produces alteration in the distribution of exchanges among and along the chromosome arms, often, but not invariably, accompanied by a change in overall frequency of exchange. The characteristics of these mutants suggest that the distributions of exchanges within and among cells as well as the distribution of exchanges along the length of the chromosome are all under resolvable genetic control. The second type of mutant alters the frequency but not the spatial distribution of recombinants.
3. Finally, although many meiotic mutants affecting the first meiotic division have been recovered, the mutants affecting the second division are relatively rare.

Cytogenetic Structure of Crop Plant Species

In this chapter we will see (i) chromosome number and karyptype (ii) ploidy level (2x or poly x) (iii) types of ploidy (auto or allo) (iv) Genome donors and relationship and (v) type of control of homoeologous pairing in some crop species including wheat, oats, tobacco and cotton.

11.1 WHEAT (TRITICUM AESTIVUM)-GENOMIC RELATIONSHIP

Wheat (*Triticum aestivum*) is an amphidiploid with 6x= 42 with the genome structure as AABBDD and thus there are three genomes, A,B and D. Genomic relationship(also given in Figure 11.1) is as follows.

Diploid 2x= 14	is	*T. monococcum* or similar species	AA
A genome donor	is	*T. monococcum*	
B genome donor	is	*Aegilops speltoides* or *T. speltoides*	BB(?)
Tetraploid 4x= 28	are	*T. turgidum*	AABB
		T. timopheevi	AABB
D genome donor	is	*T. tauschii*(*A. squarrosa*)	DD
Hexaploid 6x= 42	is	*T. aestivum*	AABBDD

The crosses between different species have shown different number of bivalents and univalents as follows.

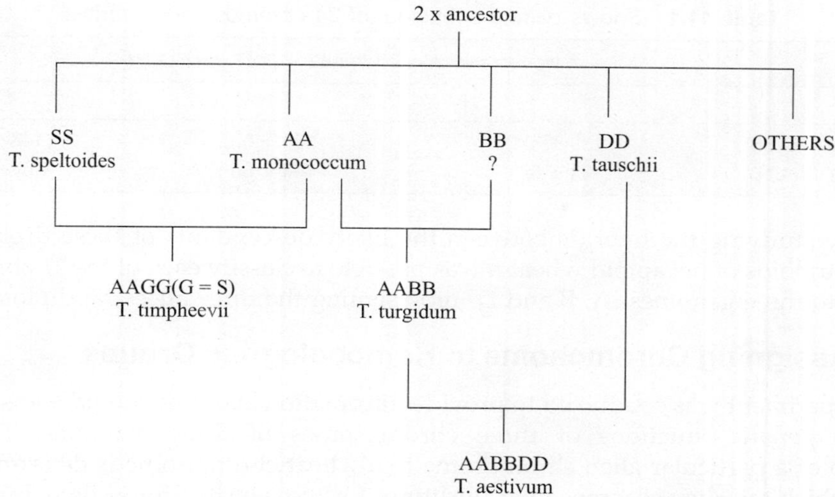

Fig. 11.1 Showing origin of cultivated wheat (*T. aestivum*) (Kimber, 1974).

Diploid(2x)	x	Tetraploid(4x)	⟶	7II + 7I's
Diploid(2x)	x	Hexaploid(6x)	⟶	7II + 14I's
Tetraploid(4x)	x	Hexaploid(6x)	⟶	14II + 7I's

When hexaploid wheat was crossed with tetraploid the hybrid had 14II + 7I and when tetraploid or hexaploid was crossed with diploid the F1 had up to 7II + 7I or 14 II + 7I. This means that three distantly related diploids, AA, BB and DD had combined to form the hexaploid and the lack of multivalent pairing was due to a high degree of differentiation among the chromosomes concerned. The above crosses show that AA x BB would have yielded AABB which upon crossing with DD would have resulted in the production of cultivated bread wheat (*T. aestivum*). Thus for A genome, the donor is *T. monococcum* and for D genome it is *T. tauschii* (*Aegilops squarrosa*). The donor of B genome is most likely *A. speltoides* or *T. speltoides*. The B genome donor is either extinct or there is polyphyletic origin, i.e. more than one species have contributed to the development of B genome. By 1952 it had become clear that each chromosome of hexaploid wheat had a close relative (homoeologue) in each of the other two genomes and the three closely related diploids had changed only slightly from those of the common parent of the three and thus the hexaploid wheat has a great amount of duplication and triplication of genes. The 21 chromosomes of wheat have been placed in seven homoeologous groups of three chromosomes each, each group comprised of one chromosome from each genome (Sears, 1965) and are designated as 1A-7A, 1B-7B and 1D-7D. For studying this three different types of studies were conducted.

11.1.1 Genome Study

In this study 21 different monos were crossed with AABB and AADD natural and artificial tetraploid and pairing behaviour in 34 chromosomes hybrids were studied. The numbers of bivalents and univalents in various crosses are presented in Table 11.1 as follows.

Table 11.1 Shows pairing behaviour of 34 chromosomes hybrids.

	Monosomic derived from 2n-1(=41, for20,21,14) X 28		
	A	*B*	*D*
2n – 1 x AABB	13II + 8I	13II + 8I	14II + 6I
2n – 1 x AADD	13II + 8I	14II + 6I	13II + 8 II

Thus by studying the hybrids between the likely descendants of these diploid species and the aneuploids of hexaploid wheat it was possible to classify each of the 21 chromosomes of wheat into three genomes- A, B and D- representing the three ancestral diploid species.

11.1.2 Assigning Chromosome to Homoeologous Groups

Here the experiment was conducted to provide information about the homoeologous groups, i.e. similar genetic functions of three chromosomes of three genomes. The genetic relationship of a particular alien chromosome to a wheat chromosome is determined by the degree to which the alien chromosome substitutes for the wheat chromosome. In case of two wheat chromosomes it consists of determining the degree, if any, to which extra dose of one chromosome compensates for the absence of the other and in practice this means synthesizing the nullisomic-tetrasomic combination and observing whether or not it is more vigorous or fertile than the simple nullisomic (Sears, 1952, 1965, 1969). By means of nullisomic-tetrasomic tests, the 21 chromosomes of wheat were placed in 7 homoeologous groups of three chromosomes each, each group comprised of one chromosome from each genome. Although this result does not exclude the possibility that minor homologies exist between certain chromosomes of different homoeologous groups, it does establish that the major relationships lie within the groups. On the basis of the seven homoeologous groups and the three genomes they belong to the 21 chromosomes of wheat are designated as 1A-7A, 1B-7B and 1D-7D, respectively (Table 11.2). For example, using tetrasomes for one chromosome crossed with nullisomes for another chromosomes information regarding possible duplications of genetic material and homology can be obtained from morphological analyses of the hybrids. In case a degree of homology exists the extra dose of one chromosome may partly compensate for deficiencies in the other chromosomes. Monosomic/ trisomic combinations were synthesized by the following crossing programme by Sears(1952).

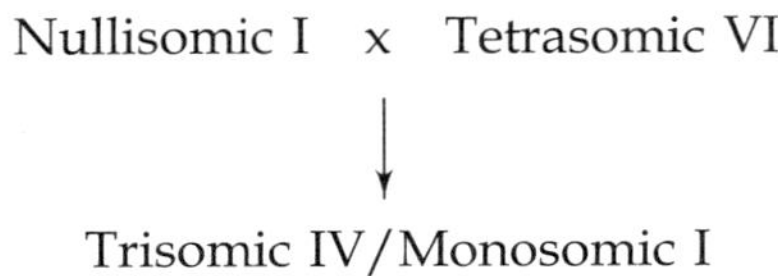

It will have normal phenotype if extra chromosome(lost) of Mono is homoeologous to one extra of Trisomic IV and thus there will be duplication/deficiency compensation. Here 1 and 6 form one *homoeologous* group and thus seven homoeologous groups were formed. Similarities between different nullisomics were the basis for the first assignment of chromosomes to homoeologous groups. When the gene, *Ph*, which suppresses pairing between homoeologous chromosomes is missing as in nullisomic 5B plants homoeologous chromosomes from the A, B and D genomes can pair with each and with the homoeologous chromosomes from the related species.

Table 11.2 Shows seven homoeologous groups of chromosomes in wheat based on their coding for similar proteins, their ability to synapse under similar conditions and their ability to compensate for each other in nullisomic stocks.

	Genomes		
Homoeologous groups	*A*	*B*	*D*
1	1A	1B	1D
2	2A	2B	2D
3	3A	3B	3D
4	4A	4B	4D
5	5A	5B	5D
6	6A	6B	6D
7	7A	7B	7D

11.1.3 Diploidization of Polyploidy Wheat

Although wheat is a hexaploid but it behaves like diploid. Wheat is a segmental allopolyploid. The A, B and D genomes are sufficiently closely related and thus some pairing is possible at meiosis between A, B and D genomes. A and D genomes are quite similar in terms of gene content and so in the cross AA x DD → AD there would be quite a bit of pairing. In tetra- and hexaploid there will be no homoeologous pairing. In hexaploid with six chromosomes(multivalents(6)) there would be formation of multivalents but then there is no homoeologous pairing as it is supposed and there is only homologous pairing. In the hexaploid wheat there is regular mitosis, high fertility and high genetic stability but when there will be homoeologous pairing these properties would be lost and there would be much more variation. The diploidization behaviour is due to action of a single gene on 5 B which suppresses homoeologous pairing. When this activity is removed or suppressed genetically, homoeologous chromosomes which are genetically and evolutionary related chromosomes of different genomes, will pair with each other.

11.1.4 Elucidation of Genetic Control of Pairing

Here we would like to know first whether or not there is a gene controlling the homoeologous pairing and secondly if there is a gene then what is its location. The genetic control of pairing can be studied in the following ways.

1. **Study of nulli haploids** - Haploids of the hexaploid wheat would have 2n=3x=21 and here there will be 21 univalents generated with 7A, 7B and 7D chromosomes. One can also synthesize nullihaploids with 2n=3x-1= 20 which will show 20 univalents and and there will be as many as 21 nullihaploids. When nulli5B-haploid with 2n= 3x-1(5B)= 20 was synthesized it showed 3-4 bivalents, 1-2 trivalents and 0-1 quadrivalents. This shows that the location of gene for control of pairing is on 5B chromosome.

2. **Study of nullisomics** - In the nullisomics, 2n=6x-2= 40 there would be 20 bivalents formed. But when nullisomic with 5B was synthesized and studied then there was observation of multivalents formation and multivalents involving up to 6 chromosomes (hexavalents) were observed which indicates presence of homologous and homoeologous pairing. In other words, nullisomy for the long arm of chromosome 5B or a recessive mutation, *ph*1b at the *Ph*1 locus in this arm restores metaphase I pairing among homoeologous chromosomes of the A, B and D genomes.
3. **Study of monosomics** - Monosomic 5B behaves like normal monosomic in pairing behaviour. Considering a gene determining the homoeologous pairing this gene is present in hemizygous condition in the monosomic5B and it is expressed as one chromosome is equivalent to two chromosomes. From the above studies it can be said that the factor controlling the homoeologous pairing resides on 5B chromosome.

In yet another study (Wall et al., 1971) crossed hexaploid wheat (6x=42) with rye (2x= 14). In the F1(2n=4x=28) there were normally 28 univalents(21 univalents of wheat and 7 of rye). They then treated the hexaploid wheat with some mutagen the pairing gene got mutated and would lack the suppression activity and so there will be homologous pairing within wheat chromosomes in the F1and thus we will get some bivalents instead of 28 univalents. This shows that there is a gene called *Ph*(pairing homoeologous) controlling the pairing. The normal dominant allele was symbolized ***Ph*** and the recessive allele *ph*. Thus the association of the mutant condition(Mutant 10/14 of *Triticum aestivum*(2n=6x=42) affecting homoeologous chromosome pairing at meiosis with chromosome 5B was determined by (i) the absence of segregation in hybrids obtained when Mutant 1o/13 monosomic 5B was pollinated by rye (*Secale cereale*, 2n=2x= 14), ie. monosomic analysis (ii) the occurrence of trisomic segregation for pairing behavior in 28-chromosome wheat-rye hybrids obtained from 5B trisomic wheat parents with two 5B chromosome from a non-mutant and one from mutant parent(trisomic inheritance) and (iii) the absence of segregation for pairing behavior in the 29-chromosome wheat-rye hybrids obtained from the same trisomic wheat(absence of segregation in 29-chromosome, 5B-disomic wheat-rye hybrids)

11.1.5 Location of Ph Gene

For locating *Ph* gene on the chromosome telocentic analysis was carried out.

Telocentric analysis - The 5Bchromosome has one short and one long arm. In this analysis Telo $5B^L$ and $5B^S$ were produced. $5B^L$ refers to 5B chromosome with long arm whereas $5B^S$ refers to 5B chromosome with short arm. Thus for all 21 wheat chromosomes telocentric chromosomes can be produced. After that nullisomics for half chromosomes are produced, i.e. nullisomics for Telo 5BL and Telo 5BS. From such type of study it was found that Telo$5B^L$ had suppression activity whereas $5B^S$ did not have and so *ph* gene is located on $5B^L$(Okamoto, 1975; Riley and Chapman, 1958b and Sears and Okamoto, 1958). Further a cross in which one of the two parents was a nullisomic for half chromosome and thus had *Ph* gene in heterozygous condition and the other was normal(*Ph*/*Ph*) was made and the individuals were studied for the pairing behaviour and it was concluded that the *Ph* gene is near the end of $5B^L$ and is one map unit from the centromere.

Nullisomic(*Ph*/*ph*) x Normal(*ph*/*ph*)

*Ph*1 plays a role in discrimination between homologous and homoeologous chromosomes in wheat. Thus in wheat breeding *Ph*1 represents a barrier to introduction of alien genes and recessive mutants at this locus are important tools for obtaining introgression of alien chromosome segments with economically important genes into wheat. Thus long arm alone of the chromosome($5B^L$), responsible for the prevention of homoeologous pairing was shown from the behavior of haploid plants which had either the complete euploid complement of 21 chromosomes or which were short arm separately (Riley and Law, 1965). Chromosome pairing at meiosis in the deficiency of short arm resembled that in euploid whereas pairing in the deficiency of the long arm resembled that in 20-chromosome haploid lacking the entire chromosome(Wall et al., 1971).

11.1.6 Mechanism of Controlling Pairing

To understand the mechanism by which *Ph*1 distinguishes homoeologous chromosomes from homologous chromosomes, it is necessary to understand (i) the nature of differentiation between homoeologous chromosomes and (ii) the physiological mode of the action of *Ph*1 gene(Dubvosky et al., 1995).

11.2 DIFFERENTIATION BETWEEN HOMOELOGOUS CHROMOSOMES

There are different hypotheses regarding this. The traditional hypothesis is that the chromosomes lose homology by structural rearrangements that alter the sequence of loci by chromosome inversions, translocations, duplications, etc. This hypothesis has led to the concept of segmental chromosome differentiation of homoeologous chromosomes (Giorgi and Cuozzo, 1980). As per this hypothesis, homoeologous chromosomes are mosaic of homosequential segments and structurally altered segments. Structural chromosome heterozygosity may not manifest cytologically (may be absent) and the absence of inversion bridges or multivalents in hybrids does not constitute evidence against this hypothesis.

Alternatively chromosomes of related genomes may be differentiated by accumulation of 'substructural' changes (Dvorak, 1981). Substructural differentiation (formerly referred to as 'non-structural differentiation) is assumed to leave chromosome homosequential but alters their homology at the level of nucleotide sequences. This can range from nucleotide substitutions to various types of nucleotide sequence arrangements including insertions and deletions. Differentiation was shown to involve all chromosomes in a genome and to occur throughout the entire lengths of chromosome arms. Comparative mapping has not revealed structural differences between homoeologous chromosomes of wheat, A, B and D genomes but then it could be said that these chromosomes do differ structurally but the differences are cryptic, i.e. too small to be detected in low density comparative maps. Chromosomes are not mosaics of homologous and homoeologous segments as assumed by segmental allopolyploidy concept but show a uniform differentiation along entire lengths. Chromosomes 1A and $1A^m$ are shown to be collinear and they are differentiated 'substructurally'.

11.2.1 Physiological Mechanisms

Different physiological mechanisms have been suggested by different workers. Riley (1966) suggested that some kind of timing mechanism works which controls pairing. *Ph* gene imposes a time limit on pairing. Feldman (1966b) suggested that regulation of pairing and suppression of homoeologous pairing is due to pre-alignment mechanism. There is sorting out of chromosomes into homologous pair before meiosis, i.e separation of homologues through the exclusion of homoeologues. This mechanism is found in all species. Pairing starts in zygotene. There is bivalent interlocking and sorting of chromosomes into homologous pair. Feldman reported that *Ph* controls homoeologous pairing through an effect on somatic association. Somatic association hypothesis is a widely accepted physiological mechanism by which *Ph*1 prevents pairing and recombination between differentiated (homoeologous) chromosomes. This hypothesis postulates that the pattern of meiotic pairing in wheat is pre-determined by the distance (association) between and homologous and homoeologous chromosomes in the nucleus which persists throughout the entire life cycle. The *Ph* reduces the forces causing somatic association enough to prevent homoeologues from associating but not enough to keep homologues apart. As a result homologues come into meiosis already effectively paired and homoeologues are excluded from pairing. He suggested that in the absence of 5B homologues and homoeologues are associated premeiotically. When 5B chromosome is present then only homologues are associated. In case of material with extra 5B chromosomes some homologues are too far apart to be able to come together for pairing; some homoeologues are closer to each other than to their homologues and therefore they pair. Since pairing begins at the ends of the chromosomes some bivalents as they come together trap one or more other chromosomes between them causing interclocking. Driscoll et al. (1979) suggested that there is no restriction on pairing and we get both homologous plus homoeologous pairing but only the homologous pairing leads to chiasmata formation. It is thought that *Ph*1 affects chromosome pairing by modifying spindle protein and primarily acts on the centromere.

11.3 OTHER GROUPS OF CHROMOSOME 5 AFFECTING PAIRING

Chromosomes other than those of group 5 are known to affect pairing. Nulli 5D leads to asynaptic condition under condition of less than 15°C temperature, while it is normal at higher temperatures. In case of Nulli5A the pairing is normal. 5A and 5D carry promoters of pairing. *Ph* gene of 5B can be mutated to act like a suppressor. Although the gene on chromosome 5BL is having the greatest effect on homoeologous pairing, it is not only the one affecting the process. There are several others (Table) which promote pairing. Similarly there are others which suppress pairing. The most strong of these promoters is the one on 5DL whose absence leads to asynapsis if the temperature is unfavorably either high or low or if the genetic background tends toward asynapsis as in case of tetrasomic-5B. There are also weak promoters on 5BS, 5DS and possible 5AS. All of the promoters are believed to operate by increasing somatic association. Genes on 2A, 3A and 3B are believed to affect chiasma pairing(Table 11.3).

11.4 ALIEN SOURCES OF PH GENE

Homoeologous pairing can be also be induced by alien species which can suppress the pairing control mechanism of wheat. In fact, two relatives, *T. speltoides* and *T. tripsacoides* (*Ae. mutica*) have such strong promoters that in hybrids with hexaploid wheat they are able to supresss the effect of 5BL and bring about homologous pairing. The suppression produced by *Ae. speltoides* has been employed to make transfer of genes from alien addition or substitution lines.

In a cross of *T. aestivum* and *T. speltoides* or *T. tripsacoides* (Riley et al., 1961) which carries 5B the *Ph* gene is inactivated or suppressed suggesting that these species carry dominant homoeoallelic gene to *Ph*. Other genes present on 3B chromosomes have effect on pairing but the effect is much weaker in comparison to gene on 5B chromosome.

A. aestivum x *T. speltoides* (or *T. tripsacoides*)

↓

Ph gene is suppressed or inactivated suggesting that these species carry dominant homoeoallele to *ph*.

Table 11.3 Location of factors affecting chromosome pairing in wheat and its relatives.

Chromosome location	*Effect on pairing*	
	Promoter	*Suppressor*
1A	Promoter	
1D	"	
2AS	"	
2B	"	
2DS	"	
3AL	"	
3AS		Suppressor
3BL	Promoter	
3BS	"	Suppressor
3DL	Promoter	
3DS	"	Suppressor
3D(Mutant)	Promoter	
3R	"	
$3H^b$	"	Suppressor
4D	"	Suppressor
5A	Promoter	
5BL	"	Suppressor
5BL(Mutants)	Promoter	
5BS	"	
5D	"	
5RL	"	Suppressor
5RS	Promoter	
5U	"	
6B	"	
7A	"	Suppressor
7DL	Promoter	

11.5 OTHER CROP SPECIES

Oats (*Avena sativa*)

The cultivated oat (*Avena sativa*) is a hexaploid with 2n=6x=42 with genomic structure as AABBDD. There has been restructuring of genome within polyploid due to translocation. The F1 of Polyploid x Suspected donor cross will provide information regarding the pairing. In case of the cross *A. longiglums*(dominant gene) x A. sativa(recessive gene), the F1 shows some homoeologous pairing in hybrid. The dominant gene of *A. longiglums* suppresses the *ph* gene activity and so there is some homoelogous pairing.

Tobacco

Nicotiana tabaccum is an allotetraploid with 2n = 2x= 48 and has the genomic structure as SSTT. S denotes genome of *N. sylvestris* whereas T genome represents *N. tomentosiformis*. There is some doubt whether or not there is gene suppressing homoeologous pairing because S and T genomes are sufficiently different. The same is true with cotton as well.

Cotton

Cotton (*Gossypium hirsutum*) is 2n=4x= 52 with genomic structure as AADD. D genome is from *G. herbaceum* (Old world cultigen) and D genome is from *G. raimondii* (from Peru, New world cultigen). There is preferential pairing of dissimilar chromosomes as A genome is larger than D genome.

Chromosome Manipulation

Chromosome manipulation can be done at the following levels:

1. Genome
2. Subgenome; whole chromosome and chromosome segments
3. DNA molecules.

The transfer of and integration of fragments of chromosomes or whole chromosome sets from one cell (genotype or species) to another cell (genotype or species) is called '**chromosome engineering**'. The subgenome refers to a subset of genomic DNA sequences that represents a fraction of the complete genome, for example, one chromosome or part of a chromosome.

12.1 GENOME MANIPULATION

Genome manipulation involves combining of two or more genomes normally present in two or more species. Thus new synthetic species combining the chromosome complements of diverged but related species can be developed. The synthetic amphidiploid will show recovery of fertility but the problems associated with the development of synthetic amphidiploid are stability and the reproductive capacity of this new species. The three steps involved in genome manipulation are:

1. Hybridization
2. Chromosome doubling
3. Adjustment of meiosis.

First, the parents, the two species with different genomes say A and B, are crossed to produce F1 (AB) which will be sterile and/or instable. Now thorough chromosome doubling using colchicine the chromosome number is doubled. The resultant amphidipoloid (AABB) is now fertile and stable as shown below.

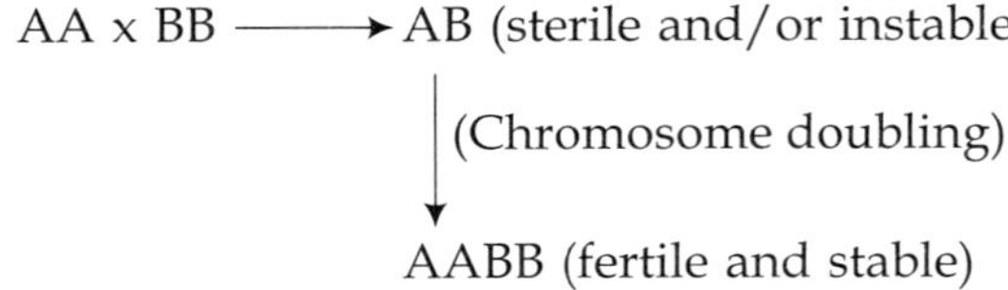

Whenever there is problem of meiotic instability, it is improved through continuous selection of genotype with improved fertility in the inbreeding generations of the newly synthesized allopolyploid. Where two species can cross easily and F1 produced is fertile, genome manipulation is easy but where crossing is not possible, i.e. where there is incompatibility barrier or the frequency of F1 produced is very low there different methods for increasing hybridization rate that can be employed.

Applications of somatic cell genetics: There are many applications of somatic cell genetics. Embryo culture can be used to rescue hybrid embryos in the event of endosperms failure. Mass culturing of higher plant cells can be done to obtain special plant products just like antibiotic synthesizing micro-organisms. In case of asparagus super males(YY) can be generated through microspore culture followed by differentiation of seedlings. Super males when crossed with normal females(XX) will yield progenies which are male(XY). These males give high yield(about 20% more spears than female as they do not produce seed). Cell and callus culture can also be used to show the genetically determined reaction of plants to pathogens. In other words, it can be used to screen for resistance to pathogens. It has generally been observed that the extreme tip (about 0.1mm) of plant apices are free from viruses even in infected individuals. Thus when minute tips are isolated and cultured under aseptic conditions, virus-free plants can be obtained. This apical meristem culture has been successfully used in producing virus-free foundation stocks in several species such as green pepper, strawberry, cassava, potato, geranium, chrysanthemum, orchard trees, etc.

12.1.1 Methods for Increasing Hybridization Rate

The different methods for increasing the rate of hybridization are as follows.

1. Somatic cell fusion
2. Embryo culture
3. Bridging cross
4. Bud pollination
5. Cutting styles or removing stigma
6. Application of growth substances
7. *in-vitro* pollination- unfertilized ovule culture
8. Mentor pollen or mixed pollen
9. Immunosuppressive drugs

For details of methods 1, 2, 7 see Roy (2000). **Bridging cross**- Suppose the two species with different genomes say A and B do not cross then in that situation species with A genome is crossed with a different species say with C genome with it crosses and F1 (AC) is obtained. This F1 (AC) upon chromosome doubling produces anamphidiploid (AACC). This amphidiploid (AACC) now crosses with species, BB. Thus B genome is combined with A genome through the use of C genome and here species with C genome is called a bridging species.

AA x BB ⟶ No F1 produced

AA x CC ⟶ AC ⟶ AACC x BB ⟶ ACB ⟶ AABBCC

In case of tobacco, nematode resistance from *N. repanda* was transferred to *N. tabaccum* with the help of bridgine species, *N sylvestris*.

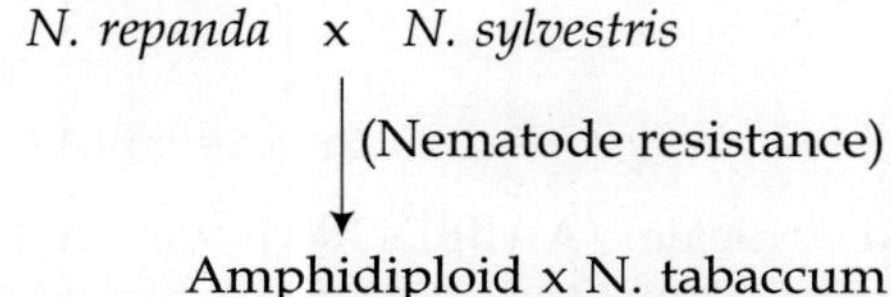

Bud pollination - The principle underlying bud pollination is that the plant determining incompatibility coincides with the opening of flower, i.e. , the enzyme/protein determining incompatibility is present in adequate quantity at this stage of development to inhibit self-pollination and so pollination at either bud stage or at the end of flowering could result in the setting of seed.

Cutting styles and removal of stigma - The substances responsible for incompatibility is present in either stigma or style or in both, and thus compatibility can be brought about by removing or scratching the stigmatic surface or by cutting the style.

Application of hormones - The application of hormones such as IAA (Indole acetic acid) which inhibits flower abscission thereby enabling the slow growing pollen tube to reach the ovary before flower dropping.

Use of mentor pollen - Mentor pollen may either be contributing the protein necessary for the compatibility reaction to occur or providing the growth substances essential for the growth of the incompatible pollen. Dead pollen has also been shown to show mentor effect.

12.1.2 Synthetic Allopolyploids

Although there are an number of crops such as cotton, wheat, Nicotiana, Brassica, banana, etc which are allopolyploids in nature, new allopolyploids can be synthesized and tested experimentally for its adaptability and usefulness. The problems associated with synthetic allopolyploids are: 1. Hybridization 2. Developmental problems. There is problem of combining useful features of two species besides fertility and stability. Fertility is affected by the meiotic behavoiur which changes because of the result of multivalents and univalents formation. There are many crops where synthetic allopolyploids have been developed.

1. **Triticale** - Triticale was developed using wheat (*Triticum aestivum*) and rye (*Secale cereale*). Triticale combines the winter hardiness of rye and yield and nutritional quality of wheat. Triticale had high number of spikelet number per spike and higher grain lysine content derived from rye and had large number of fertile florets per spikelet and greater grain size derived from wheat. It is developed by crossing hexaploid wheat with rye, the diploid which produces F1 which upon colchicines treatment produces the triticale with genomes from

wheat and rye as follows.

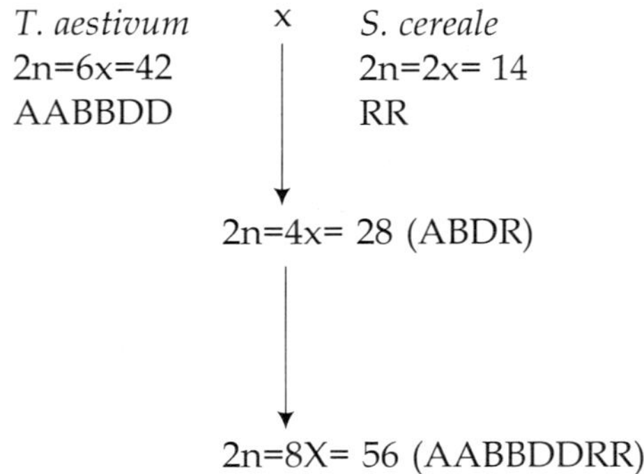

The raw octoploid triticale (AABBDDRR) was not adaptable, had inadequately developed grain endosperm, premature sprouting of grain, low frequency of tillering , had reduced fertility and low yield level because of aneuploids (chromosomal instability) which upon prolonged selection reduces to AABBRR (hexaploid triticale) which is stable. Hexaploid triticale was better than octoploid triticale as it had fewer aneuploids.

Forage grasses (*Lolium festuca*) - In this forage grass both interspecific crosses and intergeneric crosses have been attempted. In case of first interspecific cross, *L. multiflorum* is crossed with *L. perenne*, the closely related species and in the second interspecific cross, *L. perenne* is crossed to *L. italicum* and allopolyploids have been produced as follows.

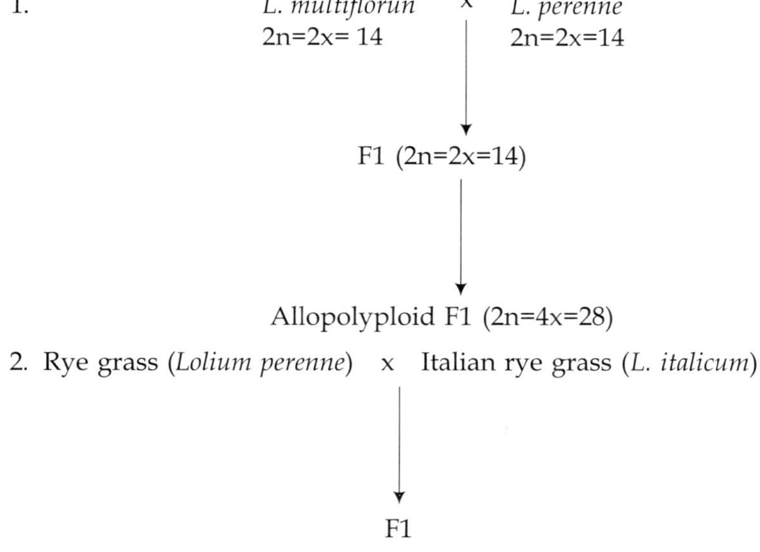

Amphidiploid (Tetraploid)

This hybrid rye grass has early rapid growth and high nutritive value derived from Italian parent, persistence of perennial rye grass and greater genetic stability (Breese et al, 1975)

The F1 was fully fertile and F2 showed wide range of variation. There were production of multivalents, because of segmental allopolyploid. Thus recombination and segregation results in the production of unbalanced gametes thereby resulting in unfertility. In case of allopolyploids because there is preferential pairing in p with p and m with m, the number of bivalents found was very high enough. In another allopolyploid developed from interspecific cross of *L. perenne* with *L. temulentum*, the presence of B chromosome, the supernumerary chromosome in the amphidiploid suppressed the homoeologous pairing to regularize and stabilize this particular hybrid combination (Evans and Macefield, 1972). But they found that the transmission of B-chromosome is irregular and for B-chromosomes to be effective they would have to be incorporated into A chromosomes. 2. **Intergeneric cross**- In case of intergeneric cross between *L. multiflorum* (more vigorous, nutritionally better) and *F. arundinacea* (winter hardiness), (Lewis, 1966) the allopolyploid developed showed partial restoration of sterility, meiosis was irregular and there is production of multivalent and univalent.

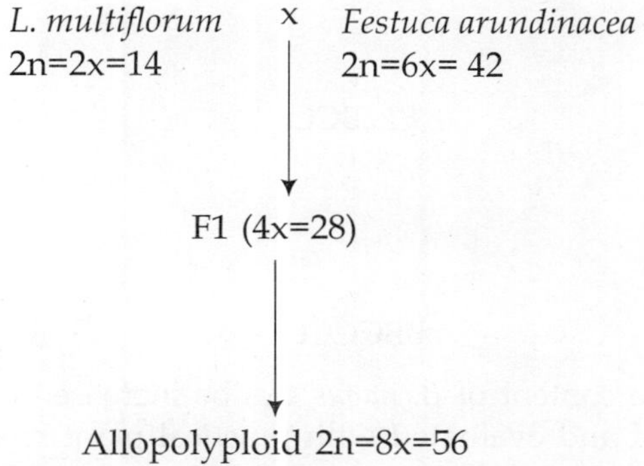

Here intergeneric cross pairing results in the recombination of specific characters but at the expense of regular assortment of the two sets of chromosomes at meiosis. Irregularity of meiosis results in loss of chromosome in subsequent generations of the amphidiploid and thus leads to aneuploids. The ideal amphidiploid is one in which no interspecific pairing occurs. In case of oat cultivated hexaploid oat can be crossed to wild hexaploid and gene transfer can be accomplished through backcrossing but when diploid and tetraploid Avena are crossed with hexaploid cultivated oat the F1 is sterile and so the regularity of meiosis does not necessarily ensure that the amphidiploid combinations will be successful breeding material since such combinations are not always physiologically harmonious. Genetic interactions brought about by combining genomes can be deleterious. This is happening in some combinations of Lolium/Feastuca group.

Brassica (Cruciferae) - Artificial *B. napus* can be constructed by crossing *B. oleracea*, cabbage and *B. campestris*, turnip in the following way.

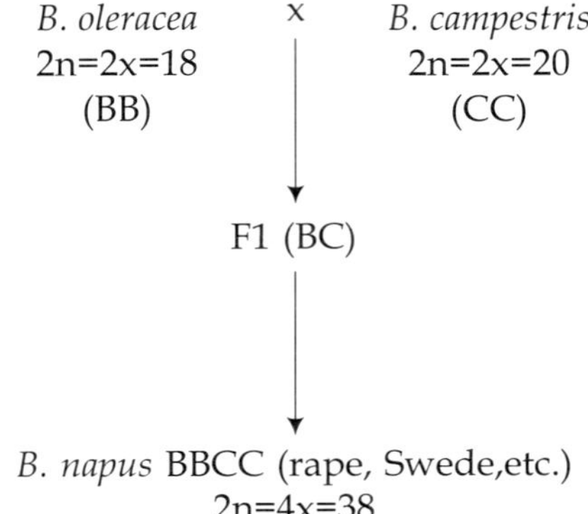

Thus resynthesis of B. napus can be done and which gives an opportunity for creating novel genome combinations, e.g. BBCCCC in the following manner.

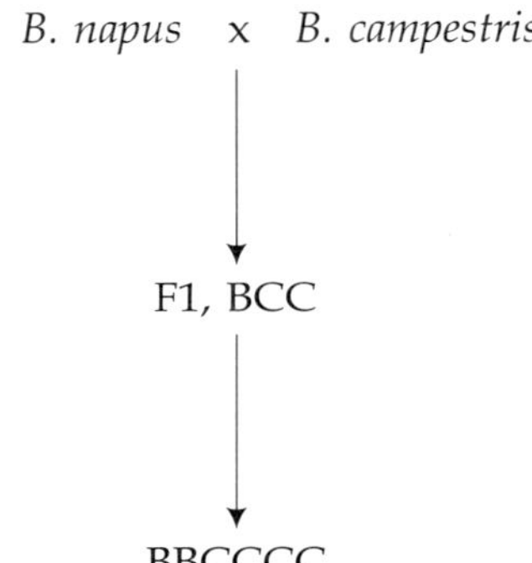

In this way C genome content of *B. napus* can be increased and thus different types of *B. napus* can be developed and evaluated with practical plant breeding objective of coming with a new crop. Further crossing between natural *B. napus* and synthetic *B. napus* can be made to introduce natural variation from diploid to tetraploid species.

Raphanobrassica - In another example of wide intergeneric cross *B. campestris*, kale has been crossed with *Raphanus sativus* (forage radish). Allopolyploid has been obtained in the following way which combines hardiness of kale with the quick growth and disease resistance of the fodder radish.

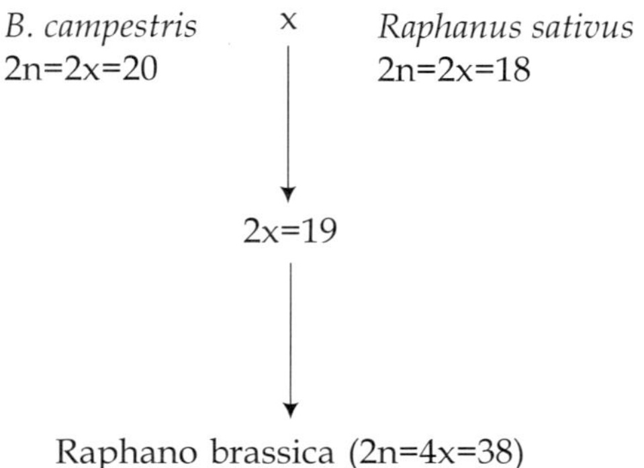

In case of amphidiploid, Raphano-brassica the chromosome pairing is regular, fertility is restored but the embryos formed abort before maturity.

12.1.3 Allotetraploidization

Allotetraploidization is the creation of artificial allotetraploids. It can be achieved by restructuring a genome so that its chromosome will not pair with those of normal chromosome (genome).It can be done in self-fertilizing species like maize, sorghum, rice, millets and other diploids. Allotetraploids will be superior to autotetraploids as 1. it would be a true breeding hybrids and 2. as no aneuploid gametes are formed, an allotetraploids population will not have the reduction in vigor and fertility due to aneuploidy found in autotetraploid populations.

12.1.4 Differential Pairing Affinity

When two species, say A and B show enough differential pairing affinity (DPA) to prevent the pairing of their homoeologous chromosomes in the autotetraploid (AABB) produced by the doubling of the chromosome number of the hybrid. DPA is the result of many factors. Chromosome pairing consists of three distinct but sequentially dependent processes. 1. Congressional pairing 2. Synapsis 3. Chiasmatic pairing. Congressional pairing is the movement towards each other of homologous chromosomes (or segments thereof) across intracellular distances. This probably occurs premeiotically. Synapsis is the very close alignment of homologous chromosomes which takes place at zygonema. The synaptic forces appear at the start of diplonema and the chromosomes are held together by chiasmata which were produced by crossingover. This chiasmatic pairing disappears at the start of anaphase. DPA factors can affect any or all of these processes. DPA factors fall into three different groups.1. Structural non-homology resulting from chromosome aberration such as deficiencies, duplications, inversions or reciprocal translocation. Chromosomal aberrations so small that they are not visible using the light microscope are called **cryptic structural changes,** are possibly common. 2. Qualitative or quantitative differences in the pairing code resides in certain palindromic segments of DNA (Sobell, 1975). The validity of this theory is not established. 3. The third type of DPA involves genes which control and modify chromosome pairing, the metabolic process is under genetic control. There are a large number of these genes known. One gene, *Ph,* is known in wheat that prevents the association of homoeologous chromosomes (Riley and Chapman, 1958).

The restructuring can be done by concentrating induced or naturally occurring visible and cryptic chromosomal aberrations and qualitatively different genetic material into a single line by recurrent selection type of breeding programme. There will be differential pairing affinity between normal and restructured chromosomes. Differential pairing factors occur naturally in exotic crosses and in standard corn belt inbred lines and that they may be readily induced by X-ray irradiation and chemical mutagens.

12.1. 5 Use of Unreduced Gametes in Plant Breeding

Many diploid plants from natural populations of the *Medicago sativa-falcata-coerulea* are good producers of 2n gametes. Similarly, the diploid hybrids obtained from crossing of haploids of the cultivated *Solanum tuberosum* (2n=2x 24) and wild Solanum (e.g., *S. phureja*) showed higher frequency of 2n gametes from male or female side. The production of unreduced 2n gametes by certain species (alfalfa, potato) can be exploited to develop heterotic tetraploids populations through BSP (Bilateral Sexual Polyploidization) and/USP (Unilateral Sexual Polyploidization) in practical plant breeding. The use of 2n gametes allows breeding at the diploid level which makes it possible to achieve highly heterotic combinations of genes whether they are due to multiple allelic interactions or to accumulation of favourable linkages, more rapidly than would breeding at the tetraploid level (Mc Coy and Walker, 1984). In addition to the production of tetraploid populations via BSP , 2n gametes may be useful for other breeding purposes such as production of highly heterotic populations and manipulations of certain qualitative traits such as disease resistance. The 2n gametes may also be used for gene mapping by half-tetraploid analysis and to explain the origin of polyploidy complexex like alfalfa. In alfalfa FDR is thought to be more suitable than SDR for transferring the selected parental genotype to the tetraploid progenies. The screening techniques for isolation of plants producing 2n gametes includes use of pollen morphology (pollen diameter) and seed set of hybrids (average seed set is used as a measure of 2n gametes production) and a combination of cytological, morphological and molecular analyses. Sexual polyploidization has the advantages of providing heterosis, variability, minimal inbreeding, high fertility and maximum heterozygosity which leads to new intra and inter genic interactions. (Peloquin, 1981). For detail on application of 2n gametes see Chapter 13.

12.1.6 Use of Haploids

Haploids can be produced in crops such as potato, tobacco, wheat and improvement can be made using plant breeding methods and again polyploids can be resynthesized. For details see Chapter 7.

12.1.7 Trisomics

Ramage (1964, 1965) and Khush and Rick (1976) described a method called **balanced tertiary trisomic system** by which one could use tertiaty trisomics with recessive male sterile genes (ms) for male sterile production for producing hybrid barley and tomatoes, respectively.

12.2 SUBGENOMIC MANIPULATION

12.2.1 Whole Chromosome Manipulation

Aneuploidy, both intra and interspecific, has proved to be of great value in crop species for chromosomal allocation and mapping of genes and of unknown-function genetic marker. Chromosome manipulation involving whole chromosome requires backcrossing programme. Whole chromosome transfer can occur in one of the following two ways. Either one whole chromosome from one species can be transferred to another recipient species or the whole chromosome from one species can replace the one whole chromosome of the recipient species and thus the resultant will be the production of either addition line or substitution line. The former will be aneuploid whereas the later will be euploid but unbalanced.

12.2.2 Addition Line-Method for Producing Alien Addition Lines (Wheat/Rye)

This method was developed in 1940 by 0′ MARA and by 1958 at PBI, Cambridge all rye addition lined were developed (Riley and Chapman, 1958). Hexaploid wheat was crossed to diploid rye and F1 was obtained which upon chromosome doubling by colchicines produced an amphidiploid (the octoploid triticale) which was backcrossed to wheat. Backcrossing is continued until BC6 and selection is practiced for plant with all wheat chromosomes plus one rye chromosome which upon selfing produces plants with all the wheat chromosomes plus one pair of rye chromosomes or alien chromosomes in disomic condition (see Figures 12.1A&B). The developed alien chromosome addition line is stable and true breeding. Meiosis is regular because pairing is normal. The alien chromosome addition line has both advantages and disadvantages and as yet no alien chromosome addition line has been used as a commercial variety. All 7 chromosmes of *Agropyron elongatum* were added to wheat (Dvorak and Knott, 1974). In barley out of 7, 6 addition lines have been produced (Islam, Shepherd and Sparrow, 1975). Using the same technique addition lines have been produced in oats, cotton and tobacco.

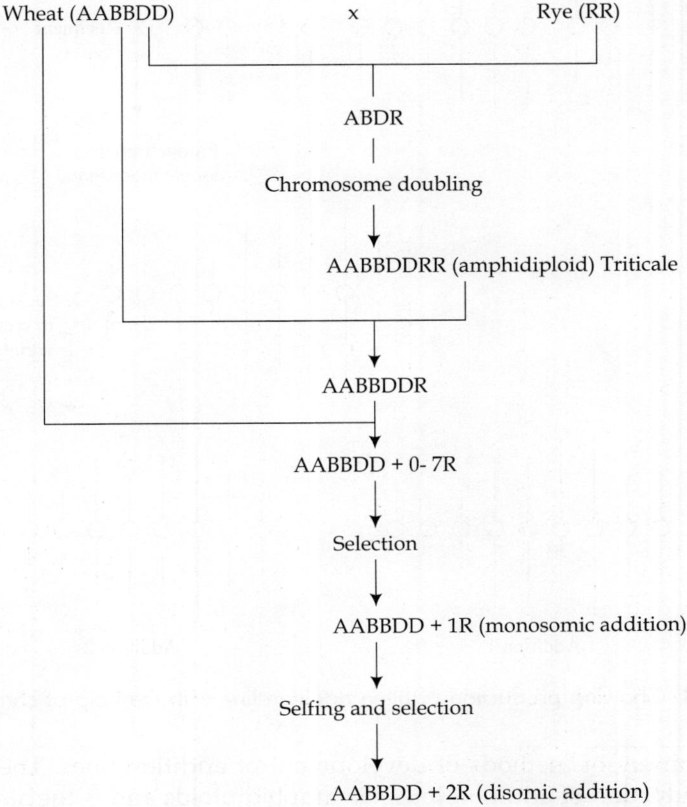

Fig. 12.1A Showing procedure for the development of alien-addition line.

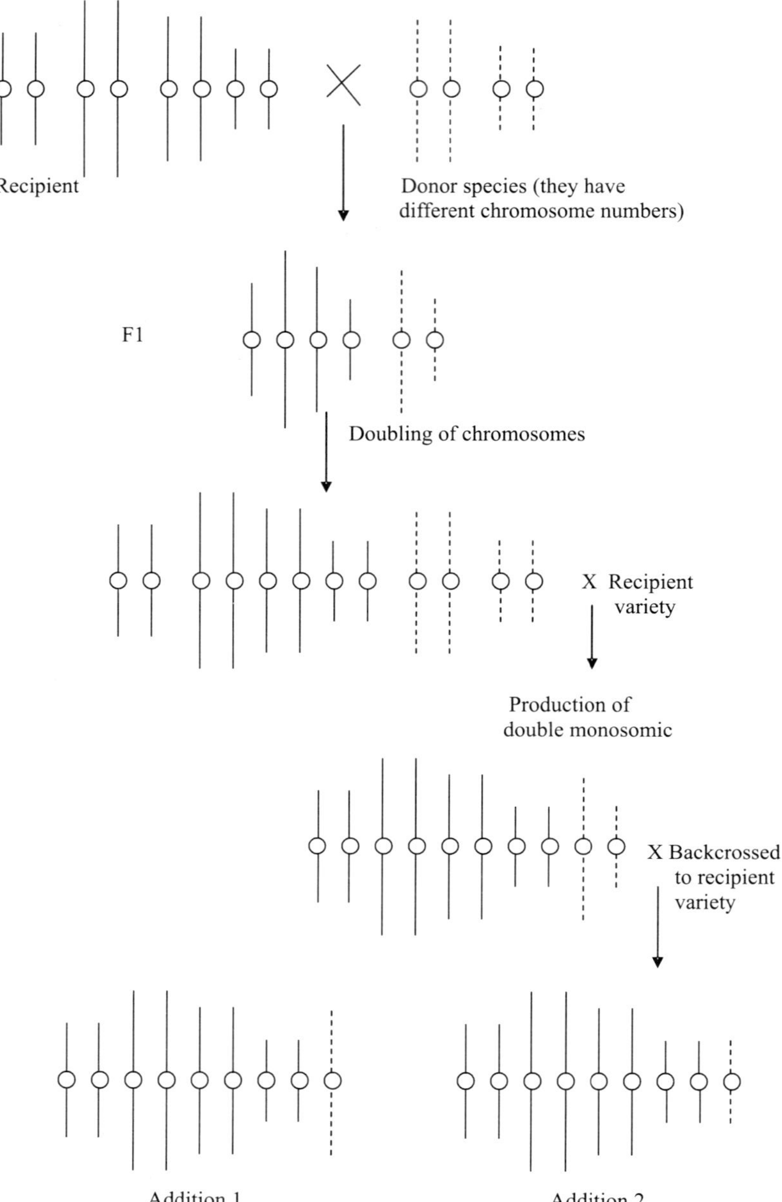

Fig. 12.1B Showing production of alien addition line with the help of chromosomes.

There are a variety of methods of development of addition lines. The above discussed method is the production of addition lines via amphidiploids and is the standard method of production of addition lines. Addition lines can also be developed through F1 hybrids, (e.g. wheat-rye (Florell, 1931); wheat-H. vulgare (Islam, Shepherd and Sparrow, 1975), via a bridging species (wheat-*Haynaldia villosa*- Hyde, 1975) and using the *H. bulbosum* induced chromosome elimination technique (Islam, Shepherd and Sparrow, 1975) and are shown in Figures 12.2a-c.

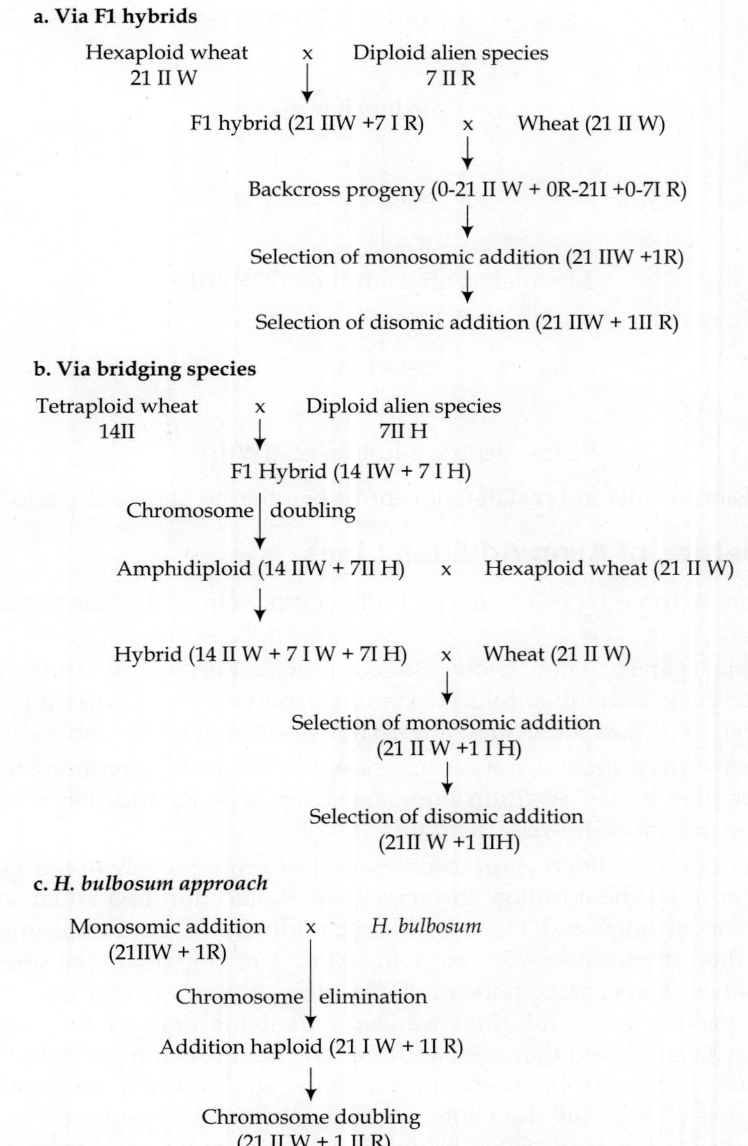

Fig. 12.2 Showing development of addition line through a. F1 hybrid b. bridging species and c. H. bulbosum technique.

12.2.3 Development of Detelocentric Addition Line

The ditelocentric addition line can be produced from the monosomic addition as given below.

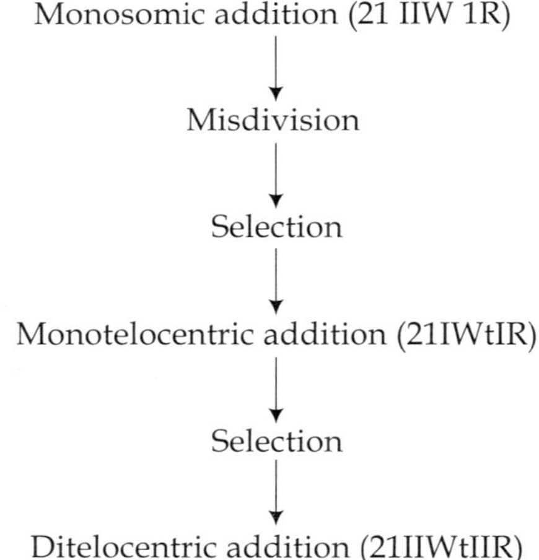

The misdivision results in breakage of centromere therby generating telocentrics.

12.2.4 Properties of Alien Addition Lines

Addition lines show three types of effects. I. Phenotypic effects II. Side effects III Addition decay.

I. **Phenotypic effects**- They show altered morphology, show character of disease resistance. They show manifold effects of phenotype. Sometimes they are similar to morphology of wheat tetrasomic lines. They also show undesirable effects.

II. **Side effects**- They show some serious side effectes such as reduced fertility. Among the wheat plus barley addition lines, most fertile is R_4 addition with 80% fertility whereas R_3 addition showed 2% fertility.

III. **Addition decay**- Addition chromosomes are lost progressively in few generations and thus we loose all the addition chromosomes. R_4 addition line when selfed produces 85% disomic addition and 15% monosomic addition. When monosomic addition line is selfed then it produces 75% euploid. After 4 or 5 generations one looses all the addition lines. The characteristic of all the alien addition is that they cause failure of chromosomes to pair and thus we get univalents and so they are lost. In the conclusion it can be said that alien addition line has little or no practical significance in Agriculture but then it is an intermediate step formed to produce desirable '**substitution**' lines. But then interspecific aneuploid types such as addition and substitution lines of complete chromosomes or chromosome arms can be used in genetic mapping studies for location of genes on chromosome. The alien addition and substitution lines can serve as donor parents in crossing programme aimed at transferring into a cultivated genome the smallest possible amount of alien chromatin exceeding a desired target gene.

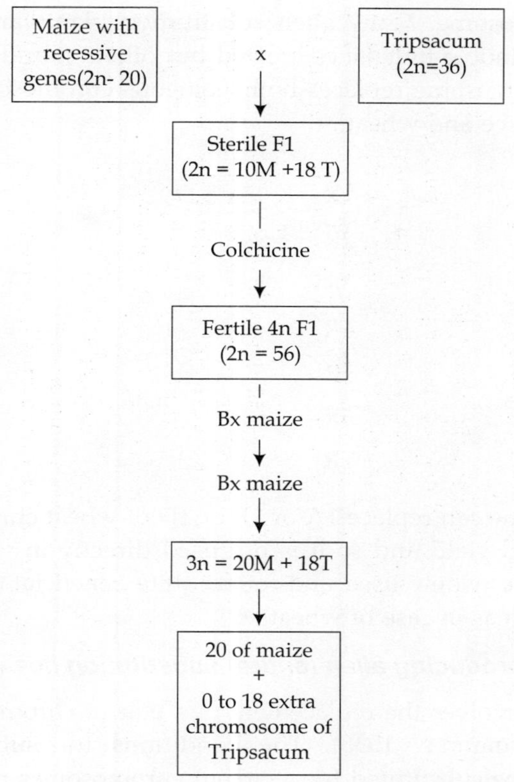

Fig. 12.3 Showing development of addition monosomic stock in corn.

In case of maize the various addition monosomic stocks of corn carrying chromosomes from Tripsacum were developed following Galinat, 1974 (Figure 12.3).

The addition bivalent is produced by self-pollination in which a 10+1 microspore fertilizes with a 10+1 megaspore producing 20+2 condition in sporophyte.

12.2.5 Chromosome Substitution

There are two uses of chromosome substitution. 1. Chromosome substitution can be developed to determine the effects of individual chromosomes in a variety when transferred to a common background generally through the use of monosome.2. Substitution of whole chromosome carrying desirable traits like disease resistance or insect resistance into an otherwise desirable variety can be made which requires a set of monosomic lines.

12.2.5.1 Properties

Many of chromosomes of *Secale*, *Agropyron* and *Triticum* (*Aegilops*) species are homoeologous with wheat chromosomes and each is able to compensate in the pollen for the absence of a particular wheat chromosome. Many alien substitution plants are poor plants with low fertility because of the double imbalance created but others found good, normal and have high fertility. Alien chromosome replaces homoeologous chromosome of wheat genome as shown below in case of rye and wheat.

1R	1 A B D
2R	2
3R	3
4R	4
5R	5
6R	**6**
7R	7

1R of rye chromosome can replace 1A or 1B or 1D of wheat chromosome. There is some reduction in fertility and yield and so it is not used directly in wheat but particularly in tobacco most of them are widely used and found quite beneficial as seed does not play an important role in tobacco as in case of wheat.

12.2.5.2 *Methods for producing alien (entire) substitution lines (wheat/rye)*

Alien-substitution line involves the replacement of a pair of chromosomes by a pair from a foreign species (Kattermann's, 1938). The conditions for substitution are: 1. Alien chromosomes can only be substituted for recipient chromosomes to which they are related and 2. the substituted and the substituting chromosomes will have evolved from the same chromosome of the common ancestor of the donor and recipient species.3. Much of their gene content must be conserved, i.e., similar in activity, linkage order and dispersion of introns. The technique involves development of an addition line (with a pair of alien chromosomes added to the complete wheat complement) and crossing this systemically with wheat monosomics.1. *controlled substitution* (Figure 12.4) 2. *uncontrolled substitution*

Weakness of method 2 It also produces other progeny in backcross with 21 bivs + 1 *univ* (21 W II + 1R I). 40W chromosome + 2 rye chromosome = 42 can arise spontaneously at very low frequency in allopolyploid particularly in the backcrosses. To overcome this problem one can use telocentric marker chromosome (t) in original cross. Critical progenies of backcross now have 1 het. bivalent. Further there is low probability of substitution. Wheat variety with replacement of 1B pair of wheat by 1R pair of rye variety has been developed that has been used most widely. Disomic substitution lines are more stable than disomic addition line. Substitution of whole chromosomes from related genera and species possessing disease resistance and other characteristics for chromosomes in economically important species may be difficult. Sears substituted an Aegilops chromosome carrying leaf rust resistance through backcrossing as follows.

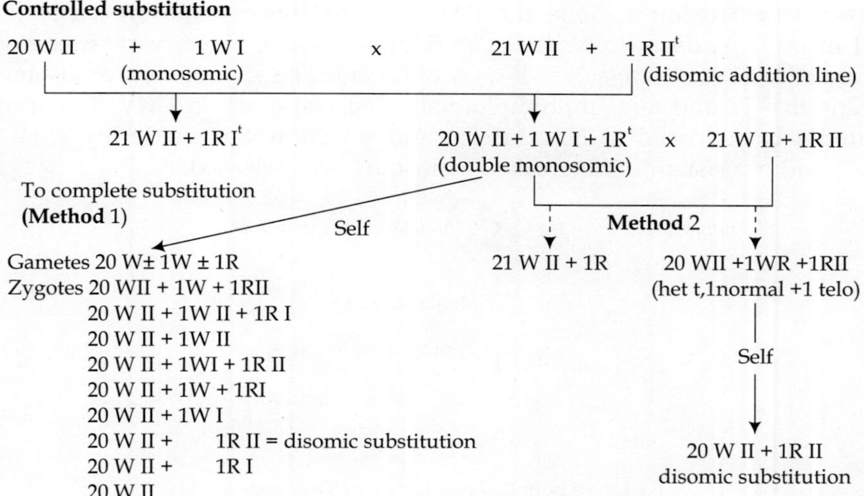

Fig. 12.4 Showing methods for producing alien (entire) substitution lines. 't' denotes telocentric chromosome.

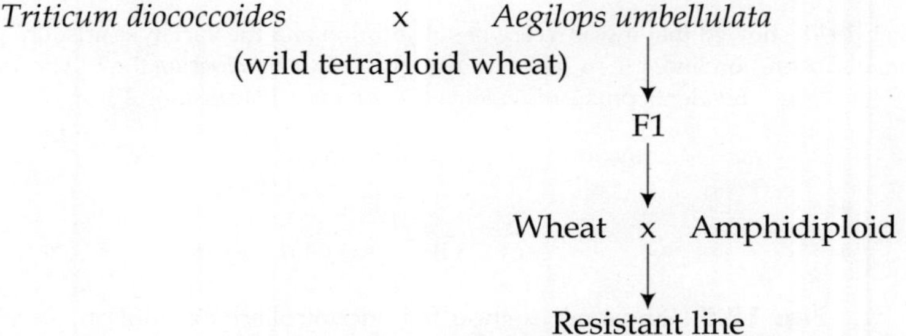

However the resistant backcrossed line also showed reduced vigor and fertility as a result of this alien chromosome.

Oat - In case of oats the *A. barbata* chromosome is substitution for *A. sativa* chromosome of the same homoeologous group. The *barbata* chromosomes are cytologically more close to their equivalents in Avena than rye chromosome to wheat. The substitution line is obtained in the following way.

Monosomic (2n-1) x Disomic addition line (2n +2) (20 II S + 1IIB)

↓ Pollen (20 IIS + 1IB)

F1 (20 II + 2 Is

↓ selfing

Substitution line (true breeding)
(20 IIS + 1 II B)

Uncontrolled substitution - Gene for resistance to tobacco mosaic virus (TMV) was transferred from *N. glutinosa* to *N. tabaccum* (Holmes, 1938). *N. glutinosa* was crossed to N. tabaccum and the F1 with 2n = 3x= 36 was obtained. The F1 upon chromosome doubling produced 2n= 6x= 72 and this amphidiploid showed resistance to TMV. This amphidiploid was repeatedly backcrossed to *N. tabaccum* and selection was made for TMV resistance (Figure 12.5) and a resistant variety called 'samsoun' was released.

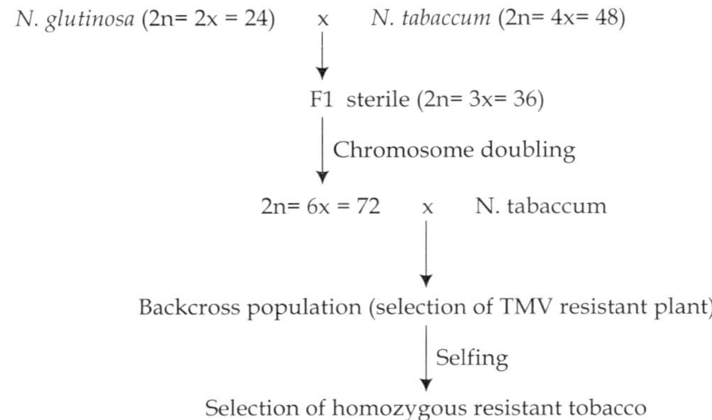

Gerstel (1940) showed that it was a case of substitution and the variety 'samsoun' was a disomic substitution line. When samsoun was crossed to *N. tabaccum* the F1 showed 23 bivalents plus 2 uinvalents (I *glutinosa* + I *tabaccum*).

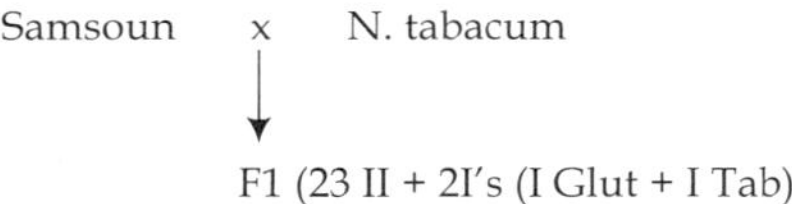

Fig. 12.5 Showing method for uncontrolled substitution.

3. Method for producing inter varietal (entire) chromosome substitution - It involves the transfer of a single unchanged pair of chromosomes of a donor variety into a recipient variety where it replaces the homologous pair. Development of inter varietal substitution lines is helpful in the study of hybrid vigor. What is required to have is the reciprocal substitutions for all the chromosomes whose hybrid has a substantial amount of hybrid vigor. Each chromosome can then be tested for its effect when either in two doses, in one dose monosomic or in one dose accompanied by its homologue from the other variety. Tests can be made against the background of each variety, also of their hybrids. Genes on individual chromosomes can be mapped and their effects assessed against the background of each parent. There are two methods of chromosome substitution. 1. Use of nullisomic 2. use of monosomic line.

1. Use of nullisomic - In this method disomic is crossed to nullisomic and the F1 obtained if monosomic which is backcrossed to nullisomic. The backcross population consists of nullisomic (4%) and monosomic (96%) from which monosomic is selected and again backcrossed to nullisomic and the backcrossing is continued until BC5. In the BC5 population

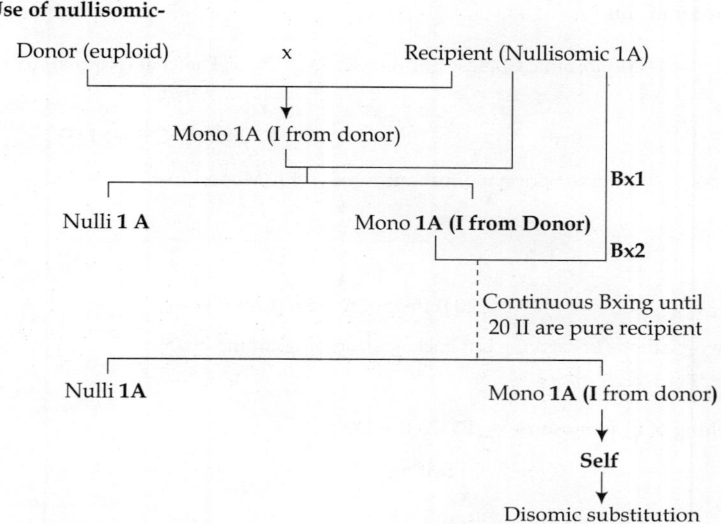

Fig. 12.6 Showing method for producing inter-varient chromosome substitution.

monosomic is selected and selfed which results in disomic, monosomic and nullisomic of which monosomic is selected and again selfed and thus disomic substitution is obtained (Figure 12.6).

2. Use of monosomic line - The monosomic line of recipient variety as recurrent seed parent is crossed with a donor euploid parental variety used as male. The resulting F1 monosomic offspring will have the donor monosomic chromosome but will lack the recipient homologue which upon continuous backcrossing will result in monosomic derivatives which upon selfing produces offspring from which a small proportion of disomic substitution lines can be selected (discussed in section). In actual practice the non-recurrent variety or the variety to be analysed chromosome by chromosome is generally crossed to each of the series of monosomes. The Figure 12.7 shows the development of chromosome substitution of 'Chinese Spring' chromosome by chromosome from 'Thatcher'.

Use of mono-telocentric or monoisochromosome - Substituting a particular chromosome from one variety for that of another is possible provided a monosomic series exists in one of the varieties and the substituted chromosome or the chromosome it replaces is marked by a readily identifiable gene. Because there is a lack of readily identifiable gene markers on most of the chromosomes of wheat, cytological chromosome markers which can be easily identified under the microscope can be used. There are two types of chromosome markers. 1. Telocentric chromosomes which are easily recognized in both mitotic and meiotic cell preparations and 2. isochromosome which can only be identified with any accuracy during meiosis. The combination of either of these cytological markers with monosomic condition, to give monotelocentric or monoisosomic lines, enables homologous donor chromosomes to be followed and maintained intact throughout a backcross programme. Use of telocentric or isochromosome in the monosomic condition instead of normal chromosome saves a selfed generation. The critical chromosome now derived from the non-recurrent parent is a normal

Use of monosomic line-

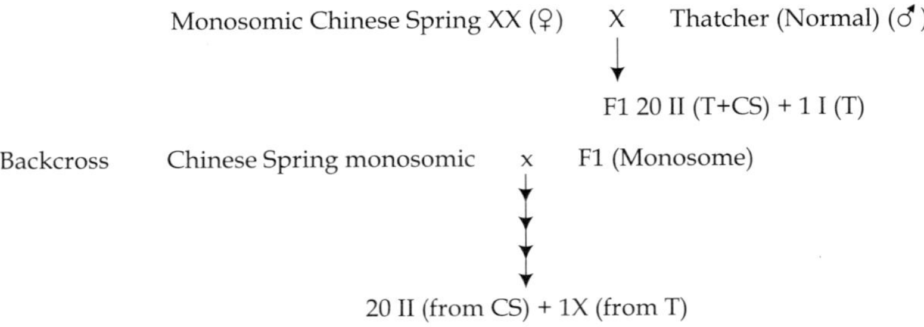

There are two methods of carrying out backcrossing programme

Method 1

Selfing of F1 monosome F1 (20 II + 1X)

↓ Selfing

Monosomic, disomic, nullisomic

(Selection of disomic homozygous for chromosome XX of Thatcher)

and Backcrossing Chinese Spring monosome (♀) x Disomic (♂)XX

↓

By alternating selfing and backcrossing 6 backcrosses can be achieved in 12 generations.

Method 2 This method is less precise and requires only 6 generations to achieve the same goal. In this method the Chinese Spring monosomic line is backcrossed to F1 hybrid used as male parent as shown below.

Chinese Spring monosome XX (♀) x Monosomic hybrid (20 II (C +S) + 1I (T))

↓ 6 backcrosses

¾ th progeny will be monosome but occasionally one in 75 of these monosomes will have CS chromosomes XX rather than the desired one from Thatcher and the probability of its occurrence at each backcross is only one in 12 (Elliott, 1958).

Fig. 12.7 Showing development of inter-varietal chromosome substituion using monosomic line.

chromosome whereas the same chromosome from the recurrent parent is a telocentric or isochromosome. After meiotic analysis of the progeny the plants with telo or isochromosome in the monosomic condition are discarded and only the monosomic plants with normal chromosome is selected and used in the next backcross. Where it is difficult to distinguish between recurrent and non-recurrent parents morphologically, telo and isochromosomes help in detecting selfs in the substitution programme. The various steps involved in the development of inter-varietal chromosome substitution line using telocentric chromosome is given in Figure 12.8. The recurrent parent is a mono-telocentric line and the donor variety is an euploid. The selection is practiced for monosomic plants in each hybrid generation for use as pollen parents in each generation of the backcross programme. The hemizygous chromosome in each of these plants will be the donor or substituted chromosome. After a

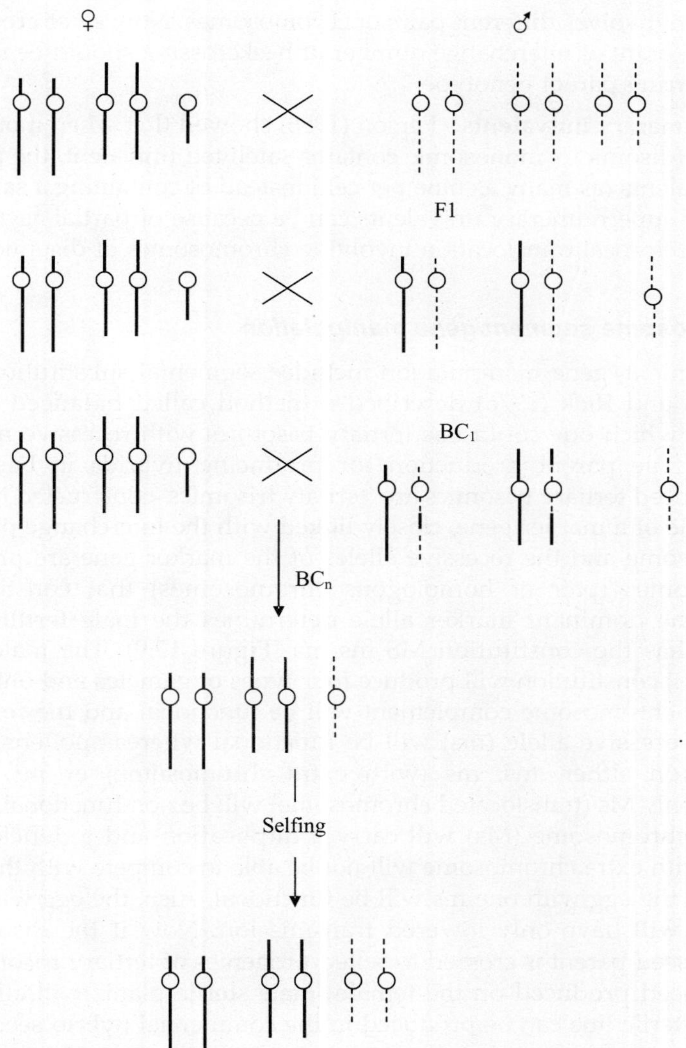

Fig. 12.8 Showing procedure for developing inter-varietal substitution line using as recurrent parent a mono-telocentric line.

number of backcross generations the selected monosomic plants are selfed and disomic plants are obtained.

Effects of reciprocal translocation in substitution programme - Wheat genotypes often differ in their karyotype by one or more reciprocal translocations. Reciprocal translocation has been detected in the intervarietal hybrids (F1's) as rings or chains of four when a single translocation is involved or as multivalents containing six or more chromosomes when more than two chromosome pairs are involved in the translocation or two rings or chains of four when translocation involves different pairs of chromosomes. Now since crossingover may be restricted near the point of interchange number of backcrossing should be increased in order to obtain the recurring parent genotype.

Effect of supernumerary univalents - Person (1956) showed that when monosome in case of the cross, normal (disomic) x monosomic contains satellited univalent, the progenies contain a number of univalents (as many as nine per cell) instead of containing a satellited univalent. The occurrence of supernumerary univalents can be because of partial asynapsis. It can also arise because of reciprocal translocation involving chromosomes of disomic and monosomic parents.

12.2.5.3 Chromosome segment/gene manipulation

Chromosome segment/gene manipulation includes segmental substitution. Ramage (1964, 1965) and Khush and Rick (1976) described a method called **balanced tertiary trisomic system (BTT)** by which one could use tertiaty trisomics with recessive male sterile genes (ms) for male sterile parent production for producing hybrids in barley and tomato, respectively. Balanced tertiary trisomics are 'tertiaty trisomics' constructed in such a way that the dominant allele of a marker gene, closely linked with the interchange point, is carried on the extra chromosome and the recessive alleles of the marker gene are present on the two normal chromosomes (pair of homologous chromosomes) that constitute the diploid complement. If the dominant marker allele determines the male fertility (Ms) then the tertiary trisomic has the constitution MS ms ms (Figure 12.9). The male plant with this genetic (Ms ms ms) constitution will produce four types of gametes and only the pollen with a normal haploid chromosome complement will be functional and the rest will abort. The pollen with one recessive allele (ms) will be functional whereas pollens with gametes of genetic constitution either, ms, ms (with extra chromosome) or ms, Ms (with extra chromosome) or only Ms (translocated chromosome) will be non functional. Pollen with only the translocated chromosome (Ms) will carry a duplication and a deficiency and will be inviable. Pollen with extra chromosome will not be able to compete with the normal type. In the female parent the egg with one ms will be functional. Also, the egg with Ms, ms will be functional but it will have only lowered transmission. Now if the disomic, male fertile (MS,MS) used as seed parent is crossed to selfed progenies of tertiary trisomic female parent then the hybrid seed produced on the female (male sterile plants)will all be diploid male fertile. The male sterile line can be produced in the commercial hybrid seed production plot using the selfed seed of balanced tertiary trisomic. Thus the female parents are the selfed progenies of trisomic plants which contain about 30% male fertile balanced tertiary trisomics (MS, ms, ms) and 70% male sterile (ms,ms) diploids as shown in Figure . Now if the extra translocated chromosome contains a second marker, say for example, a dominant gene for red plant color then all balanced tertiary trisomics (male fertile plants) will have red color and all diploids (male sterile) will have green plant phenotype and thus male sterile plants can be separated from the male fertile plants. Even when there is no extra color marker gene hybrids can be produced as the male fertile trisomics are weak, shorter and late flowering in barley and thus female parents produces almost pure stands (95 to 100%) of male sterile diploids. In order to guarantee complete crowding of the male fertile trisomic plants in the female rows the seeds from BTT plants are sown with very high seed rate (25 to 30kg/ha).

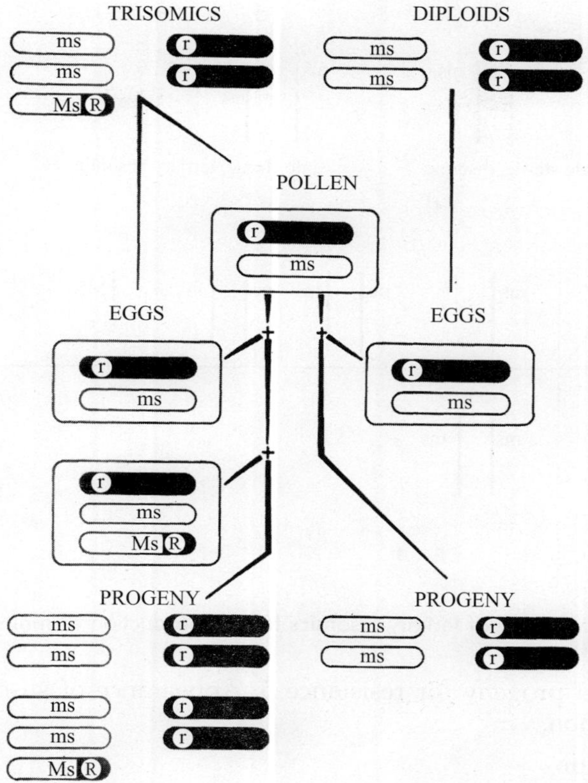

Fig. 12.9 Showing breeding behavior of a BTT marked with a dominant marker (Ms) (adapted from Ramage, 1965).

The Figure 12.10 shows the use of teritiary trisomics for the production of male sterile line is tomato.

2. Segmental substitution If there is imbalance due to substitution of one chromosome than one should go for 'segmental substituion'. Introduced alien chromosomal material should be limited to only that restricted segment necessary to incorporate in the recipient the single desired phenotypic modification from the donor. There are two methods for producing segmental substitution. 1. **controlled** and 2. **uncontrolled** (spontaneously occurring in backcrossing programme). Controlled way- Segmental substitution can be produced in controlled way. In wheat and oat they have been produced. There are two ways in which they can be produced. 1. Induced "somatic" translocation between wheat and alien chromosome 2. Induced homoeologous meiotic recombination. The process involved is as follows.1. Production of alien addition line 2. Production of translocation between wheat and alien addition and increasing the probability of translocation. Various steps involved in the somatic translocation are as follows (Driscoll and Jensen, 1963). 1. Synthesis of disomic addition 2. Irradiation of dry seed 3. Growing irradiated seed to maturity and selfing

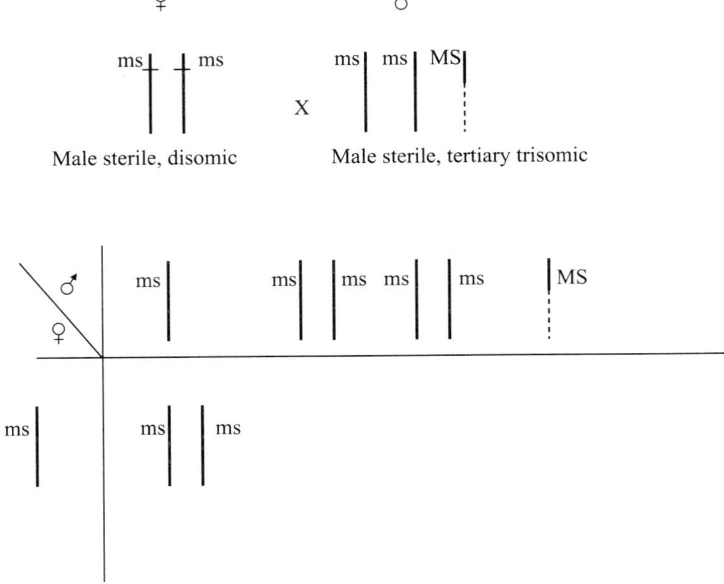

Fig. 12.10 Showing use of tertiary trisomics for the production of male sterile in tomato.

4. Screening of selfed progeny for resistance 5. Appearance of susceptible segregants is evidnce for translocation.

Controlled substitution

1. Induced ' somatic' translocation between wheat and alien chromosome

Sears' (1956) scheme for transfer of brown rust resistance from *Ae. umbellulata* (Figure 12.11) - Chromosome breakage may be induced by ionizing radiation. The breaks may then rejoin to produce novel chromosomes (translocation). In most cases there is replacement of a terminal segment of a wheat chromosome with a segment of an alien chromosome and the probability of replacement of an intercalary segment of a wheat chromosome by a similar segment of an alien chromosome is very less as it requires several simultaneous breakage and reunion events. This is a random process and the resulting translocations are likely to be deleterious. There is evidence of preferential occurrence of reciprocal translocation (Knott, 1968) in which a desired alien chromosome segment replaces an equivalent segment of a homologous wheat chromosome and this type of reciprocal translocation is likely to be advantageous. The chromosome carrying the gene (s) for leaf rust was first added to wheat. Sears made use of X-ray induced translocation in monosomic addition of an alien chromosome causing a qualitative change in the phenotype (plants carrying the extra chromosome were irradiated before meiosis and their pollen used in the crosses) and crossed it with euploid (the brown rust susceptible variety) and in the F1 progeny individual with brown rust resistance was selected. This selected plant will carry the chromosome translocation (resistant F1 plants were studied cytologically for the presence of translocations) in which a segment of the alien chromosome carrying the useful gene had

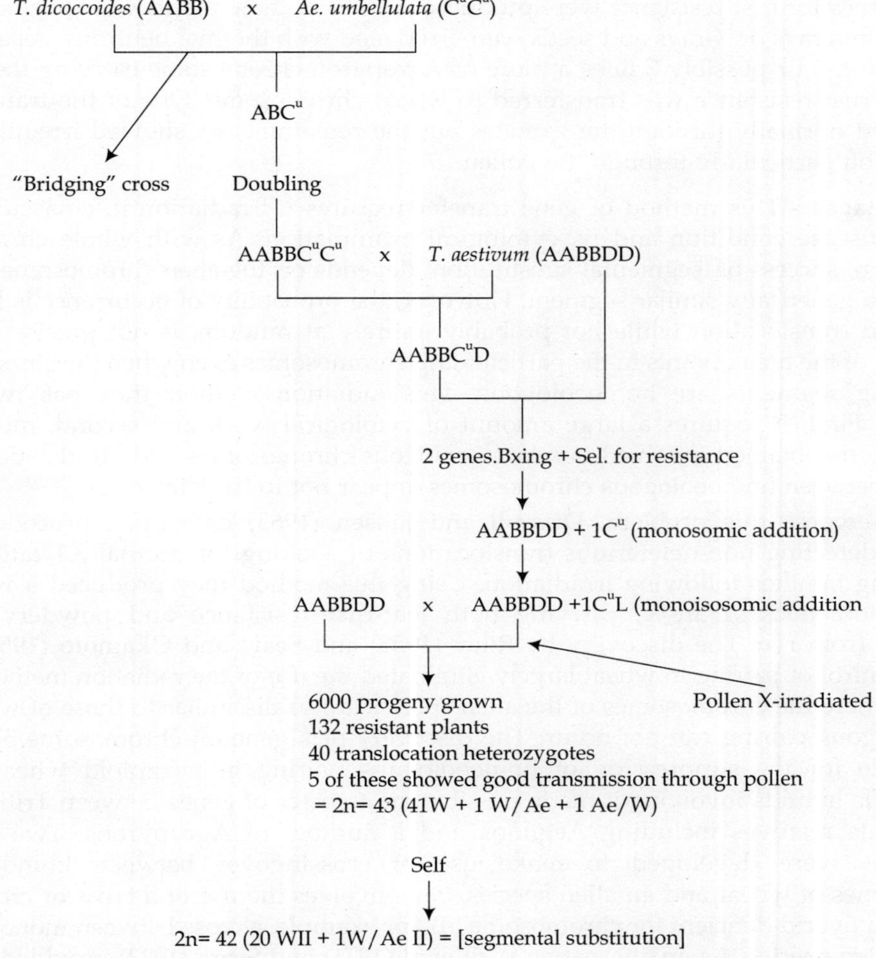

Fig. 12.11 Showing method for transfer of resistance from Ae. umbellulata.

been incorporated in a wheat chromosome. The selected plants are selfed and in the progeny selection is practiced for obtaining translocation chromosomes in the homozygous condition. Thus he transferred gene for brown rust resistance resistance from *Ae. umbellulata* to wheat. This translocation homozygote can be used in hybridization programme in order to transfer of this resistance into commercial variety. The feature of this method is the elimination of the unchanged Aegilops chromosome. This occurred through the loss of the univalent at meiosis and because pollen carrying the unchanged chromosome was at a competitive disadvantage. Thus very few of the functional male gametes carried the Ae. chromosome and the frequency of translocation was high in the resistant plant. Knott (1961) used the same method for transferring leaf rust resistance from an Agropyron chromosome to a wheat chromosome. Resistant wheat plants having 21II W plus a single added Agropyron chromosome having gene or genes for rust resistance were produced. Spikes of these plants were irradiated with either gamma rays or X-rays and seeds were irradiated with thermal neutrons. As a result of irradiation in 5 or possibly 7 lines a piece of Agropyron chromosome carrying the gene or genes for rust resistance was transferred to wheat chromosome. One of the translocation transmitted normally through the gametes but the remaining six showed irregularities in transmission particularly through the pollen.

Disadvantages - This method of gene transfer requires i. irradiation ii. emasculation iii. creating disease condition and iv. cytological examination. As with whole chromosome substitution, success of 'segmental substitution' depends on the alien chromosome segment replacing a genetically similar segment. However, the probability of occurrence is low since introduced translocation while not probably entirely at random, is not precise as to the homology of the break points in the participating chromosomes even when the chromosomes exchanging segments are homoeologous. This radiation method thus has two major problems. First, it requires a large amount of cytological work and second, most of the translocations obtained involved non-homoeologous chromosomes and are deletions. Only transfers between homoeologous chromosomes appear not to be deleterious.

To overcome this problem, Driscoll and Jensen (1963) developed procedures that involved detecting non-deleterious translocations by looking for normal 3:1 ratios in the segregating families following irradiation. Using this method they produced a wheat-rye translocation stock, Transec, carrying both leaf-rust resistance and powdery mildew resistance from rye. The discovery by Riley (1958) and Sears and Okamoto (1958) of the genetic control of pairing in wheat largely eliminated the use of the radiation method except in cases where the chromosomes of the alien species are so dissimilar to those of wheat that homoeologous pairing can not occur. The discovery of a gene on chromosome 5B largely responsible for the suppression of homoeologous pairing in hexaploid wheat and in interspecific hybrids involving it made possible the transfer of genes between Triticum and many of its relatives including Aegilops and a number of Agropyrons. Two types of procedures were developed to make use of crossingover between homoeologous chromosomes of wheat and an alien species. One involves the use of a cross or crosses that results in a hybrid deficient for chromosome 5B, for example, a cross between monosomic 5B and the alien species. As an alternative Wall et al (1971) and Sears (1972) provided mutants of the gene on 5B and these allowed homoeologous pairing in interspecific hybrids. The second method involved the use of a genotype of *Aegilops speltoides* which suppresses the effect of chromosome 5B (Riley and Kimber, 1966)

2. Induced homoeologous meiotic recombination

The discovery that lack of pairing between homoeologous chromosomes depends on pairing suppressors present in wheat chromosomes and that certain diploid relatives of wheat (*Aegilops speltoides*) have genes that are dominant to *Ph* provides opportunity to transfer genes from alien species to cultivated wheat. The problem with transfer of desirable traits from alien species to cultivated hexaploid wheat through addition or substitution of an entire alien chromosome is that almost always there are agronomically undesirable effects. This problem can be solved by transfer of desired alien chromosome segments through radiation as exchange of parts between chromosomes can be induced by ionizing radiation. But the problem again is that alien segment can not replace a random segment of wheat as it might contain essential genes and so in practice the satisfactory transfer strongly tends to involve substitution of the alien segment for a homoeologous wheat segment although with a very low frequency of transfer. Further with the discovery of *Ph* gene it became obvious that transfers from alien chromosomes might be effected through induced homoeologous pairing. Various techniques for using homoeologous pairing ranging from simply producing an interspecific or intergeneric hybrid lacking monosome 5B to synthesizing plants monosomic for a single alien chromosome, monosomic for a homoeologous wheat chromosome and nullisomic for 5B are possible (Sears, 1972a). In case of inter specific or generic hybridization there is lack of meiotic chromosomes pairing and recombination between chromosomes of crop species and those of other species from which a useful gene is to be sought. This meiotic barrier is due to a single gene at the *Ph* locus on 5B in wheat which acts to prevent genetically related chromosomes that are fully homologous from pairing together at meiosis and so the removal of *Ph* allele will cause homoeologous chromosomes from different species to pair which will lead to recombination. By this method segments of wheat can be replaced by genetically corresponding segments of alien chromosomes and thus size of alien chromosome segment can be minimized. There are two ways in which the pairing between homoeologous chromosomnes can be induced. One way will be to remove the 5B chromosome which carries the *Ph*1 gene which prevents homoeologous pairing and the second way will be to suppress the expression of this gene. Thus genes from alien species can be transferred to cultivated wheat. The homoeology exists between the genomes within the Triticeae such as those found in Aegilops, Agropyron, Secale and Hordeum which are possible donors of useful variation to wheat. This approach of transferring alien genes into the cultivated hexaploid wheat using *ph* gene is referred to as '**chromosome engineering approach**'.

Alien gene transfer by promotion of homoeologous recombination

Example 1 shows the transfer of alien genetic variation from alien species to wheat through the removal of 5Bchromosome.

Transfer of alien genetic variation (non-specific) from *T. bicorne* by **removal of 5B.**

The *Triticum aestivum* monosomic 5B, 2n-1 (5B) is crossed to *T. bicorne* (2n=14) and selection is made for plants with 27 chromosomes (20W plus 7A) but minus chromosome 5B and so there will be pairing between homoeologous chromosome of wheat (W)and alien chromosome (A), allowing recombination between alien genome and wheat genome. This particular plant is backcrossed to euploid wheat and from the progenies plant (s) with 35 chromosomes are selected which is again backcrossed twice to euploid wheat. In the backcross population plants with different number of chromosomes will be there and from which plant with chromosome number approaching 42 is selected and selfed for three generations and finally plants with 42 chromosomes are selected which are meiotically stable. These plants will contain segment of chromosome transferred from *T. bocorne* and will differ in agronomic traits in comparison to wheat euploid. The different steps involved in the segmental substitution are shown in the Figure 12.12.

Example 2 In this example segmental substitution induced by suppression of 5B is shown.

Transfer of yellow rust resistance from *Ae. comosa* by **suppression of 5B**.

This method uses monosomic addition line. Here monosomic addition (21WII + 1 *Ae. comosa*) is crossed to *T. speltoides* and in the progenies selection is made for plant having 29

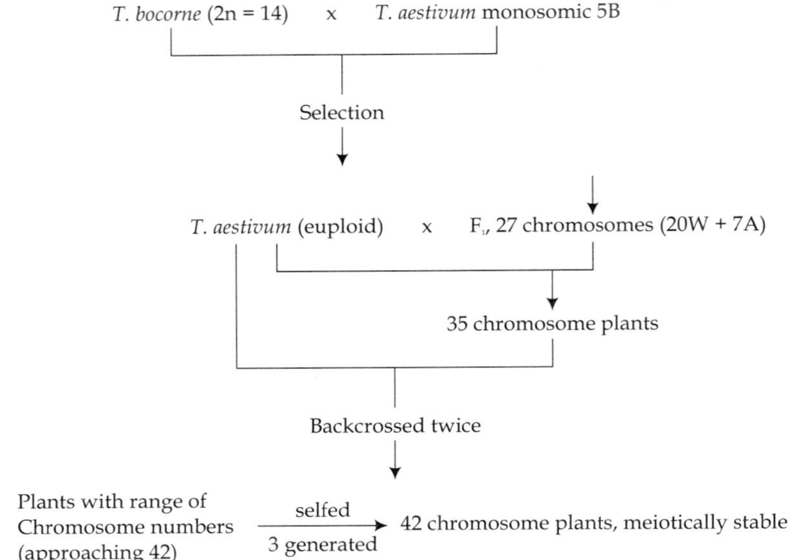

Fig. 12.12 Showing transfer of variation by removal of 5B.

chromosome (21 W + 7 *T. speltoides* = 1 *Ae. comosa*). The *T. speltoides* genome suppresses the expression of 5B and thus there is pairing between wheat and *Aegilops comosa* chromosome. This particular plant is backcrossed to euploid *T. aestivum*. In the backcrossed population selection is carried out for yellow rust resistant plants which are then selfed and finally a plant with 42 chromosomes having yellow rust resistance gene is selected. The different steps involved in the transfer of gene for yellow rust resistance from *Ae. comosa* are sown in the Figure 12.13. In the addition line the *comosa* chromosome does not pair with any wheat chromosome although it will substitute for chromosomes 2A, 2B and 2D.

The resistant 42-chromosome line was designated 'Compair'. In Compair, the resistance proved to be on wheat chromosome 2D showing that homoeologous pairing and crossingover had involved 2D and the *comosa* chromosome.

The second method, first, is producing directly good translocation at higher frequency in comparison to irradiation and secondly, undesirable linkage can easily be recognized again in this method. Both methods have been used sucessessfuly (Sears, 1972; Riley et al., 1968) but both have disadvantages. The most desirable method is a simple deletion of the *Ph* locus and this type of mutant has been obtained. The induced mutant appears to be a deletion of the gene (s) on the chromosome 5B that suppress homoeologous pairing (Sears, 1977). The mutants have several advantages over nullisomy for chromosome 5B: (i) reduced aneuploidy in the progeny (ii). higher fertility than nulli-5B, tri-5D and (iii). recombinants between chromosome 5B and its homoeologues can be obtained.

phI mediated gene transfer - The approach of transferring desired chromosomal segments between different Triticeae using phI mutants is referred to as **'Chromosome engineering'** and the greatest number of successful alien introgressions has been achieved is by far the

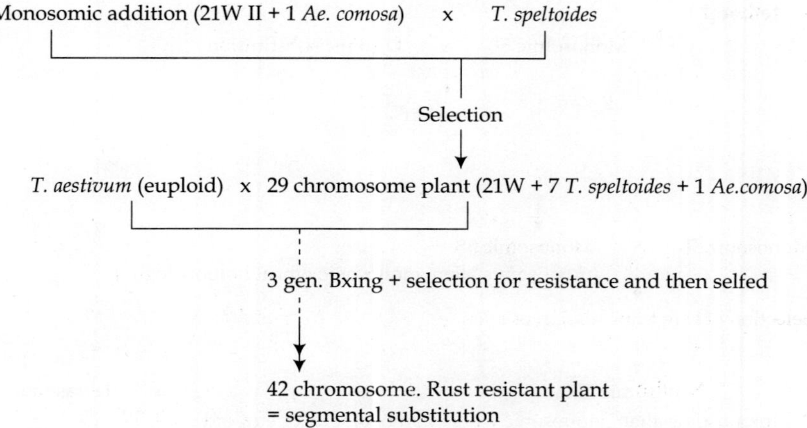

Fig. 12.13 Showing transfer of resistance by suppression of 5B.

hexaploid wheat, *Triticum aestivum*. Then attempt were made to introgress genes from *A. longissima* chromosome segment containing gene for powdery mildew resistance, Pm13 into tetraploid durum wheat which can in comparison to hexaploid wheat less tolerate chromosomal and genic unbalances. In this method homoeologous transfers involving A or B genome chromosomes are obtained first at the hexaploid level and subsequently passed into a durum background simply by homologous recombination. This method called 'second step' transfers may have less adverse effect on the recipient durum wheat genotype than those directly induced at the tetraploid level. Following this method Ceoloni et al. (1996) incorporated the Pm13 gene into the 3BS arm of durum wheat. The advantage of the *ph*I mediated approach in comparison with other transfer methods lies in the possibility to limit the amount of alien chromatin flanking the desired gene (s) up to a segment size that can well be tolerated even at the tetraploid level. Thus both the phIa mutant obtained by Sears in common wheat cv. "Chinese Spring" and the phIc mutant isolated by Giorgi in durum wheat have been employed. The transfer of Pm13 gene and other genes, namely Glu-DI and tightly linked genes, Gli-DI/Glu-D3, coding for storage proteins have been incorporated into durum wheats and introgressions have been confirmed by FISH analyses.

Use of substitution line - Sears (1972) suggested a more elegant method (Figure 12.14a and b) of the use of alien chromosome substitution in overcoming the effect of 5B. By use of a nullisomic 5B-tetrasomic 5D line plants can be obtained which lack 5B but are monosomic for both an alien chromosome and a wheat homoeologue. In this way the induced recombination can be directed from a single alien chromosome, known to carry a required gene, to a specific wheat chromosome. The extra 5D doses are required to improve fertility.

Although a number of alien addition and substitution lines have been developed in wheat but very few have proven to have any potential as commercial varieties. Alien chromosomes can generally only be substituted for a homoeologous wheat chromosome. Such substitutions are generally stable and occasionally show some promise. A number of European wheats carrying mildew and stripe rust resistance from rye have been shown to be either substitutions of rye chromosome 1R for wheat chromosome 1B or spontaneous 1B/1R

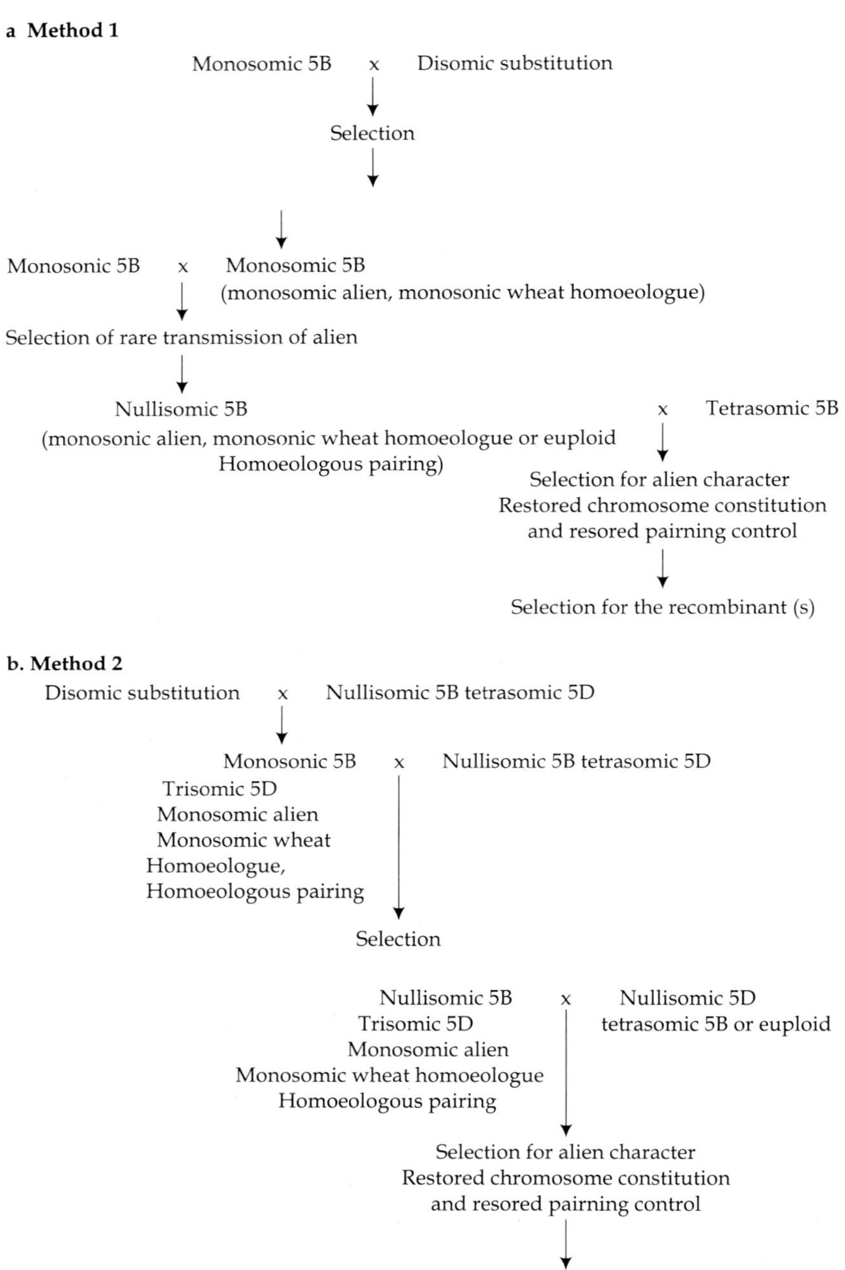

Fig. 12.14a-b Showing use of alien chromosome substitution in overcoming the effect of 5B.

translocation (Mettin et al., 1973). Sears (1972) suggested a method in wheat plant which is monosomic for two chromosomes, both monosomes may misdivide and rejoin to give a chromosome combining an arm from each. Such a method could provide a useful means of transferring an alien chromosome arm to a wheat chromosome. As wheat and rye chromosomes show little homoeologous pairing, it is likely that the spontaneous wheat-rye translocations mentioned above occurred by joining of alien chromosome arms (Knott and Dvorak, 1976)

Other methods of transfer of gene (s) from alien species into cultivated wheat-

Robertsoninan translocation - Besides the above discussed methods for transferring gene (s) from alien species to cultivated wheat the Robertsonian translocation can also be used.

Centric fusion - When the chromosomes occur as univalents at meiosis, occasional misdivision results in breakage of chromosome at the centromere thereby producing telocentrics. However if more than one chromosome misdivides simulataneously then arms of different chromosomes may join (Sears, 1973). If chromosomes with an alien chromosome of monosomic addition line are in univalent state then spontaneous whole arm Robertsonian translocation may occur between them. Robertsoninan translocation has been found between chromosome 1B of wheat and 1R of rye (Zeller, 1973; Mettin, Bluthner and Weinrich, 1978). This method of producing wheat-alien translocated chromosome for transferring gene (s) from alien species is not an ideal method as large amounts of undesirable and possibly detrimental alien genetic material are also introduced. But at present this is a useful technique in case of species where it is not possible to bring about pairing and recombination with wheat chromosome.

Use of cytoplasmic variation Alloplasmic wheat in which the cytoplasm has been derived from alien sources show considerable variation and the characters affected by introduction of an alien cytoplasm includes fertility, biomass, height, days to heading, ear characteristics and haploid production. The cytoplasms of *T. timopheevi* and *Ae. variabilis* confer male sterility.

12.3 DNA MOLECULE MANIPULATION-GENETIC ENGINEERING TECHNIQUES

For transfer of alien disease and insect resistance into crop species with least possible disturbance of genotype of the recipient species one should go for genetic engineering (see Roy, 2009). There are four gene pools, GP1, GP2, GP3 and GP4. Land races along with various wild ancestors of domesticated plants constitute Primary Gene Pool (GP1). Plants of related species constitute Secondary Gene Pool (GP2). Distantly related plants (from different genera) constitute Tertiary Gene Pool (GP3) and plants from other crop species or other organisms constitute GP4. Crossing between GP1 x GP1 produces fertile F1s. Also GP1 x GP2 (different species) produces F1 which is fertile. Genes transfer from either GP1 or GP2 to GP1 can be made through back crossing. GP1 x GP3 (different genus) produces F1 which is sterile and here use of tissue culture technique and embryo rescue can be made. In case of transfer of genes from gene pool of all other organisms such as other crop species, bacteria and other organisms (GP4) to GP1 which can not possibly cross with GP4 genes from GP4 can only be transferred through genetic engineering techniques.

2n Gametes and Its Applications

13.1 2N GAMETES

There are examples of plant species such as Datura, maize, potato and alfalfa where meiotic modifications result in the systematic production of 2n gametes (gametes with the sporophytic chromosome number). The 2n gametes could be either pollen or egg. Further, there are genotypes which either produce 2n pollen or both 2n pollen and 2n egg. Thus there can be either 'Unilateral Sexual Polyploidization' or 'Bilateral Sexual Polyploidization'.

13.1.1 Unilateral Sexual Polyploidization

When F1 hybrids obtained from the crossing Tuberosum dihaploid (2n=2x=24) produced through either parthenogenesis or anther culture and Phureja (2n=2x=24), the cultivated diploid potato from Columbia and Peru is crossed to the common tetraploid (2n=4x=48) potato, many tetraploid families are produced which are highly vigourous, showing heterosis (high yielder than the parents) and are also much more uniform than would be expected from the crosses between highly heterozygous parents. The high yields in the progeny from particular tetraploid x diploid crosses can be attributed to 2n pollen produced by the diploid parent(Mok and Peloquin, 1975a).

13.1.2 Bilateral Sexual Polyploidization

The F1 hybrids from Phureja x haploid Tuberosum upon intercrossing produced tetraploid as well as haploid individuals. Tetraploids from crosses between diploids can only be obtained when the diploid parents produce 2n pollen and 2n eggs. Mendiburu and Peloquin(1976) gave the term diplandroid and diplogynoid for 2n pollen and egg, respectively. The tetraploids obtained were more vigorous, highly heterotic and outyielded their parents by 30-40%. The highly heterotic response of diploid hybrids can be due to maximum intra and intergenic interaction

13.2 MECHANISMS FOR DIPLOID 2N GAMETES PRODUCTION

A number of mechanisms during the premeiotic, meiotic and postmeiotic stages of gamete formation may induce 2n gametes (Veilleux, 1985). The six distinct possible modes of 2n gamete formation are: 1. Pre-meiotic doubling II. First Division Restitution(FDR) III. Chromosome replication during meiotic metaphase IV. Second Division Restitution(SDR) V. Postmeiotic doubling and VI. Apospory. In potato FDR and SDR are the two important modes of 2n gamete formation. The meiotic restitution nucleus refers to a single nucleus with unreduced chromosome number, produced in place of two nuclei owing to failure of the first or second meiotic division (Rosenberg, 1927). The mitotic restitution nucleus refers to a single nucleus with tetraploid chromosome number generated in place of two diploid nuclei due to failure of mitosis(as is found in case of C-mitosis).

13.2.1 First Division Restitution(FDR)

In case of FRD in many cells the Met II spindles are parallel rather than oriented as they are normally where the poles of the two spindles define a tetrahedron. Ramanna(1979) and Veilleux et al. (1982) attributed FDR to fused spindles during the second meiotic division. The consequence of parallel spindles following one cleavage furrow at TeloII is production of two 2n microspores instead of the normal four n microspores (Figure 13.1). The Figures (13.2 and 13.3) show the genetic consequences for a pair of chromosomes of a diploid heterozygous(Aa) individual. The FDR leads to the production of two 2n microspores. The genotypes of the meiotic products will be determined by the presence or absence of a cross over between the locus and the centromere. In the absence of crossingover between chromatids carrying A and a there will be formation of 'No exchange tetrads' (NET) and all the gametes are of the genotype Aa (Figure). In case of crossingover 'Single exchange tetrads' (SET) are formed and two types of chromosomal orientation are possible. The first orientation will result in production of gametes with genotypes AA and aa whereas the second orientation will lead to the production of only one type of gametes with genotype Aa. Assuming β being the frequency of the SET the ratio of AA, Aa and aa gametes will be $\beta/4$: $(1- \beta /2)$: $\beta /4$.). Approximately 75-80% of the heterozygosity is transmitted from parent to offspring with FDR and also a large fraction of the epistatic interactions are maintained. The meiotic mutant, parallel spindles(*ps*) is inherited as a simple recessive(Mok and Peloquin, 1975c). The synaptic mutant, *sy*-3, in combination with 'parallel spindles' can produce FDR 2n pollen with no crossingover, i.e. for any gene $\beta = 0$. The mutant, sy-3 was found in *S. phureja*-dihaploid hybrids.

13.2.2 Second Division Restitution(SDR)

The first division followed by premature cytokinesis with occurrence of no second division leads to formation of two 2n microspores(Peloquin, 1975a). SDR mechanism is a simply inherited phenomenon(pc-1 and pc-2). The genetical consequences of SDR are shown in Figure 13.4. With NET the 2n gametes of genotypes AA and aa will be produced in equal

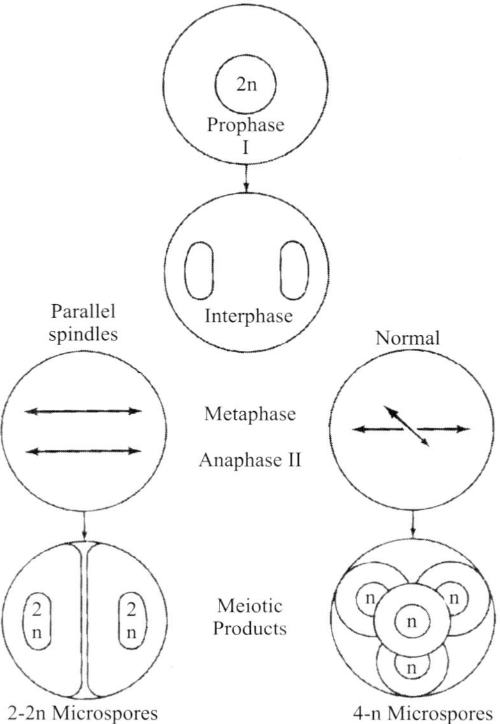

Fig. 13.1 Showing the comparison of meiotic product from normal and parallel orientation of second division spindles. Parallel spindles signifies the FDR mechanism.

frequency. Again two types of orientation are possible in case of SET (Figure) but in both cases the gametes produced are of genotype Aa. Assuming β being the frequency of SET the ratio of gametes AA, Aa and aa will be (1-β)/2 : β : (1-β)/2. In contrast to FDR all heterozygous loci from centromere to the first cross over will be homozygous in the gametes and with SDR only 40-45% of the heterozygosity of the parent is transmitted to the progeny. Thus in potato FDR hybrids showed 40-60% superiority over SDR hybrids in yield. The coefficient of β ranges from 0 to 1 for both FDR and SDR. When β = 0, chromosomal segregation occurs whereas randon chromatid segregation occurs when β = 2/3. Random chromatid segregation leads to the production of gametes with AA, Aa and aa genotypes in the ration of 1:4:1 with both FDR and SDR.

Parallel spindles, premature cytokinesis and synaptic mutants are the mechanisms for the production of 2n pollen whereas omission of the second meiotic division is the predominant mechanism of 2n egg formation. It is determined by s single recessive gene(*os*). Another mechanism is the failure of cytokinesis after the second meiotic division which is under control of a single recessive gene(*fc*). Both these mechanisms of 2n egg formation are genetically equivalent to second division restitution(SDR). The desynaptic gene, ds-1 has been identified in diploid clones which produce FDR 2n egg through a direct equational division of univalent chromosomes at anaphase I.

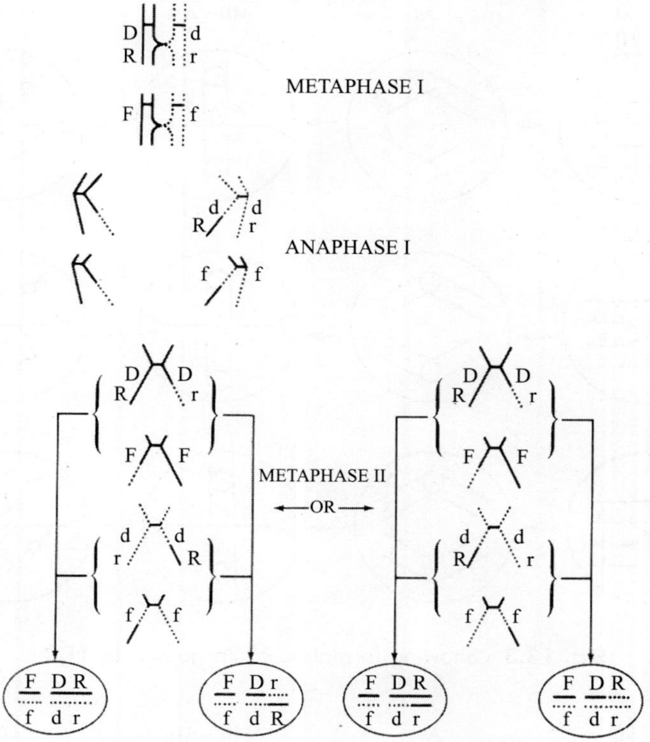

Fig. 13.2 Genetic consequences of FRD as a result of parallel orientation of second division spindles.

Diploid species and dihaploid- species hybrids in general produce both 2n and n gametes. Matings between 4x and 2x parents(4x-2x and 2x-4x crosses) produce almost entirely 4x progenies due to a 'triploid block mechanism'(Marks, 1966) whereas matings between 2x parents(2x-2x crosses) produce both 2x and 4x progenies. Frequencies of 2x and 4x progeny vary between interdiploid crosses. Finally, the 4x progenies from 2x-2x crosses can be identified on the basis of chromosomal counts or some other means such as establishing the mean number of chloroplasts in stomata(Wagenvoort and Zimnoch-Guzowska, 1992). In potato only the seeds that survive are mainly tetraploids. Thus the number of seeds obtained following a 2x-4x cross is a measure of egg production in the diploid.

Thus meiosis can fail in either the first or the second division. If pairing is poor and the divisions delayed, the chromosomes may not move far from the plate and may return to be included in a restitution nucleus(Walters, 1958; Wagenaar, 1968a). One can have situation in which first division is completed but it is followed by failure of cytokinesis in second division, then the chromosomes included in each daughter cell of MI are doubled in MII. Most unreduced gametes arise from failures in either the first or second meiotic division. The effects of two kinds of meiotic failure or restitution can strikingly different. If MI fails completely then a whole nucleus is formed with no chromatin excluded and if this is

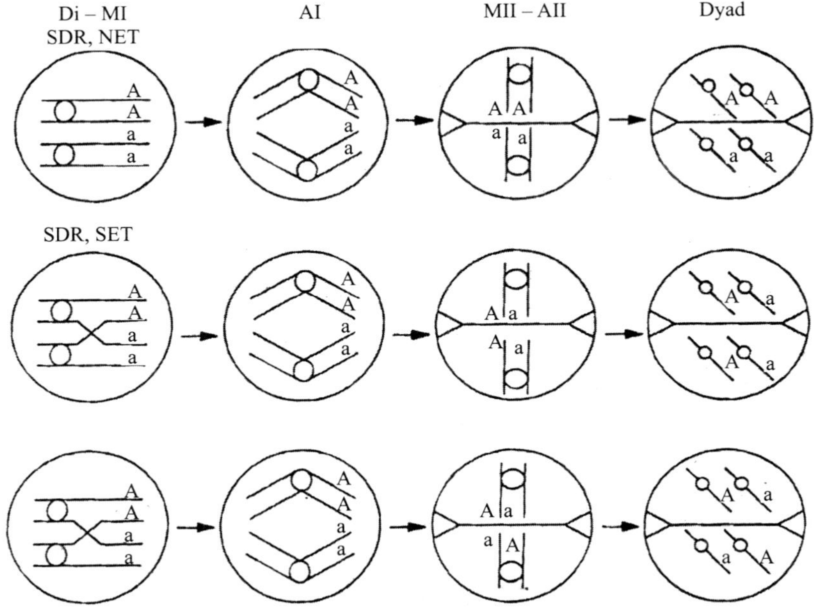

Fig. 13.3 Showing formation of 2n gamtes by FDR.

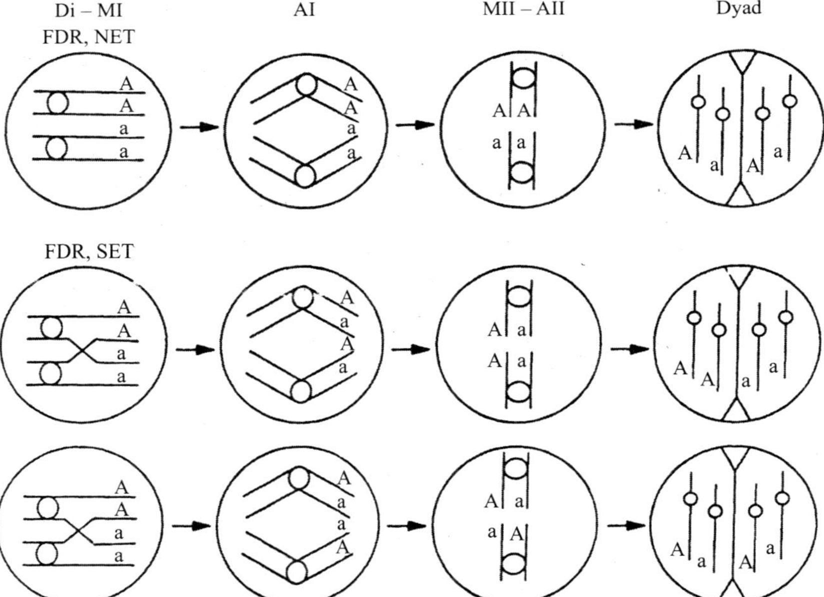

Fig. 13.4 Showing formation of 2n gametes by SDR. Di=Diakinesis; M=Metaphase; A= Anaphase.

followed by an equational division then all gametes will be similar and there will be no segregation. If MI is completed and cytokinesis fails in MII then there may be considerable chromosome segregation and each chromosome should have an exact partner. The consequences may be important in the next generation as the gametes will contribute full autosyndetic pairing capabilities (Harlan and deWet, 1975). One consequence is the stabilized 'triploid' line reported by deWet.

13.3 CONSEQUENCES OF SEXUAL POLYPLOIDIZATION

The genetic consequences of 2n gametes produced by 2x(FDR and SDR) and 4x parent are shown in Table 13.1 in terms of frequency and type of heterozygous 2n gametes produced by a heterozygous parent.

Table 13.1 Shows 2n gametes produced by 2x and 4x parents.

		2n gametes			
		Homozygotes		*Heterozygotes*	
Parent	*Genotype*	*Frequency*	*No.*	*Frequency*	*No.*
Tetraploid					
Monoallelic	$A_1A_1A_1A_1$	1	1	0	0
Unbalanced diallelic	$A_1A_1A_1A_2$	$(1+\alpha)/2$	2	$(1-\alpha)/2$	1
Balanced diallelic	$A_1A_1A_2\,A_2$	$(1+2\alpha)/3$	2	$2(1-\alpha)/3$	1
Triallelic	$A_1A_1A_2\,A_3$	$(1+5\alpha)/6$	3	$5(1-\alpha)/6$	3
Tetra-allelic	$A_1A_1A_3\,A_4$	α	4	$1-\alpha$	6
Diploid					
Monoallelic	A_1A_1	1	1	0	0
Diallelic(FDR)	A_1A_2	$\beta/2$	2	$1-\beta$	1
Diallelic(SDR_	A_1A_2	$1-\beta$	2	β	1

Hermsen (1984a) calculated that the percentage of the parental heterozygosity in FDR and SDR gametes for four types of chromosomes with known sites of centromere and crossovers are 80.2% and 39.6%, respectively. The mean and variance of FDR progenies are expected to be larger and smaller, respectively than those of SDR ones when a locus is close to the centromere. This is just the reverse when the locus is distal to the centromere. The 2n gametes influence much more in 2x-2x crosses than in 2x-4x crosses. Further the performance of FDR x SDR crosses falls between those of FDR X FDR and SDR x SER in 2x-2x crosses. Computer simulation study of Watanabe et al., (1991) showed that 2n gametes transmit about 80%(FDR) and 40%(SDR) of the parental heterozygosity to the 4x offspring whereas somatic doubling transmits all of the parental heterozygosity. Tetra-allelic and tri-allelic genotypes can arise in 4x progenies produced by 2n gametes. As in case of somatic doubling all of the loci are duplex or balanced diallelic genotypes are produced in 4x progenies it results in more inbreeding and fewer allelic interactions at a locus than in sexually produced 4x progenies. Amongst the three mechanisms of sexual polyploidization the FDR X FDR has less inbreeding and more genotypic diversity than SDR X SDR in 4x progenies when the locus is close to the centromere. SDRx SDR has in 4x progenies a lower degree of inbreeding than

FDR X FDR when the locus is distal to the centromere. SDR x FDR is the most stable mechanism against extreme recombination events and produces a high proportion of tri and tetra-allelic genotypes. Tri- and tetra-allelic interactions appear important in generating a heterotic effect in polyploid with tetrasomic inheritance.

In case of different types of bilateral sexual polyploidization, i.e. FDR x FDR, FDR x SDR and SDR x SDR crosses lead to the production of diallelic, triallelic and tetraallelic hybrids in different proportions as shown in Table 13.2. The FDR X FDR crosses produce more tetra-allelic progenies than the other two when the locus is within 30cM of the centromere. The FDR x FDR cross is favored over the other two types of crosses as there is higher probability of producing more genetically uniform hybrids because it has a higher frequency of tetra-allelic genotypes for loci located up to 30cM and a low frequency of diallelic genotypes for loci located anywhere in a chromosome (Tai, 1994).

Table 13.2 Shows types of crosses and production of hybrids.

Types of 2x-2x crosses	*Diallelic*	*Triallelic*	*Tetrallelic*
FDR X FDR	$\beta^2/4$	$\beta(1-\beta/2)$	$(1-\beta/2)$
FDR X SDR	$\beta(1-\beta/2)$	$(1-\beta)(1-\beta/2) + \beta^2/2$	$\beta(1-\beta/2)$
SDR X SDR	$(1-\beta)^2$	$2\beta(1-\beta)$	β^2

13.3.1 Significance of Sexual Polyploidization

Although meiotic modifications in uneven polyploids such as triploids, pentaploids lead to unbalanced gametes production the mechanism of parallel spindles endow plants with uneven ploidy levels with the ability to produce balanced gametes in a systematic fashion. The unique capacity of FDR to overcome numerical nondisjunction provides triploids with a mechanism to significantly increase male fertility (Mok et al., 1975) which otherwise would have reached the evolutionary dead ends. Mann and Sasakuma(1977) reported another mechanism of increase of fertility in which the F1 hybrid (2n=3x=21) from the cross, *Aegilops heldreichii*(2n=2x=14) x *T. durum*(2n=4x=28) showed high frequency of 2n male and female gametes and produced amphidiploid progeny. In about one-third of the microspores21 univalents were formed at Met I and divided equationally at Ana I. Telo I was followed by cytokinesis with no second division occurred and 21 chromosome gametes were produced.

13.3.2 Application of 2n Gametes

The production of 2n gametes provides new opportunities for genetic analysis with tetrasomic inheritance and for breeding highly heterotic tetraploid varieties.

13.3.2.1 Development of tetraploid hybrids

The development of tetraploid hybrids involve i. development of high quality 2n gametes producing parents, assessing their merits and ii.production of tetraploid hybrids from 4x x 2x crosses or 2x- 2x crosses and their evaluation. In case of potato the 2n producing parent is the dihaploid *S tuberosum*- diploid species hybrid. As *S. phureja* based 4x-2x hybrids represent a distinct group of genetical material in comparison to the *S. tuberosum* varieties the *S. phureja* based 2x parent should be improved. The diploid parent here serves as a bridge for introduction of resistances to biotic and abiotic stresses from other diploid species into the breeding material. In case of banana, Musa accessions chromosome doubling occurs naturally as a result of failure of first or second meiotic division and thereby production of 2n gametes. Chromosome doubling during megasporogenesis results in the production of 2n eggs and that is why 4x offspring are produced in the cross, 3x-2x(Vuylsteke, et al, 1993). Similarly chromosome doubling during microsporogenesis results in the production of 2n pollen(Asker, 1980; Ortiz, 1997a) and this suggests that 3x hybrids(2n= 3x= 33) can be derived via 2x -2x crosses where one of the 2x parents produces 2n pollen and thus is a case of unilateral sexual polyploidization. The USP derived triploid hybrids prove to be agronomically superior to diploid hybrids derived from the same cross.

13.3.2.2 Genetical study

The 4x-2x cross aaaa x Aa can be used for mapping study (Mendibure and Peloquin, 1979). The map distance is determined by measuring the frequency of nulliplex genotype(aaaa) in the tetraploid progenies. The 2x parent used for mapping should produce 2n gametes by either FDR or SDR but not a mixture of both. Wagenvoort and Zimnoch-Guzowska(1992) suggested the use of 2x-4x cross in the gene-centromere mapping study as in certain diploids SDR 2n eggs occur exclusively. Further mapping results for a gene from 2x-4x crosses can be used to distinguish between FDR and SDR pollen in 4x-2x crosses when the gene is located close to the centromere.

Cytoplasmic Manipulation

14.1 EMBRYO DEVELOPMENT

Endosperm is an important and unique tissue. It is a major source of food and feed. In 99% of the angiosperm it is necessary for normal seed development. It occurs only in angiosperm. It is triploid in genetic constitution and is a product of second fertilization and formed as a result of fusion of diploid polar cells and haploid sperm cell. As seed failure occurs in interploid crosses, interspecific crosses and during extraction of haploids it is worthwhile to study the basis of endosperm failures in interploid and interspecific crosses and the genetic mechanism of endosperm heterosis. Many hypotheses were put forward explaining the basis for normal seed development and its failure. Earlier hypotheses suggested the need for a particular balance of chromosome sets between maternal tissue, endosperm and embryo and Muntzing(1933) suggested a 2:3:2 ratio of the ploidy levels for the normal development of seed. Other research workers suggested either 3:2 ratio of chromosome sets in the endosperm and embryo (Watkins, 1932) or 2:3 ratio of maternal tissue to endosperm(Valentine, 1956). Researches from potato(von Wagenheim et al., 1960) and maize(Chase, 1964) suggested that the problems related to seed development and its failure are intrinsic to the endosperm and consequently Nishiyama and Inomata(1966) hypothesized the necessicity of a 2:1 ratio of maternal to paternal genomes in the endosperm for the normal development of seed. This 2:1 genome ratio hypothesis appears applicable to intraspecific, interploid crosses in most angiosperm (Lin, 1975). This hypothesis, however, does not explain the results obtained with interspecific crosses in potato and Avena and abnormal endosperm development occurs with a 2:1 balance of maternal to paternal genomes. In both groups the ploidy level of parents does not precisely reflect crossability, that is, some tetraploids behave as diploids in *Solanum* and some diploids as tetraploids in *Avena*. The hypothesis that has been proposed now envisages qualitative genetic factors rather than chromosome sets to be in a 2 maternal :1 paternal ratio for the normal seed development. These hypothetical quality factors have been labeled as endosperm balance numbers(EBN) in Solanum(Johnston, et al., 1980) and activation index(AI) in Avena(Nishiyama and Yabuna, 1978). When the EBN from female and male are in 2:1 there occurs normal development of tuber in potato and the deviations result in abnormal endosperm development. The AI in *Avena* is expressed as the ratio of the activating value(AV) of a male nucleus to the response value(RV) of the two polar nuclei. When AI is 0.5(AV/2RV) normal endosperm results and if this ratio deviates from 0.5 then endosperm is often abnormal, arrested or aborted.

14.1.1 Endosperm Heterosis

How the heterosis in the endosperm can be explained? The imprinting of gene action (see Roy, 2009) in maize demonstrated that the particular expression of a gene in the endosperm can be dependent on whether it came from the female or the male parent and thus different genes or gene combinations are derepressed in the polar nuclei and the sperm nucleus (Kermicle, 1978). Works with interspecific hybrids of wheat showed a factor which affected endosperm development only when transmitted paternally(Lin, 1975; Gill and Waines, 1978). Thus one can envisage endosperm heterosis in self-pollinated crop where both polar and the male gamete have identical genotype. **Gametophytic modifications**- There are genes which can modify the post-meiotic events in the formation of female gametophyte. The intermediate gametophyte(ig) gene in maize when in homozygous plants used as the female shows pleiotropic effect on seed development such as defective seeds due to increased ploidy level in the endosperm, polyembryony, heterofertilization, high frequencies of paternal haploids with both types of haploids occurring non-preferentially among seed with either single or multiple embryos(Kermicle 1971). The gene results in continued cell and nuclear division and instead of gametophyte containing one egg and two polar nuclei there is proliferation of both eggs and polar nuclei.

14.2 PRODUCTION OF MONOSOMIC DIPLOID ANGIOSPERM

Maize plants heterozygous for *r-X1* locus when used as female produces monosomic progeny. When gametes carrying *rX1* deficiency are fertilized by haploid pollen around 11% monosomic and 11% trisomic plants occur (Plewa and Weber, 1973). The *rX1* induces a high rate of chromosomal non-disjunction during the mitotic divisions when the female gametophyte is formed from a megaspore. As the developing gametophyte is either 2, 4 or 8 nucleate, the n-1 chromosome condition in one or more nuclei is compensated for genetically by the presence of one or more nuclei with n or n+1 chromosomes. Thus eggs with the n-1 chromosomes are functional although ordinarily n-1 (nullisomic) condition is lethal to the gametophyte and monosomics are not transmitted.

14.3 USES OF MONOSOMICS

Monosomics have been used to detect genes determining lipid content and composition of fatty acids in embryos by comparing plants monosomic for specific chromosome with their diploid relatives and the detected differences will be due to the dosage effects of gene(s) on that chromosome. Monosomic can also be used for analysis of nucleolar formation in microspore. Microsporegenesis in monosomic results in 2 of the 4 cells a microspore quartet being nullisomic. All microspores from plants monosomic for chromosome 1,2,4,7,8,9,10 contain a nucleolus but microspore nullisomic for chromosome 6 lack nucleoli and it never contains a nucleolus in more than two of the four cells. This shows that the factor(s) determining nucleolar formation resides on that chromosome 6. The availability of monosomic in maize provides opportunity to develop and use chromosome substitution lines similar to those used in wheat where it can be used for transferring desirable genes from one line to another. Finally the contributions of different chromosomes to specific quantitative traits can be determined and biometrical analysis appears possible (Law, 1966).

14.4 TRIPLOIDS IN SOYBEAN

In soybean a recessive gene, ms1 determines male sterility which results due to failure of cytokinesis following microsporogenesis resulting in four nuclei in each microspore. The male sterile plants produce more pollen mother cells than the male fertile sibs. The other traits associated with male sterile plants include reduced pollen fertility, polyembryonic seeds nad haploids and polyploids in the progeny. Haploids ,triploids and tetraploids were found among polyembryonic seeds and haploids, triploids, tetraploids, pentaploids and hexaploids among monoembryonic seeds(Beversdorf and Bingham, 1977). Cytological analysis showed presence of extra cells and nuclei(up to four egg cells and as mmany as extra nuclei) in the female gametophyte. The gene *ms1* appears to be similar to *ig* in action in that the eight nuclei of the female gametophyte continue to divide rather than differentiate into a seven celled mature gametophyte. Further this gene likes *ig* changes the level of ploidy in the embryo and endosperm. 2x-4x or 4x-2x crosses have never yielded triploid progeny because of triploid block mechanism which appears to be associated with endosperm failure in the tetraploid and pentaploid endosperm in the progeny of interploidy crosses. However, triploids occur regularly in the above crosses because of prevention of triploid block. Prevention of triploid block occurs when a 2n egg occurs in the same female gametophyte as two n polar nuclei double fertilization gives rise to a normal triploid endosperm and an abnormal triploid embryo.

14.5 SOMATIC INSTABILITY

Somatic instability refers to diverse phenomena such as intraplant variations in chromosome number, somatic sectoring, complement fractionation, reductional groups, somatic reduction and chromosome substitution (Peloquin, 1981). Spindle abnormalities are involved in all these phenomena. Although these abnormalities occur in low frequency but the frequency can greatly increase either by genetic factor (genotype dependent) or environmental factors such as temperature and chemical treatments. Somatic instability can also arise spontaneously. Sorghum (*Sorghum bicolor*), a diploid(2n=2x=14), is an example of somatic instability. Somatic instability has also been found in grasses. Sorghum has provided better evidence for somatic reduction than has any other higher plant. A true breeding mutant, a dwarf white-seeded sorgo type was obtained following colchicine treatment of seedlings, sorghum line experimental 3. The hypothesis put forward that this diploid, complex true breeding mutant arose through chromosome substitution during reduction of colchicines induced tetraploid cells back to diploid cells(Franzke and Sanders, 1964; Sanders and Franzke, 1969). The following observations were made in this study. I. Mutants differ from the parental lines in many traits (thus suggesting chromosome rather than gene changes) II. There was repeated occurrence of the same mutant type from a true breeding parental line III. There was recovery of parental-like types from each of the three colchicines treated F1 hybrids seedlings IV. Similar diploid true breeding mutants were obtained from diploid and tetraploid cytotypes of sorghum. The hypothesis of chromosome substitution is based on the following assumptions. I. Sorghum is a disomic polyploid in which homoeologs can substitute for each other genetically II. Bi-valent pairing is genetically controlled III. Somatic reduction occurs. When plants heterozygous for two independent reciprocal translocations were treated with colchicine to induce mutation, all true breeding mutants obtained were homozygous for the nontranslocated chromosome in the four pairs of chromosome involved in the translocations (Simantel and Ross, 1963). Such mutants can be obtained only if reduction and doubling or doubling and reduction of chromosome number occurs and thus it is an evidence for occurrence of somatic reduction. Intraplant morphological variations have also been found in several grass species, interploid hybrids, interspecific hybrids and an intergenreic hybrids such as timothy (*Phleum pretense*), bromegrass (*Bromus inermis*), sudangrass (*Sorghum vulgare* var. sudenense) ans sorghum-sudangrass hybrids. Same clone upon asexual propagation produced plants with striking differences and both sorghum and sudangrass types were obtained from asexual propagation of somatically unstable sorghum-sudangrass hybrids(Nielsen et al., 1969) and these results can also be explained by chromosome substitution but unlike sorghum it occurred without treatment in grasses. Spontaneous somatic instability has occurred in Rubus and Hymenocallis. It is likely that many of the so called' **bud sports**'(some of which have been very valuable) that have occurred in varieties of many species arose via chromosome substitution. Thus chromosome substitution can be used as a breeding tool with particular species particularly the polyploidy crops where one could obtain homologous substitutions in polysomic polyploids and homoeologous substitutions in disomic polyploids. It provides an alternative breeding scheme for improving varieties of species where sexual alteration of genotype is either undesirable or impossible. Finally, success in obtaining desirable variants in plants regenerated from tissue culture of a clone where no variation has been induced may be dependent primarily on chromosome substitutions.

14.6 HETEROFERTILIZATION

It is the occurrence of embryos and endosperms of differening genetic constitution(non-correspondence) results from the fertilization of the egg and the polar nuclei by genetically differeing sperm from different pollen tubes(Sprague, 1932). In pollinations with artificially mixed pollen samples or with segeregationg pollen marked with endosperm and embryo markers it is possible to identify kernels from embryosacs that have beeen penetrated by at least two pollen tubes and to analyze fertilizations or other events accordingly. In recombination analyses involving an endosperm trait with an embryo or plant trait segregating markers transmitted through the male must be verified by progeny tests whenever heterofertilization might confound the results. Heterofertilization frequencies of 1 to 2% are common and Sprague(1932b) found up to 25% heterofertilization in some strains. Another type of heterofertilization event due to regular nondisjunction of the centromere of B chromosomes in the male gametophyte is recognizable by noncorrespondence and deficiencies if a section of an A chromosome is attached to the B centromere by translocation (Roman and Ullstrup, 1951). In spontaneous sectors trisomic for chromosome 3, tetraspory (all the four products of meiosis contributing to the eight nuclei of the embryo sac) rather than monospory is found causing noncorrespondence and anomalous classes (Neuffer, 1964). Gynogenesis or androgenesis (maternal or paternal haploidy or diploidy) can be recognized in the embryos of dry seeds by genetic marking. In crosses involving a particular strain, Coe's stock the frequency of gynogenetic haploids is substantially increased over normal(Sarkar and Coe, 1966) while the frequency of androgenetic haploids is greately increased in seed progeny from plants bearing gene, *ig*(Kermicle, 1969). Monosomy may occur spontaneously or under the influence of *r-x 1*(Weber, 1973) and can cause noncorrespondence. Newly occurring polysomy also can cause noncorrespondence and other anomalous classes. Gynogenetic diploids from crosses 2n x 2n or 2n x 4n appear to be derived by the fusion of meiotic products to form diploid sporophytes rather than by the doubling of reduced products and thus may not be homozygous. Tetraploids arising from crosses of 2n x 4n, however, may be derived by doubling of the reduced products to give diploid gametes through the female parent while tetraploids arising from 4n x 2n appear to be derived by fusion of meiotic products in the male. Plants homozygopus for *el* produce diploid eggs whose origin appears to be from incompletely reduced meiotic products (Rhoades and Dempsey, 1966b). The genetics of autotetraploids taking into account all known phenomena that affect chromosome behaviour and transmission is discussed by Doyle (1973).

14.7 LOCATION OF FACTORS TO CHROMOSOME AND TO MAP POSITION

Besides conventional method the following methods can be used for mapping. 1. Chromosomal aberrations as markers, aberrations linked to gene markers, hemizygosity, transmission defects and trisomy. In the conventional method with marker stocks with the unplaced factor is made heterozygous with aid of a series of markers chosen to survey the linkage map and the F1 plants are testcrossed(if possible) or self-fertilized. There is no standard set of markers and the common procedure is to use whatever strains bearing contrasting markers are available. The marker strains should carry two well separated markers on the same chromosome or segment and markers that affect the seed or seedling are generally preferred as classification of progenies can be made in the laboratory and in the early stage. The markers can be dominant or recessive. Although dominant markers are more efficient in that coupling heterozygote produce a mathematically more informative array in the F2(Hanson, 1959), few suitable dominant markers are available and most are older plant traits and many have low productivity combined with variable penetrance. F2 progenies or testcross progenies can be used to study linkage and epistasis between two factors using 2 x 2 chi-square test. Data involving interactions are inefficient for the estimation of recombination percentages (Allard, 1956).

Cytological Techniques

15.1 MITOSIS IN ROOT TIP CELLS (E.G. CROCUS SPP.)

Corms of Crocus species (and other bulbous species) are generally obtainable from bulb and seed merchants in late summer. Crocus corms are normally dispatched in September , and when received they should be planted in moist vermiculite and kept outdoors in an unheated frame .Roots about 2cms long are produced within 4 weeks and these should be removed and fixed in 3:1 acetic alcohol fixative . Fixed roots can be stored for several months in the fixative, provided they are refrigerated (0-4°C)

A. Feulegen/Orcein squash procedure

1. Remove roots from fixative, blot dry and place in 5N HCl at room temperature for 40 minutes (hydrolysis +maceration).
2. Rinse briefly in distilled water, blot dry and transfer to Feulgen reagent for 10-15 minutes or until root tips are intensely stained to a magenta colour.
3. The root tips can be squashed immediately at this stage, but if they are to be stored for use more than 30 minutes later, it is preferable to wash the roots in S02 water (2 x 5 mins) and store them in distilled water in a refrigerator. They can be kept in this condition for several days.
4. Squashing. Cut off the stained meristematic root tip on a slide and discard the rest of the root.
5. Add one or two drops of acetic-orcein or lact-proprionic orcein. Feulgen staining is specific for DNA and as such will only stain nuclei and chromosomes. Orcein is amore general stain which intensifies the chromosomal stain and also stains cytoplasm and cell walls.

6. Break up the root tip in the stain by tapping with a brass rod or other suitable blunt instrument.
7. Place a coverslip on the suspension of cells.
8. Warm the slide gently on a spirit burner -do not boil the stain!
9. Sandwich the slide between layers of blotting paper. Press gently at first to remove excess stain and then apply firm pressure to flatten the cells. Use thumb pressure applied vertically. Do not move coverslip sideways while pressing or the cells will be destroyed.

B. A rapid squash method for tip cells

1. Place fresh root tips into 5N HCl for 15 minutes at room temperature.
2. Rinse in distilled water and use immediately or store in distilled water at 0-4°C for up to 24 hours.
3. Remove tips on a clean slide and discard rest of root.
4. Add one or two drops of stain (0.5% toluidine blue in Mcllvaine's buffer pH 4(12.29 ml0.1M citric acid +7.71 ml 0.2M Na_2HPO_4))
5. Break up tissue in stain by tapping, add coverslip and squash between layers of filter paper

15.2 MITOSIS IN LOCUST EMBRYOS

1. Take locust eggs 6-7 days old (containing approximately half grown embryos).
2. Tear chorion under insect saline using a pair of needles, to release contents. The embryo should be visible as a whitish structure among yellow yolk.
3. Carefully transfer embryos, using a pipette, to3:1 acetic alcohol fixative. Allow to fix for a minimum of 1-2 hours, but they can be stored in fixative at 0-4°C for weeks or months.
4. Place single embryos on clean slides, blot dry, add a drop or two of lacto-porpionic orcein or acetic orcein. Tap to break up the tissue in the stain, add a coverslip, warm on a spirit burner, and squash between layers of filter paper.

15.3 MEIOSIS IN LOCUST OR GRASSHOPPER TESTES

1. Dissect testes from last in star nymphs or young adult males. Transfer quickly to a dish containing insect saline. Using a pair of needles remove most of the yellow/ orange fat body surrounding the testis
2. Transfer to 2:1 acetic alcohol fixative. Testes should be fixed at least 1-2 hours, but preferably a few days. Testes can be stored in 3:1 fixative in a refrigerator (0-4°C) for weeks or months.
3. To prepare a squash, take 3-4 testis follicles on a clean slide and blot excess fixative.
4. Add a drop or two of acetic orcein or lacto-propionic orcein.
5. Tap to break up follicles in stain, add coverslip, warm on spirit burner and squash between layers of filter paper.]

15.4 MEIOSIS IN PLANT POLLEN MOTHER CELLS

1. Fix young flower buds in Carnoy's fixative (6ethanol: 3 chloroform: 1 acetic acid). Take care to fix a range of bud sizes or influroescences with florets covering a range of developmental stages. Preliminary work may be necessary to establish the approximate stage at which meiosis occurs.
2. Store buds in fixative at 0-4°C until required.
3. If aceto-carmine stain is to be used a few drops of saturated ferric chloride (sufficient to give the fixative a pale straw colour) should be added to the fixative as a mordant.
4. Dissect anthers from individual buds or florets.
5. Blot off excess fixative and add one or two drops of stain (carmine or orcein).
6. Tap lightly to release p.m.c.s. into stain, remove another debris, add coverslip, warm slide on spirit burner, squash between layers of blotting paper.
7. Check the stage of development of p.m.c.s. If too young –use a larger bud, if too old- use a smaller bud.

15.5 RAPID TECHNIQUES FOR PLANT ROOT TIP CHROMOSOMES (DYER, 1978)

1. Remove root tip (after pre-treatment if necessary) and fix for 5 mins. minimum in the following mixture (Dyer's fix)

   Absolute ethanol - 10
   Chloroform - 2
   Glacial acetic acid - 2
   Formalin - 1
2. Wash briefly in 70% ethanol.
3. Macerate in 1N HCl at 60°C for 5 minutes.
4. Transfer to water (use muslin over test tube).
5. Tap to break tissue in a drop of lactopropionic orcein on slide.
6. Place on coverslip, warm gently and squash to flatten.

15.6 C-BANDING TECHNIQUE FOR PLANT ROOT TIP CHROMOSOMES (MARKS, 1974)

1. Fix roots in 3:1 alcohol: acetic acid after pretreatment with colchicine or other .
2. Macerate in 45% acetic acid at 60°C for 20 minutes.
3. Transfer to water (use muslin over test tube).
4. Tap to break tissue in a drop of 45% acetic acid.
5. Place on coverslip and squash to flatten. Preparations can be checked for presence of chromosomes at this stage.

6. Remove coverslip after freezing slide in liquid nitrogen. Place slide directly in absolute ethanol (2 changes) and allow to air dry.
7. Place slides in 5% Barium hydroxide at 20°C for 5 minutes.
8. Wash in running tap water for at least 15 minutes.
9. Place slides in 2 x 55C at 60°C for 30 minutes.
10. Stain in 2% Giemsa in phosphate buffer pH 6.8 for 5 minutes.
11. Rinse slides in distilled water and check progress of staining under microscope. Return to stain if necessary and repeat.

15.7 PREPARING SEMI-PERMANENT OR PERMANENT SLIDES

1. Ordinary squash preparations can be preserved semi-permanently by sealing the coverslip edges e.g. with rubber solution, or other substances. The advantage of rubber solution is that it can be easily removed if it is desired to make a permanent preparation.
2. The basic operation in making permanent preparations is removal of coverslip, without disturbing the underlying cells, so that a permanent mountant can be introduced.
   (a) The older method is to invert the slide in 10% acetic acid (for aceto-carmine stained slides) or in 3:1 acetic alcohol (for orcein- stained slides) until the coverslip drops off. Slides and coverslips are then passed through 2 changes of absolute ethanol and mounted in Euparal (or in DPX after air-drying).
   (b) A quicker (and better) method is to freeze the slides on a block of dry ice (or in liquid nitrogen) and lever the coverslip off with a safety razor blade. The slides are hen dehydrated in 2 changes of ethanol and mounted directly in Euparal- or air dried in DPX.

15.7.1 Recipes for Stains

Feulgen reagent (leuco-basic fuchsin)

Dissolve 0.7g basic fuchsin and 3.8g Na_2S_2O5 in 200ml of 0.15N HCl. Shake or stir for at least 2 hours. Add 3g activated charcoal (decolouriser), shake well for about 1 minute and filter rapidly through coarse filter paper. The solution can be adjusted to pH 3.6 by adding 1N NaOH, but this is not essential. Store in a dark tightly stoppered at 0-4°C.

Caution: leuco basic fuchsin is a carcinogenic substance.

Aceto-carmine

Add 4 gms of carmine to 100ml 45% acetic acid. Boil gently in a reflux condenser for 1-2 days. Filter and store in a refrigerator. This stock solution can be diluted with 45% acetic acid, if required.

Lacto-propionic orcein

Substitute a 1:1 mixture of 45% lactic acid and 45% propionic acid for acetic acid in the above recipe.

Note: All carmine and orcein stains are inclined to precipitate in storage and they should therefore be filtered from time to time.

SO_2 water

5ml 1N HCl
5ml 10% $Na_2S_2O_5$
100ml distilled H_2O

Insect saline

0.7g NaCl
100ml distilled water
0.2ml 10% aqueous $CaCl_2$

Tables 15.1 and 15.2 shows the plant and animal species, their chromosomes numbers, parts to be used for mitotic and meiotic studies and the seasons in which they would be available.

Table 15.1 Showing plant and animal species suitable for study of mitosis.

Species	*Common name*	*Chromosome number*	*Material*	*Season*
Plants				
Crocus balansae	crocus	2x = 6	corms	September/October
Crocus candidus		2x = 6	corms	September/October
Crocus oliveri		2x = 6	corms	September/October
Allium ascalonicum	shallots	2x = 16	bulbs	August-March
Vicia faba	Broad beans	2x = 12	seeds	All year round
Tradescantia paludosa	Spiderwort	2x = 12	cuttings	All year round
Tulip species	Tulips	2x = 24	bulbs	Autumn
Hyacinths	Hyacinths	2x = 16*	bulbs	Autumn
Scilla campanulata	Garden blueball	2x = 16**	bulbs	Autumn
Trillium	Wood lillies	2x = 10	rhizomes	April-September
Secale cereale	Rye	2x = 14	seeds	All year round
Hordeum vulgare	barley	2x = 14	seeds	All year round
Pisum sativum	Garden pea	2x = 14	seeds	All year round
Crepis capillaris	hawksbeard	2x = 16**	seeds	All year round
Animals				
Locusta migratoria	locusts	2x = 23 ♂	embryos	All year round
Schistocerca gregaria				
Dugesia lugubris	Turbellarian flatworms	2x = 8	Cocoons	Spring and early
Dendrocoelum lacteum		2x = 16	Young animals	summer
Polycelis nigra		2x = 16	Regenerates	

*Different varieties have different chromosome numbers

**Has a useful karyotype, all chromosomes identifiable

Table 15.2 Showing animal and plant species suitable for study of meiosis.

Species	*Common name*	*Chromosome number*	*Material*	*Season*
Animals				
Locusta migratoria	Locusts	2x = 22 + x ?	testes	All year round
Schistocerca gregaria		2x = 22 + x ?		All year round
Chorthippus brunneus	grasshoppers	2x = 16 + x ?	testes	June-October
Chorthippus parallelus				
Omocestus viridulus				
Myrmeleolettix maculatus				
Triturus spp.	newts	2x =	testes	June-July
Plants				
Lilium spp.	lillies	2x = 24	anthers	Spring/Summer
Allium cepa	Common onion	2x = 16	anthers	Spring/Summer
A.schoenoprosum	Chives	2x = 16	anthers	Spring/Summer
A.ursinum	Ramsons	2x = 14	anthers	Spring
Vicia faba	Broad bean	2x = 12	anthers	Spring/Summer
Pisum sativum	Garden pea	2x = 14	anthers	Spring/Summer
Secale cereale	Rye	2x = 14	anthers	Spring/Summer
Hordeum vulgare	Barley	2x = 14	anthers	Spring/Summer
Paeonia	Paeony	2x = 10	anthers	Spring/Summer
Crepis capillaris	Hawksbeard	2x = 6	anthers	Spring/Summer
Tradescantia palludosa	spiderwort	2x = 12	anthers	All year round

15.8 HUMAN LEUCOCYTE CHROMOSOME

1. Put into a sterile 10ml bottle (inject through rubber seal) 4ml medium (Eagles or 199) +60 units/ml streptomycin and penicillin

   0.5 ml Calf Serium

   0.05ml Phytohaemagglutinin (Welcome reagent ampoule)

   0.05ml Heparinized blood

   (Blood can be dripped directly into a tube of dried heparin, enough for 5ml; or smaller amounts can be sucked into a hypodermic which contains a few drops of heparin. The final concentration of heparin should be 25 units/ml).

1. Mix the ingredients by rocking or swirling but do not agitate or get bubbles. Incubate without shaking at 37°C, preferably in a covered water bath, for 48 hours. This should yield 1st round of mitosis; 2nd and 3rd appears by 72 hours.
2. Accumulate metaphases by injecting 0.25ml colchicine 0.1 mg/ml 1-3 hours before fixation, i.e. at 45-47 hours incubation. Arrested metaphases have very contracted chromosomes but can be counted easily.
3. After 48 hours, mix and transfer to a conical glass centrifuge tube. Spin 1000x g 5mins. Decant supernatant. Resuspend pellet by agitating as next solution is added slowly, drop by drop. This is important in prevevting the aggregation of cells.
4. Slowly add 5ml warm 0.075M KCl, incubate at 37°C10 mins. This swells the cells. Repeat 4.

5. Slowly add fixative while mixing 95ml methanol-acetic acid, 3:1). Leave 5 minutes then repeat 4.
6. Replace with fresh fixative. Leave 10 minutes then repeat 4.
7. Add 0.2-0.5 ml fresh fixative.
8. Take slides from acid alcohol, wipe them dry on dark surface. Use a pasteur pipette to put 2 drops of the final mixture cell suspension onto the centre of each slide. Allow to dry completely. Mark the slide with a diamond or crayon.

Conventional squash or air-dried preparations of fixed chromosome stained with basic dyes or with the Feulgen procedure specific for DNA reveal little of their substructure. Apart from the primary constriction (centromere), secondary constrictions including nucleolar organizers and under certain conditions, some of the more obvious heterochromatin blocks, metaphase chromosome appears as a tightly coiled structure with little in the way of longitudinal differentiation (Evans, 1973).

A. Standard Giemsa Staining

1. Place air dried slides in 5% Giemsa solution for 15 minutes.
2. Rinse in distilled water and allow to air dry.
3. Mount in DPX, do not examine with oil-immersion objective until DPX has dried-overnight.

The Giemsa procedure comprises a maximum of six steps, of which one or more could be omitted with some plants (Schweizer, 1973): 1. fixation 2. maceration 3. preparation of squashes 4. pretreatment 5. incubation 6. staining.

B. C-banding

1. Place air dried slides in 5% barium hydroxide at 50°C for 3-4 minutes.
2. Wash thoroughly for 5 mins., rinse in distilled water.
3. Place slides in 2 x SSC at 62°C for 1 hour.
4. Rinse with distilled water.
5. Stain in 5% giemsa (pH 6.8) for about 10 mins.
6. Rinse with distilled water, air dry and mount in DPX.

C. G-banding (1)

1. Place slides in 2 x SSC at 62°C for 1 hour.
2. Rinse briefly with distilled water.
3. Stain in 5% giemsa (pH 6.8) for about 10 mins
4. Rinse in distilled water, check staining under microscope, air dry and mount in DPX.

G-banding (2)

1. Place slides in 1% Lipsol in tap water for 10-15 seconds.
2. Rinse in distilled water.
3. Stain in 5% giemsa (pH 6.8) for about 3 mins.
4. Rinse in distilled water, check staining under microscope, air dry and mount in DPX.

Recipe for phosphate buffer

A. 1/15M (KH_2PO_4) 8.16 g/L
B. 1/15M (Na_2PO_4) 8.52 g/L

15.9 THE DIFFERENTIAL STAINING OF CHROMATIDS

1. Add 0.05ml of 1mM BUdR to culture bottles to give a final concentration of 10μM$^{\pm}$.
2. Incubate cultures in the dark for 65 days.
3. Harvest cells and prepare air dried slides in the usual manner.
4. Stain slides in a 0.5 μg/ml solution of Hoechst 33258 in phosphate buffer pH7.0 for 10 mins.
5. Rinse in distilled water.
6. Place slides in moisture chambers (coplin jars with about ¼ inch water in bottom). Expose to UV for 1 hour.
7. Incubate slides in 2 x SSC at 60° C for ½ hour.
8. Rinse in distilled water, stain in 5% Giemsa (pH 6.8) for 5 mins.
9. Rinse in distilled water, air dry and mount in DPX.

15.10 AUTORADIOGRAPHY

N.B. All operations to be conducted in a darkroom illuminated only by P904 (Ilford) Safelights.

1. Melt 25ml Ilford K2 emulsion by gentle stirring in a glass vessel in a water bath at 43°C.
2. Dilute melted emulsion with an equal volume of distilled water and stir gently to mix. Leave for a few minutes for bubbles to rise.
3. Pour mixture into a dipping jar (e.g. Coplin jar).
4. Dip slides into mixture one at a time, allow to drain and wipe off excess emulsion.
5. Place dipped slides in metal slide racks and store in a light proof box containing silica gel at 4°C.
6. After a period of exposure (anything from a few days to several months) the autoradiographs should be processed as follows.
7. Develop in KodakD19B developer for 5mins at 20°C.
8. Stop bath and rinse in distilled water.
9. Fix in Amphix fixative for 2 minutes.
10. Wash in running water 5-10 minutes (this can be done in light conditions).
11. Stain slides in 5% Giemsa solution for 15 mins. Rinse in distilled water and allow to air dry.
12. Mount with a coverslip using DPX mountant.

15.11 TRANSFORMATION OF *ESCHERICHIA COLI* WITH PLASMID DNA

1. Dilute a fresh overnight culture 0.1ml into 5ml LB.
2. Grow at 37°C for 120 min with shaking in 25ml conical flask.
3. Pour bacterial suspension into sterile plastic centrifuge tube and harvest bacteria by spinning at full speed in bench centrifuge for 5 min.

4. Resuspend bacterial pellet in 2ml 10mM-Nacl, ice cold; harvest bacteria as above.
5. Resuspend bacterial pellet in 2ml 100mM-$Cacl_2$, ice cold; incubate on ice for 20 min.
6. Harvest bacteria and resuspend in 0.5ml 100mM-$Cacl_2$, ice cold; bacteria are now competent.
7. Use 100µl of bacterial suspension and 1µl DNA per transformation mixture in small sterile glass test tube.
8. Incubate on ice for 60 min.
9. Heat shock for 2min at 42°C.
10. Add 0.5ml LB; incubate at 37°C for 120 min.
11. Plate 0.2ml samples on L agar plates containing 300µg penicillin/ml

A **competent** bacteria, yeast or higher eukaryotic cell is one which is capable of taking up and integrating foreign nucleic acid molecules. It occurs naturally at a particular time of growth cycle or may be induced by calcium (in *E. coli*) or lithium ions (in yeast and higher eukaryotes). For **molecular cytogenetic techniques** see Roy (2009).

Problems in Cytogenetics and Genetics

CYTOGENETICS PROBLEMS

1. Suppose that we discover Drosophila male which is heterozygous for a reciprocal translocation between the second and third chromosomes, each break having occurred near the centromere (which for these chromosomes is near the centre.
   (a) Draw a diagram showing how these chromosomes would synapse at meiosis.
   (b) We find that this fly has the recessive gene *bw*(broad eye) and *e*(ebony body) on the nontranslocated second and third chromosomes, respectively. It is crossed with a female having normal chromosomes that is heterozygous for *bw* and *e*. What type of offspring would be expected and in what ratio? (Zygotes with an extra chromosome arm or deficient for one do not survive and there is no crossing over in Drosophila male).
2. Suppose that we are studying the cytogenetics of five closely related species of Drosophila and we find the following gene order(the letters indicate genes identical in all five species) and chromosome pairs in each species. Show how these species probably evolved from each other, describing the changes that occurred at each step.

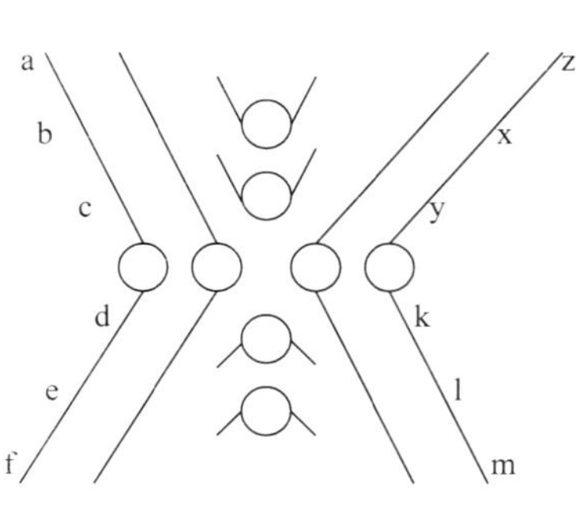

Species A

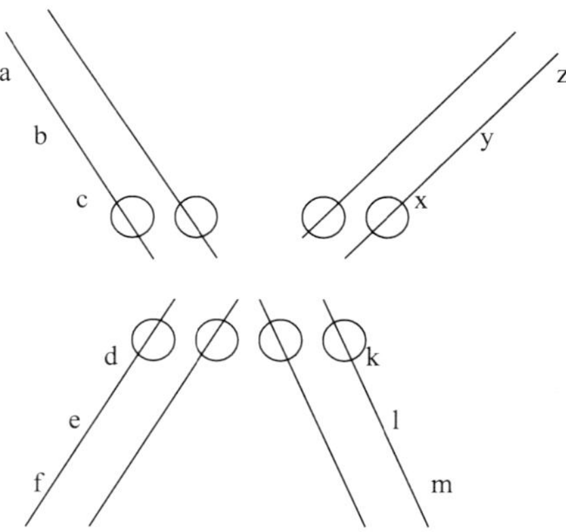

Species B

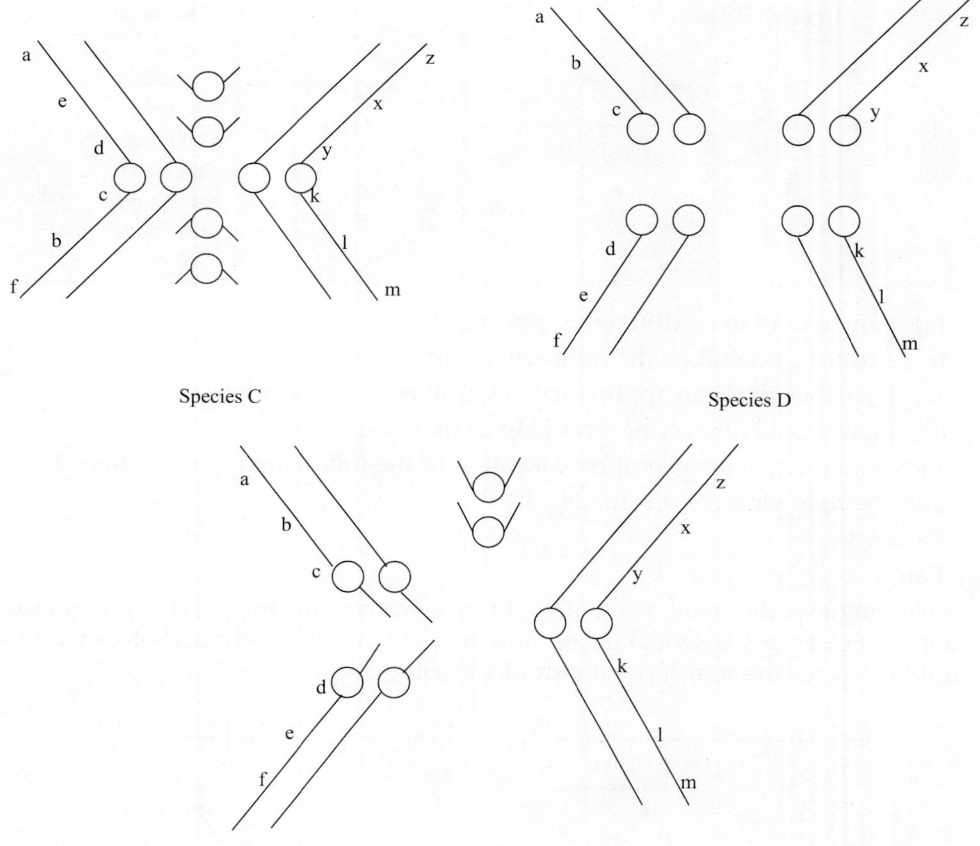

Species E

3. In Neurospora, the genes *a* and *b* are on separate chromosomes. In a cross of a standard ab strain with a wild type obtained from nature, the progeny are as follows: ab 45%, ++ 45%, a + 5%, + b 5%. Interprete these results and explain the origin of all progeny types under your hypothesis.
4. Assume that in a study of hybrid cells, three genes in humans have been assigned to chromosome 17. These genes are a, *b*, and *c* and are concerned with making the compounds *a*, *b* and *c*- all of which are essential for growth. If $a^- b^- c^-$ mouse cells are fused with $a^+ b^+ c^+$ human cells, assume that we find a hybrid in which the only human component is the right arm of chromosome 17(17R), translocated by some unknown mechanism to mouse chromosome. The hybrid can make the compounds *a*, *b* and *c*. Treatment of cells with adenovirus causes chromosome breaks. Let's assume that we can isolate 200 lines in which bits of the translocated 17R have been clipped off. These lines are tested for ability to make *a*, *b* and *c* and the results are given below.

Number of lines	Can make
0	*a* only
0	*b* only
12	*c* only
0	*a* and *b* only
80	*b* and *c* only
0	*a* and *c* only
60	*a*, *b*, and *c*
48	nothing

(a) How would these different types arise?
(b) Are *a, b,* and *c* all on the right arm of chromosome 17
(c) If so, then draw an approximate map showing relative positions.
(d) How would quinacrine dyes help us in this ?

5. In Neurospora, a reciprocal translocation of the following type is obtained: and the following cross is made.
Parent 1
Parent 2
Assuming that the small, wavy piece of the chromosome involved in the translocation does not carry any essential genes, how would you select products of meiosis that are duplicated for the translocated part of the solid chromosome?

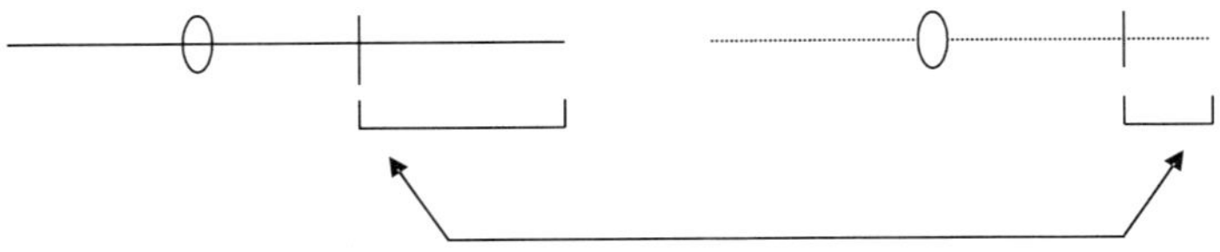

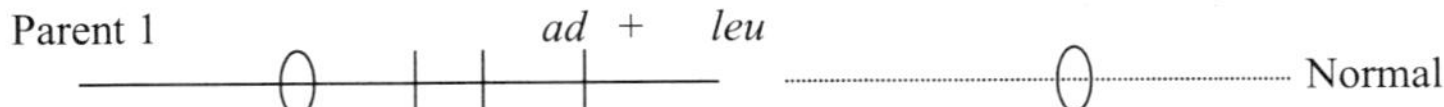

6. Several kinds of sexual mosaics are well documented in humans, some examples of which are given below. Suggest how each may have arisen.
XX/XO, XX/XXYY, XO/XXX, XX/XY and XO/XX/XXX

7. The table given below shows the meiotic pairing observed in various hybrids between genotypes of *Avena sativa* (2n= 6x = 42) and *Avena longiglumis* (2n = 2x = 14). What conclusions do these results allow concerning the control of pairing in *Avena sativa.* CW57 and Cc 4751 are accessions of *Avena longiglumis*.

	2n	I	II	III	IV	V	Proportion of complement paired
CW 57 x Pendek	28	6.83	4.83	3.11	0.41	0.11	-.75
Cc 4751 x Manod	28	16.40	5.73	0.03	-	-	0.41
Cw 57 x Cc 4751	14	-	7.00	-	-	-	1.00
Pandek x Manod	42	-	21.00	-	-	-	1.00
Pandek x (CW57 x Cc 4751)-1	28	7.82	5.12	2.91	0.32	-	0.72
Pandek x (CW57 x Cc 4751)-2	28	8.31	4.75	2.98	0.10	0.18	0.70
Pandek x (CW57 x Cc 4751)-3	28	15.30	5.26	0.58	0.11	-	0.45

8. A barley plant homozygous for a reciprocal translocation between chromosomes 1 and 5 was crossed to a plant with untranslocated chromosomes but homozygous for the recessive allele n, (naked caryopsis). All the F1 plants showed the wild-type phenotype(covered caryopsis) and were semi-sterile due to heterozygosity for the 1/5 translocation. Upon backcrossing this F1 to the homozygous recessive parent the following phenotypes were recovered:

Naked caryopsis Fertile	Naked caryopsis Semi-sterile	Covered caryopsis Fertile	Covered caryopsis Semi-sterile
119	12	17	105

What can you deduce from these data concerning the chromosome location of the n allele? Suggest further crosses you might perform to determine its location more precisely. What is the map distance of the n locus from the translocation breakpoint?

9. Dry seed of a disomic alien addition line of wheat, carrying a dominant resistance gene to a fungal pathogen on the alien chromosome, were treated with X-rays. The seeds were germinated and the resulting plants grown to maturity and self-pollinated. Seed harvested from these plants were grown out in families(one family = selfed progeny of one irradiated parent plant). Many families consisted of entirely resistant plants but a few families contained some susceptible plants. The resistant sibs of these susceptible plants were screened cytologically for 42 chromosome resistant plants. In a few cases, the backcrossed(to euploid wheat) progeny of these plants segregated resistant and susceptible progeny in a 1;1 ratio. Selfing resistant plants yielded some 42 chromosome resistant plants which bred true for resistance. With the help of diagram , explain the chromosomal changes accompanying this breeding programme.

10. Hybrids between various diploid and tetraploid species of *Brassica* gave the following pairing arrangements at meiosis:

    B. nigra (2n=16) x *B. carinata* (2n = 34) = 8 II + 9 I
    B. oleracea (2n= 18) x *B. carinata* (2n = 34) = 9 II + 8 I
    B. oleracea (2n = 18) x *B. napus* (2n = 38) = 9 II + 10 I
    B. campestris (2n= 20) x *B. napus* (2n = 38) = 10 II + 9 I
    B. campestris (2n= 20) x *B. juncea* (2n = 36) = 10 II + 8 I
    B. nigra (2n=16) x *B. juncea* (2n = 36) = 8 II + 10 I

    Produce a diagram summarizing the phyolgenetical relationships of these species. Suggest a method for testing your conclusion.

11. Assume that gene x_ is a new mutant in corn(Zea mays). An xx_ plant is crossed with a triplo-10 individual(trisomic for the chromosome 10) carrying only dominant alleles at the x_ locus. Trisomy progeny are recovered from this cross and backcrossed to a homozygous recessive(xx_) parent plant. What phenotypic ratios can be expected if the locus x_ locus is (a) not on chromosome 10(b) on chromosome 10? Assuming that locus x_ could be definitely assigned to chromosome 10 by this method, how would you locate the x_ locus within chromosome 10 and what strains would you require for this purpose.
12. Two inversions in chromosome A were investigated in the plant species *Crepis capillaries*. One inversion (x) had both breakpoints in the long arm while the second inversion (y) had one breakpoint in the long arm and one in the short arm as follows:
    (i) Show by means of diagram the pairing configurations you would expect to see at zygotene of meiosis in heterozygotes for (a) inversion x and (b) inversion y.
    (ii) State the expected fertility of crosses of inversion heterozygotes x and y to normal plants using inversion heterozygotes (a) as male parents and (b) as female parents. Give a detailed explanation of your answer.
    (iii) Show by means of diagrams the chromosome configurations you would expect at anaphase I and anaphase II of meiosis in inversion x heterozygotes and estimate the relative frequencies of these configurations using the chiasma frequencies given above. Assume in your answer that there is never more than one chiasma in each region of bivalent A, that chiasma formation occurs independently in the different regions and that there is no chromatid interference.

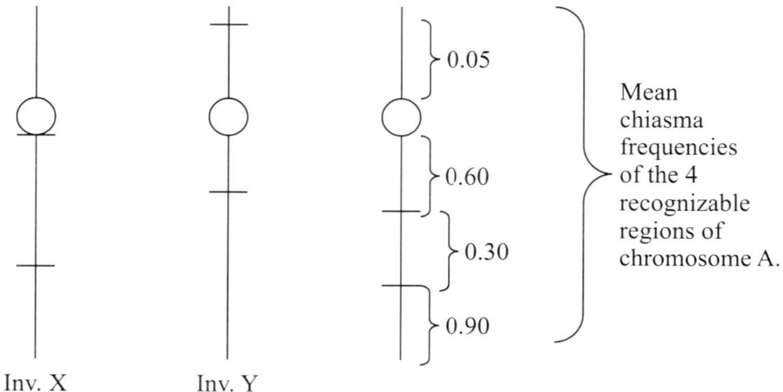

13. Chromosome 3A of *Triticum aestivum*(2n= 6x=42) shows 25% transmission when monosomic 3A plants are crossed as female parents to euploid wheat and 4% transmission when monosomic 3A plants are used as male parents. Suggest reasons for these relatively low transmission rates and account for the difference between male and female transmission. What percentages of monosomic and nullisomic 3A plants are expected following self pollination of monosomic 3A plants.

    A plant of *T.aestivum* monosomic for chromosome 3A (female) was crossed to a euploid plant(male) homozygous for the recessive gene st_. The F1 progeny of this cross consisted of plants of two distinct phenotypes; 212 were phenotypically st_ and

73 were wild type. When this cross was repeated using other monosomics, all F1 progeny were phenotypically wild type. A further cross was performed between a euploid st _ st_ plant and a ditelosomic $3A^L$ plant(disomic for the telocentric of the long arm of chromosome 3A) and the progeny were classified according to phenotype and cytology with the following results.

	Wild type	*St*
Euploid	9	95
Monotelosomic $3A^L$	102	6

What can you deduce from these results concerning the chromosomal location of st_? Give a detailed explanation of your conclusions.

14. A program for the transfer of leaf rust resistance from *Agropyron elongatum* to wheat involved the construction of a genotype simultaneously nullisomic 5B, monosomic 3D and carrying an added *AGROPYRON ELONGATUM* CHROMOSOME 3Ag (homoeologous to wheat group 3 chromosomes). This genotype was crossed with euploid wheat and progeny were selected which were resistant to leaf rust but lacked an entire Agropyron chromosome. Explain the origin of these resistant progeny plants

In order to characterize further the resistance carrying chromosomes of these plants, four different plants were crossed to wheat stocks carrying the following 'marker' chromosomes;

3Ag- Entire *Ag. Elongatum* chromosome 3

$3Ag^L$- Telocentric long arm *Ag. elongatum* chromosome 3

$3Ag^S$- Telocentric short arm *Ag. elongatum* chromosome 3

$3D^L$ - Telocentric long arm *Ag. elongatum* chromosome 3D

$3D^S$ Telocentric short arm *Ag. elongatum* chromosome 3D

The pairing behaviour of the resistance-accrying chromosomes with the various 'marker' chromosomes was observed at meiosis in the resulting hybrids, with the following results:

% pairing of resistance-carrying chromosomes with 'marker chromosomes

Resistant plants	*3Ag-*	*$3Ag^L$*	*$3Ag^S$*	*$3D^L$ -*	*$3D^S$*
1	100	100	70	100	3D
2	100	100	15	100	3D
3	100	100	1	100	3D
4	100	100	0.25	100	3Ag

The resistance- carrying chromosomes could also be classified by electrophoretic methods according to the source of the GOT (glutamate oxaloacetate transaminase0 alleles they carried, as shown in the table. Chromosome 3D of wheat and chromosome 3Ag have essentially similar physical and genetical maps. Chromosome morphology and location of the GOT allele on these chromosomes are as follows.

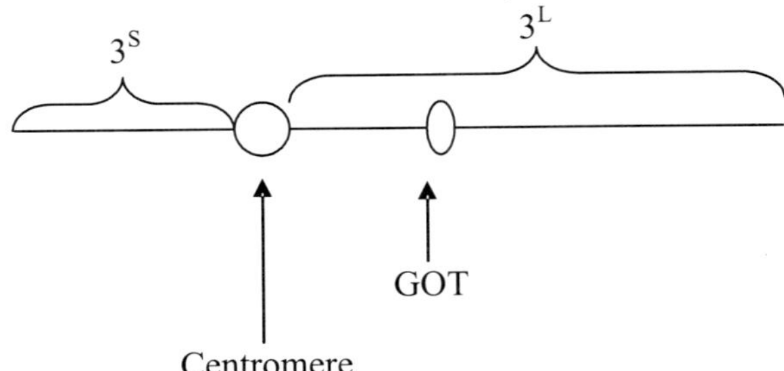

15. A mixed culture of strains A and B of homothallic Ascomycetes was established under conditions which favored sexual reproduction.(Note: A homothallic fungus is capable of producing selfed A x A and B x B asci as well as the hybrid A x B asci). Stain A carried the markers *his* and *ade*, and strain B the markers *pyr* and *pro* where *his, ade, pyr* and *pro* govern requirements for histidine,, adenine, pyridoxine and praline, respectively. A mass ascospore suspension was prepared from this mixed culture following the formation of mature fruiting bodies and , suitably diluted, 0.1 ml aliquots were plated onto (a) fully supplemented medium, (b) minimal medium(MM) supplemented with histidine and pyridoxine and (c) MM supplemented with adenine and praline. The ascospore dilution plated and the number of colonies that developed on the three media are given below.

Medium	*Dilution plated*	*Mean number of colonies per plate*
(a) Fully supplemented	10^{-3}	144
(b) MM + histidine + pyridoxine	10^{-1}	36
(c) MM + adenine + proline	10^{-2}	108

Given that the *his* and *pyr* markers are on different chromosomes determine the following.

(i) The proportion of hybrid asci in the mixed culture(i.e. those produced by the fusion of a strain A nucleus with a strain of B nucleus).

(ii) The linkage relationship of the *ade* and *pro* marker.

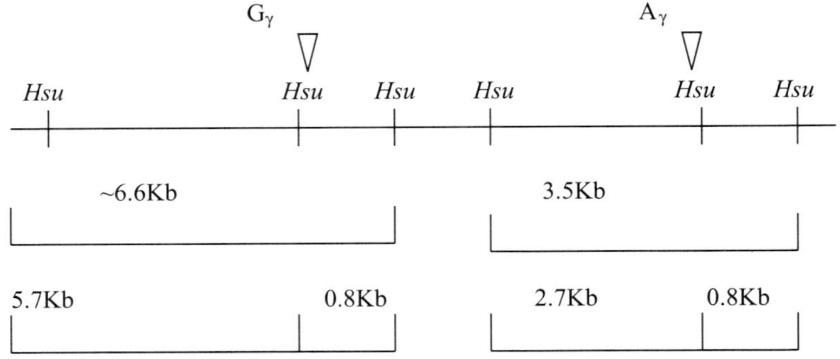

16. In Neurospora crosses *ab* x ++(in which *a* and *b* represent different loci in each cross), 100 linear (ordered) asci were analyzed from each cross to give the results shown in the Table given below.(Note that only spore pair genotypes are listed and further, Neurospora having an eight spored ascus).

Cross number	Ascus type and number						
	ab ab ++ ++	a + a + +b +b	ab a+ ++ +b	ab + b ++ a+	ab ++ ++ ab	a+ + b +b a+	a+ + b ++ ab
1	34	34	32	0	0	0	0
2	55	3	40	0	2	0	0
3	9	6	24	22	8	10	20
4	31	0	1	3	61	0	4
5	95	0	3	2	0	0	0
6	69	0	10	18	0	1	2
7	16	14	2	60	1	2	5
8	51	49	0	0	0	0	0

For each cross, map the genes in relation to each other and to their respective centromere or centromeres.

The above questions are taken from M.Sc. (Applied Genetics) examinations of University of Birmingham, England.

GENETICS PROBLEMS

1. The diagrams reproduced below show nuclear division stages taken from the plant species *Crepis capillaries* (2n = 6). Identify and label on the diagrams the following:
   (a) The nuclear division stages shown
   (b) The different chromosomes as they appear in the different stages
   (c) Identifiable features, such as centromeres, secondary constrictions and chiasmata
2. Starting from early interphase a nucleus first undergoes mitosis and then enters meiosis. Assuming that the DNA content of the nucleus is 2c at the beginning, indicate the DNA quantity per nucleus at the following stages.

Stage	DNA content
Early interphase	
Late interphase	
Mitotic prophase	
Mitotic metaphase	
Mitotic telophase	
Early interphase	
Late interphase	

Leptotene Zygotene Pachytene Diplotene Metaphase I Telophase I Interphase Prophase II Metaphase II Telophase II	

3. The following is a sequence of bases from one polynucleotide strand of DNA, the 5′ end being to the left. In the space provided, fill in the bases of the complementary (sense) strand, the mRNA, the t-RNA (most likely sequence) and the amino acids of the polypeptide chain which is produced.

C	*T*	*G*	*G*	*T*	*A*	*C*	*A*	*T*	*DNA*
									DNA strand transcribed mRNA tRNA Amino acids of polypeptide chain

4. A double stranded DNA source has a G-C ratio(i.e. A+T/G+C) of 1.30. In an *in vitro* experiment RNA can be transcribed from this DNA. Two base ratios can then be computed from the RNA, namely
   (i) the G-C ratio(i.e. A+U /G+C)
   (ii) The purine : pyrimidine ratio (i.e. A+G / U+C)

   What would these ratios be in the following, hypothetical situations?
   (a) Only one strand of DNA is transcribed and the RNA is single stranded.
   (b) Only one strand of DNA is transcribed but the RNA is double stranded.
   (c) Both strands of the DNA are transcribed and RNA single or double stranded

RATIO	
(A +U/ G+C)	*(A +G / U+ C)*

If any ratio can not be determined from the data given then write "?".

In two bacterial species the following data were obtained.

	DNA		RNA
Species	(A + T / G + C)	(A + U / A + C)	(A + G / U + C)
E.coli	1.00	0.98	0.80
B. subtilis	1.36	1.30	1.02

For each species which of the situations a, b, c(above) might apply.

E.coli	
B. subtilis	

5. Lysine is an amino acid and in wild type Aspergillus it can be synthesized from the constituents of the minimal medium which includes inorganic nitrogen.

   Autotrophic mutants having a nutritional requirement for lysine have been found. Certain organic compounds have been shown to be capable of substituting for lysine in some mutants but not in others.

   In an experiment involving 7 newly isolated auxotrophs, the following pattern was found.

Nutritional supplement	*Auxotrophic mutant*						
	A	*B*	*C*	*D*	*E*	*F*	*G*
Hydroxynorleucine	+	+	+	+	+	+	–
No supplement	–	–	–	–	–	–	–
Lysine	+	+	+	+	+	+	+
A amino-adipic acid	–	+	–	–	+	–	–

Note: + indicates growth on minimal medium + supplement
– indicates no growth on minimal medium + supplement

From this information alone and using the diagram provided below, answer the following questions.

(i) What is the likely path way giving rise to lysine?

(ii) Indicate the point in the path way at which each mutant is blocked.

(iii) What compounds might be expected to be accumulated by mutants A, B and G?

Minimal medium

Mutants

Mutant	*Accumulates*
A	
B	
G	

6. Heterokaryons were constructed between all possible pairs of mutants shown in Q5 in order to test for complementation. The heterokaryons were inoculated onto minimal medium, with the result shown below, where as before, +/– indicates growth or no growth.

	A	*B*	*C*	*D*	*E*	*F*	*G*
A	–	+	+	–	+	+	–
B		–	+	+	–	+	+
C			–	+	+	–	–
D				–	–	+	–
E					–	+	+
F						–	–
G							–

   (i) From a consideration of your answer to Q5 fill in the circled blanks in this table.
   (ii) What can you deduce about the mutants A- G from the table above?
   (iii) What can be said about the enzyme(s) involved in the immediate production of hydroxynorleucine?

7. Three stocks of Drosophila (A, B and C) all breed true for eyes which are reduced in size when compared to wild-type stocks. Furthermore, these stocks can not be distinguished by inspection.
   (i) On crossing "A" with true breeding wild-type stock, all the F1 had reduced eyes like "A" stock. On crossing all stocks "B" and "C" to wild-type however, all the F1 had large eyes like wild-type. What do you conclude from these results?
   (ii) The stocks "B" and "C" were intercrossed and the resulting F1 were wild-type. What conclusion can be drawn from this?
   (iii) A fourth stock "D" was available which resembled "B" and "C" yet was slightly more extreme in expression. "D" behaved like "B" and "C" when crossed to wild type, yet the crosses D x B and D x C yielded F1 progeny with reduced eyes like "B" and "C". What is the likely genotype of "D"?

8. In a three point cross linkage experiment the triple recessive stock *cu_ cu_, ou_ ou_, st_ st_* was crossed to wild-type and the F1 females were backcrossed to the triple recessive stock. The frequency of the different types of offspring were as follows:

cu	*ou*	*st*		350
+	*ou*	*st*		19
cu	+	*st*		110
cu	*ou*	+		5
+	+	*st*		7
+	*ou*	+		118
cu	+	+		29
+	+	+		362

(i) Construct a linkage map of the three genes in the space below.
(ii) Estimate the coincidence value. Given that this estimate is significantly different from 1.0 , what does it mean?

9. Five boys living in the same family have the following phenotypes:

Boy	*Blood groups*		*Tasting ability*	*Colour vision*
1	AB	MN	Taster	Normal
2	O	MN	Non-taster	Colour blind
3	A	MN	Taster	Normal
4	B	M	Non-taster	Normal
5	B	N	Non-taster	Colour blind

The fathers phenotype is known to be AB, N, Taster, colour blind.

(i) Which children do you suppose might be adopted?
(ii) Assuming that all of those children of whome he could be the father are really his own by one wife, what can you say about the genotype of the husband and wife?

Trait	*Husband*	*Wife*
ABO		
MN		
Tasting ability		
Colour vision		

N.B. The A-B-O system is determined by a series of 3 alleles at a locus, of which 1^A and 1^B show no dominance to one another while 1^0 is recessive to both 1^A and 1^B. The MN system has two alleles Ag^M, Ag^N which show no dominance to one another. Tasting ability is determined by two alleles *T*(for tasting) being dominant to t(for non-tasting). Colour vision is governed by two alleles *Cb*(for normal) being dominant to cb(for colour blind). This locus is sex linked, the previous three being autosomal.

10. The temperate bacteriophage λ(lambda) and its host bacterium *Escherichia coli* were used to measure the frequency of recombination between phage genes.

Two strains of λ were used R^+C^+ and R^-C^-.

R^- is a mutant allele of λ which prevents the phage from growing on E.coli stock A, although it will grow on E.coli stock B.

R^+ phage, on the other hand, will grow on A and B. C is a plaque morphology locus, CO producing clear and C^+ turbid plaques.

A culture of *E.coli* stock B was mixed-infected with approximately equal numbers of particles of both strains. After lysis had occurred phage particles were recovered in 1ml of lysate.

A dilution series was made from this such that if the original was 10^0(= 1), the others were 10^{-1}, 10^{-2}, 10^{-3}, 10^{-4} and 10^{-5}. From each of these .1ml was taken and spread on a lawn of *E.coli* of strain A and a similar .1ml was spread on a lawn of *E.coli* B. After incubation the number of clear and turbid plaques was recorded, with the following results.

	E.coli stock			
	A		*B*	
Dilution	C^+	C^-	C^+	C^-
10^{-4}	-	-	-	-
10^{-5}	154	21	180	170

(i) What do the numbers on "B" indicate?

(ii) Estimate the number of phage particles in the original 1ml lysate.

(iii) Estimate the frequency of R-C recombinants.

In a control experiment, the R^-C^- strain alone was used with the following results.

	E.coli stock			
	A		*B*	
Dilution	C^+	C^-	C^+	C^-
10^0	0	26	-	-
10^{-1}	0	3	-	-
10^{-2}	0	0	-	-
10^{-3}	0	0	-	-
10^{-4}	0	0	-	-
10^{-5}	0	0	0	360

(iv) What does this tell you about mutation?

(v) What do the –'s signify?

11. An *E. coli* geneticist isolated a number of autotrophic mutants all of which responded to the same substance. Ten of these mutants($aux\_^{-1}$ to $aux\_^{-10}$) were found to be located between the early *argS* and later *pabB* genes in Hfr x F^- conjugation experiments. Strains were then constructed so that a pair of reciprocal crosses could be made for each pair of aux_ mutants as follows.

$$Hfr\ arg^-\ aux^-\ \text{x}\ pab^+ \text{ x } F^-\ arg^+\ aux\text{-}y\ pab^-$$

$$\text{and } Hfr\ arg^-\ aux^-\ \text{y}\ pab^+ \text{ x } F^-\ arg^+\ aux\text{-}y\ pab^-$$

In these crosses conjugation was allowed to continue beyond the time of entry of the pab_ gene and prototrophs were selected by plating similar aliquots on minimal medium. The following data give the number of colonies which grew in the reciprocal crosses for each of a number of the possible pairwise combinations. What is the relative order of the sites of the ten *aux_* mutants and of the *argS* and *pabB* genes?

12. Recent work has shown that it is possible to perform diagnosis of β-thalassaemia(defects in the β-globin gene) by detecting restriction enzyme site polymorphism segregating within the family at risk. The polymorphism is detected as variation in the band pattern obtained in a Southern blot after digestion with the appropriate restriction enzyme. For example, a probe for γ-globins detect *HsuI* site polymorphism in the γ-globin genes (Gγ and Aγ) which are closely linked to the β-globin gene. The *HsuI* restriction pattern for the γ-globin genes is shown with the polymorphic sites marked ( ) and the sizes of the fragments detected by the γ-globin probe indicated.

DNA of parents and existing children is analyzed and corrected with the β-thalassaemia state(wild type, heterozygous carrier or homozygous thalassaemic). If one of the children is either wild-type or homozygous thalassaemic then the β-thalassaemic state of a subsequent foetus may be predicted. Illustratedbelow are the gel patterns for two families A and B. Work out the genotype with respect to the restriction sites for each individual in a family assuming no intermarker recombination and draw the *HsuI* restriction map of the γ-globin genes linked to the thalassaemic β-globin gene for each parent.

Family A

Fragment Size in Kb	Mother carrier	Father carrier	Daughter carrier	Son Thalassaemic
Origin ↓				
6.6	——	——	——	——
5.7	——	——	——	——
3.5	——	——	——	——
2.7		——	——	

Family B

Origin ↓	Mother carrier	Father carrier	Daughter Thalassaemic	Son carrier
6.6	———	———		———
5.7	———	———	———	———
3.5	———	———	———	———
2.7		———	———	

References

Allard, R.W. 1960. Principles of Plant Breeding. Wiley and Toppan

Alleman, M et al. 2006. An RNA-dependent RNA polymerase is required for paramutation. Nature, 442: 295-298

Arrighi, F.E. and Hsu, T.C. 1971. Localization of heterochromatin in human chromosomes. Cytogenetics, 10: 81-86

Auerbach, C. 1951. Problems in chemical mutagenesis. Cold Spring Harbor Symp. Quant. Bio., 16: 199-214.

Baker, B.S., Carpenter, A.T.C., Esposito, M.S., Esposito, R.E. and Sandler, L. 1976. The genetic control of meiosis. Ann. Rev. Genet., 10:53-134

Baker, R.L. and Morgan, D.T., Jr. 1969. Control of pairing in maize and meiotic interchromosomal effects of deficiencies in chromosome 1. Genetics, 61: 91-106

Battaglia, E. 1964. Cytogenetics of B chromosome. Caryologia, 17: 245-299

Beadle, G.W. 1932c. The relation of crossing over to chromosome association in Zea-Euchlaena hybrids. Genetics, 17: 481-501

Benzer, S. 1959. On the topology of genetic fine structure. Proc. Natl. Acad. Sci., U.S.A., 45: 1607-20

Bernardi. 1995. The human genome: organization and evolutionary history. Ann. Rev. Genet., 29: 445-476

Blackwood, M. 1956. The inheritance of B-chromosomes in Zea mays. Heredity, 10: 353-366

Blixt, S. 1969a. Publications by Herbert Lamprecht. Agri. Hort., Genet., 27:7-18

Blixt, S. 1972. Mutation genetics in Pisum. Agri. Hort. Genet., 30:1-293

Blakeslee, A.F.; Belling, J. and Farhnam, M.E. 1923. Inheritance in tetraploid. Daturas. Bot. Gaz., 76: 329-373

Bobrow, M. 1973. Acridine-orange and the investigation of chromosome banding. Cold Spr. Harb. Symp. Quant. Biol., 38: 435-440

Brody, H. and Carbon, J. 1989. Electrophoretic karyotype of *Aspergillus nidulans*. Proc. Natl. Acad. Sci., U.S.A., 86: 6260-3

Burkholder, G.D and Duczek, L.L. 1982. The effect of chromosome banding techniques on the proteins of isolated chromosomes. Chromosoma(Berl.), 87: 425-35

Burkholder, G.D. 1975. The ultrastructure of G- and C-banded chromosomes. Exp. Cell Res., 90: 269-278

Burkholder, G.D. 1981. Ultrastructure of R-banded chromosomes. Chromosoma, 83: 473-480

Burnham, C.R. 1932a. The association of non-homologous parts in chromosomal interchange in maize. Proc. 6th Intern. Congr. Genet., 2:19-20

Burnham, C.R. 1932b. An interchange in maize giving low sterility and chain configurations. Proc. Natl. Acad. Sci., U.S.A., 18: 434-44

Burnham, C.R. 1934. Chromosomal interchanges in maize: Reduced crossing-over and association of non-homologous parts. Am. Naturalist, 68: 81-82

Burnham, C.R. 1950. Chromosome segregation in translocations involving chromosome 6 in maize. Genetics, 35: 446-481

Burnham, C.R. 1962. Discussion in cytogenetics. Burgess Publ. Co., Minneapolis, Minn.

Burnham, C.R. 1966. Cytogenetics in plant improvement, pp. 139-88. In: Frey, K.J. (ed.), Plant Breeding, The Iowa State University Press/ Ames, Iowa

Carlson, W.R. 1973b. A procedure for localizing genetic factors controlling mitotic nondisjunction in the B-chromosome of maize. Chromosoma, 42: 127-136

Carlson, W.R. 1977. The Cytogenetics of Corn in 'Corn and Corn Improvement'. G.F.Sprague(ed.) American Society of Agronomy, Inc, Publisher, U.S.A.

Caspersson, T.;Zech, L.; Modest, E.J.; Foley, G.E.; Wagh, U. and Simonsson, E. 1969a. Chemical differentiation with fluorescent alkylating agents in Vicia faba metaphase chromosomes. Exp. Cell Res., 58: 128-140

Caspersson, T.;Zech, L.; Modest, E.J.; Foley, G.E.; Wagh, U. and Simonsson, E. 1969b. DNA-binding fluorochromes for the study of organization of the metaphase nucleus. Exp. Cell Res., 58: 141-152

Caspersson, T.;Zech, L.;Johansson, C. and Modest, E.J. 1970. Identification of human chromosomes by DNA-binding fluorescent agents. Chromosoma (Berl.), 30: 215-227

Chase, S.S. 1969. Monoploids and monoploid derivatives of maize(Zea mays). Bot. Rev., 35: 117-167

Chovnick, A, Ballantyne, G.H. and Holm, D.G. 1971. Studies on gene conversion and its relationship to linked exchange in Drosophila melanogaster. Genetics, 69: 179-209

Clarke, A.E. and Anderson, E.G. 1935. A chromosomal interchange in maize without ring formation. Am.J. Bot., 22:711-716

Clausen, R.E. 1941. Monosomic analysis in Nicotaina tabacum.(abs.) Genetics, 26:145

Clausen, R.E. and Cameron, D.R. 1944. inheritance in Nicotaina tabacum. XVIII. Monosomic analysis. Genetics, 29: 447-477

Coe, E.H. 1959. A line of maize with high haploid frequency. Am. Nat., 93: 381-382

Comings, D.E. 1975. Mechanism of chromosome banding VIII. Hoechst 33258-DNA interaction. Chromosoma(Berl.), 52: 229-243

Comings, D.E. et al.1975. Mechanisms of chromosome banding V. Quinacrine banding. Chromosoma(Berl.), 50:111-145

Cowell, J.K. 1984. A photographic representation of the variability in the G-banded structure of the chromosomes in the mouse karyotype. Chromosoma(Berl.), 89:294-320

Craig, J.M. and Bickmore, W.A. 1993. Chromosome bands. BioEssay, 15:349-54

Creighton, H.B. and McClintock, B. 1935. The correlation of cytological and genetical crossing over in Zea mays. A corroboration. Proc. Natl. Acad. Sci., U.S.A., 21: 148-150

Creighton, H.B. and McClintock, B. 1931. A correlation of cytological and genetical crossing-over in Zea mays. Proc. Natl. Acad. Sci., U.S.A., 17: 492-497

Crissman, H.A. and Tobey, R.A. 1974. Cell cycle in 20 minutes. Science, 184: 1297-1298

Darlington and Bradshaw. Teaching Genetics. Oliver and Boyd. 1963

Darlington and Lacour 1976. The Handling of Chromosomes. 6th Edtn. Gorge Allen and Unwin.

Dawe, R.K 1998. Miotic chromosome organization and segregation in plants. Ann. Rev. Plant Physiol. Plant Mol. Biol., 49: 371-95

Dawe, R.K. and Cande, W.Z. 1996. Induction of centromeric activity in maize by suppression of meiotic drive 1. 1996. Proc. Natl. Acad. Sci., U.S.A., 93:8512-8517

De Witte, W. and Murray, J.A.H. 2003. The plant cell cycle. Ann. Rev. Plant Mol. Biol., 54: 195-222

Dempsey, E. and Smirnov, V.1964. Cytological location of gl-15. Maize Genet. Coop. News Letter, 38:71-73

Deutrillaux, B. and Lejeune, J. 1971. Su rune nouvelle technique d'analyse du caryotype humain. C.R. Acad. Sci., Paris 272: 2638-2640

Dooner, H.K. and Kermicle, J.L. 1971. Structure of the R-r tandem duplication in maize. Genetics, 67: 427-436

Doyle, G.G. 1979. The allotetraploidization of maize. Theor. Appl. Genet., 54: 103-112

Driscoll, C.J. and Jensen, N.F. 1963. A genetic method for detecting induced intergeneric translocations. Genetics, 48: 459-468

Driscoll, C.J., Bielig, L.M and Darvey, N.L. 1979. An analysis of frequencies of chromosome configuration in wheat and wheat hybrids. Genetics, 91: 755-767

Elliott, F.C. 1958. Plant Breeding and Cytogenetics. McGraw-Hill Book Company, New York.

Emmerling, M.H. 1858b. An analysis of intragenic and extragenic mutations of the R-r gene complex in Zea mays. Cold Spring Harbor Symp. Quant. Biol., 23: 393-407

Endo, T.R. and Gill, B.S. 1984. Somatic karyotype, heterochromatin distribution and nature of chromosome differentiation in common wheat, T. aestivum L. em Thell. Chromosoma(Berl.), 89: 361-369

Endrizzi, J.E. 1974. Alternate-1 and alternate-2 disjunctions in heterozygous reciprocal translocations. Genetics, 6: 209-240

Evans, H.J. 1973. Molecular architecture of human chromosomes. Br. Med. Bull., 29 (No. 3): 196-202.

Finch, J.T. and Klug, A. 1976. Solenoidal model for super structure in chromatin. Proc. Natl. Acad. Sci. U.S.A., 73: 1897-1901

Fincham, JR.S., Day, P.R. and Radford, A. 1979. Fungal Genetics. Blackwell Scientific Publications, Oxford.

Funaki, K; Matsui, S and Sasaki, M. 1975. Location of nucleolar organizers in animal and plant chromosomes by means of an improved N-banding technique. Chromosoma(Berl.), 49: 357-370

Fu, T.K. and Sears, E.R. 1973. The relationship between chiasmata and crossing over in Triticum aestivum. Genetics, 75: 231-246

Fussell, C.P. 1975. The position of interphase chromosomes and late replicating DNA in centromere and telomere regions of Allium cepa L. Chromosoma(Berl.), 50: 201-210

Gale, M.D. and Miller, T.E. 1987. The introduction of alien genetic variation into wheat. In Lupton, F.G.H.(ed.)(1987). Wheat Breeding. Its scientific basis. Chapman and Hall. London

Galloway, S.M. and Evans, H.J. 1975. Sister chromatid exchange in human chromosomes from normal individuals and patients with ataxia tilangiectasia. Cytogenet. Cell Genet. 15: 17-29

Gerlach, W.L. 1977. N-banded karyotypes of wheat species. Chromosoma(Berl.), 62: 49-52

Gerstel, D.U. 1945. Inheritance in Nicotian tabacum. XX. The addition of Nicotiana glutinosa chromosomes to tobacco. J. Heredity, 36: 197-206

Gilham, N.W. 1978. Organeller Heredity. Raven Press, New York.

Gilles, A. and Randolph, L.F. 1951. Reduction of quadrivalent frequency in autotetraploid maize during a period of 10 years. Am. J. Bot., 28: 12-17

Harlan, J.R. and deWet, H.M.J. 1975. On Ö. WINGE AND A PRAYER: The origins of polyploidy. The Bot. Rev., Vol. 41, 361-390

Haskell and Wills 1968. Primer of Chromosome Practise. Oliver and Boyd

Hayes, W.1968. The Genetics of Bacteria and their Viruses. Blackwell Scientific Publications, Oxford.

Hecht, F. et al. 1974. The human cell nucleus: quinacrine and other differential stains in the study of chromatin and chromosomes. In: The Cell Nucleus, Vol. 2(H. Busch ed.) pp. 33-121, Academic Press, NewYork

Hof Van't, J. 1974. The duration of chromosomal DNA synthesis, of the mitotic cycle, and of meiosis of higher plants. In: R.C. King(ed.) Handbook of Genetics. pp. 363-377. Plenum Press, New York.

Holmquist, G. 1979. The mechanism of C-banding, Depurination and â-elimination. Chromosoma, 72: 203-224

Holmquist, G.P. 1992. Chromosome bands, their chromatin flavors, and their functional features. Am. J. Hum. Genet., 51: 17-37

Horney, K.J. 1975. The exploitation of polyploidy in sugar-beet breeding. J.Agric. Sci., Camb., 84:543-557

Jack et al. 1985. The structural basis for C-banding. Chromosoma(Berl.), 91: 363-368

John and Lewis. Somatic Cell Division. Oxford Biology Readers (26). OUP

John and Lewis. The Meiotic Mechanism. Oxford Biology Readers (65). OUP

John, B. and Hewitt, G.M. 1965. The B-chromosome system of Myrmeleotettix maculates(Thunb). II. The statics. Chromosoma, 17: 121-138

John, B. and Lewis, K.R.1965. The Meiotic System (a monograph). Protoplasmatolgia Vl Fl Springer-Verlag, Vienna.

Jones, R.N. and Rees, H. 1967. Genotypic control of chromosome behavior in rye. XI. The influence of B-chromosomes on meiosis. Heredity, 22: 333-347

Kasha, K.J.(ed.) 1974. Haploids in higher plants. University of Guelph, Guelph, Canada.

Kato, H. and Yosida, T.H. 1972. Banding patterns of Chinese hamster chromosomes revealed by new techniques. Chromosoma(Berl.), 36: 272-280

Kenton, A. 1981. Chromosome evolution in the Gibasis linearis alliance(Commelinaceae). I. The Robertsonian differentiation of G. venustula and G. speciosa. Chromosoma(Berl.), 84: 291-302

Kenton. A. 1978. Giemsa C-banding in Gibasis(Commelinaceae). Chromosoma(Berl.), 65: 30-9-324

Kermicle, J.L. 1969. Androgenesis conditioned by a mutation in maize. Science, 166: 1422-1424

Khus, G.S. 1973. Cytogenetics of Aneuploids. Academic Press, New York.

Khus, G.S. and Rick, C.M. 1967. Tomato tertiary trisomics: origin, identification, morphology and use in determining position of centromeres and arm location of markers. Can. J. Genet. Cytol., 9:610-631

Kihara, H. and One, T. 1926. Chromosommenzahlen und systematische Gruppierung der Rumex-Arten. Zeitschr. F. Zellforsch. Mikro Anat. 4: 475-481

Kikudome, G.Y. 1959. Studies on the phenomena of preferenatial segregation in maize. Genetics, 44: 815-831

Kezer, J; Sessions, S.K. and Leon, P. 1989. The meiotic structure and behaviour of the strongly heteromorphic X/Y sex chromosomes of neotropical plethodontid salamander of the genus Oedipina. Chromosoma, 98: 433-42

Kikudome, G.Y. 1961. Cytogenetic behavior of a knobbed chromosome 10 in maize. Science, 134: 1006-1007

Kimber, G. 1974. A reassessment of the origin of the polyploidy wheat. Genetics, 78: 487-492

Knott, D.R. 1961. The inheritance of rust resistance IV. The transfer of stem rust resistance from Agropyron elongatum to common wheat. Can. J.Pl. Sci., 41: 109-123

Knott, D.R. and Dvorak, J. 1976. Alien germplasm as a source of resistance to diseases. Ann. Rev. Plant Pathology, 14: 211-235

Lamprecht, H. 1968a. Die neue Genenkarte von Pisum und warum Mendel in seinen Erbsen-kreuzungen keine Geneknkoppelung gefunden hat. Arb Steiermarkischen Landesbibliothek Joanneum Graz, 10: 1-29

Langlois, R.G., Yu, L.-C, Gray, J.W. and Carrano, A.V. 1982. Quantitative karyotyping of human chromosomes by dual beam flow cytometry. Proc. Nat. Acad. Sci., U.S.A., 79: 7876-80

Latt, S.A. 1973. Microfluorometric detection of deoxyribonucleic acid replication in human metaphase chromosomes. Proc. Natl. Acad. Sci., U.S.A., 70: 3395-3399

Law, C.N and Worland, A.J. 1974. Eur. Wheat Aneuploid Co-op Newsl., 4: 9-10

Law, C.N. and Worland, A.J. 1973. Plant breeding Institute Annual Report for 1972. pp. 25-65

Law, C.N.; Snape, J.W. and Worland, A.J. 1987. Aneuploidy in wheat and its uses in genetic analysis. In Lupton, F.G.H.(ed.)(1987). Wheat Breeding. Its scientific basis. Chapman and Hall. London.

Levin, L. 1969. Biology of the Gene. The C.V. Mosby Company.

Lewis and John. The Matter of Mendelian Heredity, 2nd Edtn. Churchill(Chapter 6)

Lewis, K.R. and John, B. 1963. Chromosome Marker. Churchill, London.

Longey, A.E. 1956. The origin of dimunitive B-type chromosomes in maize. Am. J.Bot., 43: 18-23

Longey, A.E.1945. Abnormal segregation during megasporogenesis in maize. Genetics, 30:100-113

Mann, T.J.; Gerstel, D.U.and Apple, J.L. 1963. The role of interspecific hybridization in tobacco disease control. Proc. III World Tobacco Sci. Congr., Salisbury, Zimbabway, pp. 201-7

McClintock, B.1931. Cytological observations of deficiencies involving known genes, translocations and an inversion in Zea mays. Missouri Agric. Exp. Stn. Res. Bull. 163: 1-30

McClintock, B.1933. The association of non-homologous parts of chromosomes in the mid-prophase meiosis in Zea mays. Z. Zellforsch. Mikroskop. Anat. 19: 191-237

McClintock, B.1941a. The association of mutants with homozygous deficiencies in Zea mays. Genetics, 29: 542-571

McLeish and Snoad 1958. Looking at Chromosomes. McMillan

Meiburn, K.J. and Misteli, T. 2007. Chromosome territories. Nature, 445: 379-381

Mettin, D.; Bluthner, W.D. and Schlegel, G. 1973. Additional evidence on spontaneous 1B/1R wheat-rye substitutions and translocations. Proc. 4th Int. Wheat Genet. Symp., pp. 179-84

Moens, P. 1973. Mechanisms of chromosome synapsis at meiotic prophase. Int. Rev. Cytol., 35: 117-134

Moses, M.J. 1968. Synaptonemal complex. Ann. Rev. Genet., 2:363-412

Muramatsu, M. 1963. Dosage effect of the spelta gene q of hexaploid wheat. Genetics, 48:469-482

Neale, M.J and Keeney, S. 2006. Clarifying the mechanics of DNA strand exchange in meiotic recombination. Nature, 442: 153-158

Nel. P.M. 1971. studies on the genetic control of recombination in Zea mays. Ph.D. thesis. Indian University.

Nilsson, N.O., Sall, T and Bengtsson, B.O. 1993. Chiasma and recombination data in plants: are they compatible? Trends Genet., 9:344-8

Nitzche, W. and Wenzel, G. 1977. Haploids in plant breeding. Adv. in Plant Breeding(Suppl. J. Plant Breeding). Vol. 8

O'Mara, J.C. 1940. Cytogenetic studies on Triticale. I. A method for determining the effects of individual Secale chromosome on Triticum. Genetics, 25: 401-108

O'Mara, J.G. 1940. Cytogenetic studies in Triticale I. A method for determining the effects of individual Secale chromosomes in Triticum. Genetics, 25: 401-408

Olmo, H.P. 1935. General studies of monosomic types of Nicotiana tabacum. Genetics, 20: 286-300

Pardue, ML. and Gall, J.G. 1970. Chromosomal localization of mouse satellite DNA. Science, 168: 1356-1358

Paris Conference 1971. Standardization in Human Cytogenetics. Birth defects: Original Article Series. Vol. 8, No. 7. The National Foundation, New York.

Paris Conference 1971. Standardization in Human Cytogenetics. Cytogenet., 11: 313-362

Patterson, E.B. 1952. The use of functional duplicate-deficient gametes in locating genes in maize. Genetics, 37:612

Peloquin, S.J. 1981. Chromosomal and cytoplasmic manipulations, pp. 117-150. In Frey, K.J.(ed.), Plant Breeding II, The Iowa State University Press/ Ames

Perry, P and Wolff, S. 1974. New Giemsa method for differential staining of sister chromatids. Nature, 251: 156-158

Person, C. 1956. Some aspects of monosomic wheat breeding. Can. J. Bot., 34: 60-70.

Peters, J-M. 2007. The checkpoint brake relieved. Nature, 446: 8680869

Peterson, P.A. 1970. Controlling elements and mutable loci in maize: their relationships to bacterial episomes. Genetica, 41: 33-56

Phillips, R.L., Kleese, R.A and Wang, S.S. 1971b. The nucleolus organizer region of maize(Zea mays L.): chromosomal site of DNA complementary to ribosomal RNA. Chromosoma, 36: 79-88

Phillips, R.L.; Burnham, C.R. and Patterson, E.B. 1971a. Advantage of chromosomal interchanges that generate haplo-viable deficiency-duplications. Crop Sci., 11: 525-528

Prinz, S and Amon, A.1999. Dual control of mitotic exit. Nature, 402. 133-135

Rabl, C. 1985. Uber Zelltheilung. Morphol. Jahrb. 10 : 214-330

Radding, C. 1973. Molecular mechanisms in genetic recombinations. Ann. Rev. Genet., 7: 87-111

Rakha, F.A. and Robertson, D.S 1970. A new technique for the production of A-B translocations and their use in genetic analysis. Genetics, 65: 223-240

Ramage, R.T. 1965. Balanced tertiary trisomics for use in hybrid seed production. Crop Sci., 5: 177-178

Randolph, L.F. 1935. Cytogenetics of tetraploid maize. J.Agric. Res., 50: 591-605

Reddy et al. 2007. Ubiquitination by the anaphase-promoting complex drives spindle check point inactivation. Nature, 446: 921-925

Rees, H. and Jones R.N. 1976. Chromosome Genetics. Arnold

Rees, H. 1947. Crossover chromosomes in unreduced gametes of asynaptic maize. Genetics, 32: 101

Rees, H. 1955b. Heterosis in chromosome behavior. Proc. Roy. Soc. B., 144: 150-159

Rees, H. 1957. Genotypic control of chromosome behavior in rye. IV. The origin of new variation. Heredity, 11: 185-193

Rees, H. 1957b. Distribution of chiasmata in an asynaptic Locust. Nature, 180: 559.

Rees, H. 1961. Genotypic control of chromosome form and behavior. Bot. Rev., 27: 288-318

Reese, H. and Ayonoadu, U. 1973. B chromosome selection in rye. Theor. Appl.Genet., 43:162-166

Rhoades, M.M. 1942. Preferential segregation in maize. Genetics, 27: 395-407

Rhoades, M.M. 1952. Preferential segregation in maize. P. 66-80. In : Heterosis, Iowa State College press, New York

Rhoades, M.M. 1955. The Cytogenetics of maize, chap. 4, 'Corn and Corn Improvement' ed. By G.F.Sprague, Academic Press, Inc., New York

Rick, C.M. and Butler, L. 1956. Cytogenetics of the tomato. Adv. Genet., 8: 267-382

Rieger, R.; Michaelis, A. and Green, M.M.(1976). Glossary of Genetics and Cytogenetics, Classical and Molecular. Springer-Verlag. Berlin.

Riley, R and Law, C.N. 1965. Genetic variation in chromosome pairing. Adv. Genet., 13: 57-114

Riley, R. 1958. Chromosome pairing and haploids in wheat. Proc. X Int. Congr. Genet., 2: 234-35

Riley, R. and Chapman, V. 1958. Genetic control of the cytologically diploid behavior of hexaploid wheat. Nature, 182: 713-715

Riley, R. and Kimber, G. 1966. The transfer of alien genetic variation to wheat. Rep. Plant Breed. Inst., Camb. 1964-65, pp. 6-36

Riley, R., Chapman, V. and Kimber, G. 1959. Genetic control of chromosome pairing in intergeneric hybrids with wheat. Nature, 183: 1244-1246

Riley, R., Chapman, V. and Kimber, G. 1960. Position of the gene determining the diploid like behavior of polyploid wheat. Nature, 186: 259-260

Riley, R.; Chapman, V. and Johnson, R. 1968. Introduction of yellow rust resistance of Aegilops comosa into wheat by genetically induced homoeologous recombination. Nature, 217: 383-84

Robinson, A.S. 1976. Progress in the use of chromosomal translocations for the control of insect pests. Biol. Rev., 51, pp.1-24

Rodman, T.C. 1974. Human chromosome banding by Feulgen stain aids in localizing classes of chromatin. Science, 184: 171-73

Roy, D. 2000. Plant Breeding-Analysis and exploitation of variation. Narosa, New Delhi.

Roy, D. 2009. Biotechnology. Narosa, New Delhi(IN PRESS).

Roy, D. 2009. Bioinformatics. Narosa, New Delhi(IN PRESS).

Schulz-Schaeffer, J. 1980. Cytogenetics Plants, Animals, Humans. Springer- Verlag, New York.

Schwarzacher, T.; Ambros, P. and Schweizer, D. 1980. Application of Giemsa banding to orchid karyotype analysis. Plant Syst. Evol., 134: 293-297

Schweizer, D. 1973. Differential staining of plant chromosomes with Giemsa. Chromosoma, 40: 307-320

Schweizer, D. 1976. Reverse fluorescent chromosome banding with chromomycin and DAPI. Chromosoma, 58: 307-324

Sears, E.R. 1953a. Nullisomic analysis in common wheat. Am. Naturalist, 87: 245-252

Sears, E.R. 1953b. Addition of the genome of Haynaldia villosa to Triticum aestivum. Am. J. Bot., 40: 168-174

Sears, E.R. 1956. The transfer of leaf-rust resistance from *Agropyron umbellulata* to wheat. Brookhaven Sym. Biol., 9: 1-22

Sears, E.R. 1958. The aneuploids of common wheat. Proc. 1st Intern. Wheat Symp. Winnipeg, Canada, pp. 221-229

Sears, E.R. 1966a in Chromosome Manipulations and Plant Genetics(eds. R. Riley and K.R. Lavis), Oliver and Boyd, Edinburgh, pp. 29-45.

Sears, E.R. 1966b. Proc. 2nd Int. Wheat Genet. Symp. Lund, 1963. Hereditas, 2(suppl.), 370-81.

Sears, E.R. 1969. Wheat cytogenetics. Ann. Rev. Genet., 3:451-468

Sears, E.R. 1972. Chromosome engineering in wheat. Stadler Genet. Symp. 4: 23-28

Sears, E.R. 1974. The wheats and their relatives. pp. 59-91. In Handbook of Genetics(ed. R.C.King). Plenum Press, New York.

Sears, E.R. 1975. An induced homoeologous pairing mutant in Triticum aestivum. Genetics, 80:S74

Sears. E.R. and Okamoto, M. 1958. Intergenomic chromosome relationships in hexaploid wheat. Proc. X Int. Congr. Genet., 2: 258-59

Shiraishi, Y. and Yosida, T.H. 1972. Banding pattern analysis of human chromosomes by the use of urea treatment. Chromosoma(Berl.), 37: 75-83

Shirayama, M., Toth, A., Galova, M amd Nasmyth, K. 1999. APC^{cdc20} promotes exit from mitosis by destroying the anaphase inhibitor Pds1 and cyclin c1b5. Nature, 402: 203-207

Srb, A.M.; Owen, R.D. and Edgar, R.S. 1965. General Genetics.2nd edition. W.H.Freeman, San Francisco

Stadler, D.R. 1973. The mechanism of intragenic recombination. Ann. Rev. Genet., 7: 113-127

Stebbins, G.L. Jr. 1968. Variation and Evolution in Plants. Oxford and IBH company Co. New Delhi.

Stegmeier, F. et al. 2007. Anaphase initiation is regulated by antagonistic ubiquitination and deubiquitination activities. Nature, 446: 876-880

Strickberger, M.W. 1970. Genetics. The Macmillan Company. New York

Sturtevant, A.H. 1915. The behaviour of chromosomes as studied through linkage. Zeitschr. Ind. Abstam. Vererbungsl. 13: 234-287.

Sumner, A.T. 1972. A simple technique for demonstrating centromeric heterochromatin. Exp. Cell Res., 75: 304-306

Sumner, A.T. et al.1971. New technique for distinguishing between human chromosomes. Nature(Lond.) New Biol. 232: 31-32

Swanson, C.P, Timothy, M and Young, J.W. 1973. Cytogenetics. Prentice-Hall of India Pvt. Ltd. New Delhi

Swanson, C.P. 1958. Cytology and Cytogenetics. The Macmillan Co.

Sybenga, J. 1972. General cytogenetics. American Elsevier Publ. Co., New York

Sybenga, J. 1975. The quantitative analysis of chromosome pairing and chiasma formation based on the relative frequencies of MI configurations. VII. Autotetraploids. Chromosoma(Berl.), 50: 211-222

Tai, G.C.C. 1994. The use of 2n gametes. In: Potato Genetics(eds) J.E. Bradshaw and G.R. Mackay. CAB International, pp; 109-132

Tatsuya, H. 2000. Chromosome cohesion, condensation and separation. Ann. Rev. Biochem., 69: 115-144

Trask, B.J. 1991. Fluorescence in situ hybridization: application in cytogenetics and gene mapping. Trends genet., 7: 149-154

Unrau, J. 1950. The use of monisomes and nullisomes in cytogenetics studies of common wheat. Sci. Agri., 30: 66-89

Unrau, J., Person, C. and Kuspira, J. 1956. Chromosome substitution in hexaploid wheat. Can. J. Bot., 34: 629-640

Van Holde, K.E. 1988. Chromatin structure and Function. Springer-Verlag, New York.

Vosa, C.G. 1970. Heterochromatin recognition with fluorochromes. Chromosoma(Berl.), 30: 366-372

Vosa, C.G. and Marchi, P. 1972. Quinacrine fluorescence and Giemsa staining in plants. Nature New Biol., 237: 191-192

Wagenaar, E.B. 1968a. Meiotic restitution and the origin of polyploidy. II. Prolonged duration of metaphase I as a causal factor of restitution induction. Can. J. Genet. Cytol., 10: 844-852

Wall, A.M.; Riley, R. and Chapman, V. 1971. Wheat mutants permitting homoeologous meiotic chromosome pairing. Genet. Res., Camb., 18: 311-329.

Wall, A.M.; Riley, R. and Gale, M.D. 1971. The position of a locus on chromosome 5B of T. aestivum affecting homoeologous meiotic pairing. Genet. Res., Camb., 18: 329-339

Walters, M.S. 1958. Aberrant chromosome movement and spindle formation in meiosis of Bromus hybrids: an interpretation of spindle organization. Amm. J. Bot., 45: 271-89

Wang, H.C. and Fedoroff, S. 1972. Banding in human chromosomes treated with trypsin. Nature(Lond.) New Biol., 235: 52-53

Weisblum, B. and de Haseth, P.L. 1972. Quinacrine, a chromosome stain specific for deoxyadenylate-deoxythymidylate-rich regions in DNA. Proc. Natl. Acad. Sci. U.S.A., 69: 629-632

Weisblum, B. and Haenssler, E. 1974. Fluorometric properties of the benzimidazole derivative Hoechst 33258, a fluorescent probe specific for AT concentration in chromosomal DNA. Chromosoma(Berl.), 46: 255-260

Westergaard, M. and von Wettistein, D. 1972. The synaptonemal complex. Ann. Rev. Genet., 6:71-110

White, M.J.D. 1973. Animal Cytology and Evolution. Cambridge University Press. 3rd Edtn.

White, M.J.D. The Chromosomes. Chapman and Hall. 6th Edtn. 1973

Wilson, G.B. and Morrison, J.H. 1966. Cytology. East-West Press Pvt. Ltd. New Delhi

Yunis, J.J. 1976. High resolution human chromosomes. Science, 191: 1268-70

Yunis, J.J. and Yasmineh, W.G. 1971. Heterochromatin, satellite DNA and cell function. Science, 174: 1200-1209

Zickler, D. and Kleckner, N. 1998. The leptotene-zygotene transition of meiosis. Ann. Rev. Genet., 32:619-697

Zickler, D. and Kleckner, N. 1999. Meiotic chromosomes: Integrating structure and function. Ann. Rev. Genet., 33:603-754.

Index